The Chinese Pleasure Book

The Chinese Pleasure Book

Michael Nylan

ZONE BOOKS · NEW YORK

2018

© 2018 Michael Nylan
ZONE BOOKS
633 Vanderbilt Street
Brooklyn, NY 11218

All rights reserved.

No part of this book may be reproduced, stored in a retrieval system, or transmitted in any form or by any means, including electronic, mechanical, photocopying, microfilming, recording, or otherwise (except for that copying permitted by Sections 107 and 108 of the U.S. Copyright Law and except by reviewers for the public press), without written permission from the Publisher.

Distributed by The MIT Press,
Cambridge, Massachusetts, and London, England

Library of Congress Cataloging-in-Publication Data
Names: Nylan, Michael, author.
Title: The Chinese pleasure book / Michael Nylan.
Description: New York : Zone Books, 2018. | Includes index.
Identifiers: LCCN 2018015511 | ISBN 9781942130130 (hardcover)
Subjects: LCSH: Philosophy, Chinese — To 221 B.C. | Emotions (Philosophy) |
    Pleasure. | Senses and sensation. | Mind and body.
Classification: LCC B127.E46 N95 2018 | DDC 152.4/201—dc23
LC record available at https://lccn.loc.gov/2018015511

To Naomi Richard, *une femme formidable*

Contents

PREFACE 9

ACKNOWLEDGMENTS 13

INTRODUCTION 17

I Coming Attractions 33

II Good Vibrations: The Allied Pleasures of Music and Friendship in the Masters 59

III Mencius 孟子 on Our Common Share of Pleasure 135

IV Xunzi 荀子 on Patterns of Brilliance 175

V Vital Matters: The Pleasures of Clear Vision in the Zhuangzi 莊子 213

   APPENDIX On the Dating and Composition of the Zhuangzi 261

VI Yang Xiong 揚雄 on the Allure of Words Well Chosen 263

   APPENDIX Brief Historical Background to the Haogu Movement 315

VII Semidetached Lodgings: The Pleasures of Returning Home in Tao Yuanming 陶淵明 and Su Shi 蘇軾 317

   APPENDIX "Matching Poems" by Tao and Su 373

AFTERWORD   427

NOTES   429

LIST OF NAMES   431

SUGGESTED READINGS   435

INDEX   441

Notes and Words Cited may be accessed online at:
zonebooks.org/books/133-the chinese-pleasure-book

Preface

When I told my childhood friend that I was writing a book about pleasure-seeking and pleasure-taking, she remarked, "Oh, I get it. You're writing a self-help book." Actually, this is not a self-help book. But readers may certainly treat it as one, if they find that limning the ideas of thoughtful people living in other times and places helps to provoke new thoughts about how to wrest more from life than mere survival. Since the serious dilemmas that people face in living out their days have not changed much, despite huge technological advances, immersing oneself in the questions and answers formulated by very smart people in the past, whether it's Zhuangzi or Montaigne (or both), is likely to afford a bit more clarity through useful perspectives. As one philosopher put it, "Our perspectives change, largely through what we immerse ourselves in."[1] Common tensions in lived experience have been articulated in different ways across cultures, and the differences are fascinating.

I suspect that most Americans associate pleasure with selfishness — or worse.[2] There is no need to go that route; indeed, every reason not to. One has only to examine carefully the visual and literary tropes in service to modern industries purveying a reckless coupling of blind optimism and individual identity as quasi-autonomy. Against such projects as the Berkeley-based "Science of Happiness" project, this book contends that a better grasp of social relatedness is basic if we are to figure out how to forge decent lives on the margins of civilizational ruin. While our postmodern preoccupations with representation, virtual reality, and highly mediated existences urge us to avert our eyes, there is profound pleasure to be had from knowing more about the communities we dwell in and how to interact most effectively in them, and there is even pleasure, I would argue, to be had from boldly confronting some unpleasant aspects of experience.[3]

Several quotations come to mind as I reread the previous sentence. The first, by Van Gogh, identifies "the great thing" as the ability to "wrest new vigor from reality." The second, by Virginia Woolf, says, "The future is dark, and that is the best thing for it to be, I think."[4] Written decades before Woolf's descent into suicidal depression, Woolf's remark reveals a rather sanguine observation: because we have no chance of foreseeing the future, despair is at once defeatist and irrational. We might as well try to influence our world in humane and sustaining ways. And as the fictional character in a P. D. James's novel, Adam Dalgleish, observes, "It isn't action but pleasure which binds one to existence."[5]

It is easy for a historian to act as anthropologist when journeying to lands so distant in time and place as early China, where, not surprisingly, they see things differently. In *Shame and Necessity*, Bernard Williams (one of the few gods in my academic pantheon) argued that the ethical thinking of the ancient Greeks was "in better shape" than that of the Kantian and post-Kantian moralists, insofar as the Ancients' system of ideas basically lacks the concept of *morality* altogether, in the sense of "a class of reasons or demands which are vitally different from other kinds of reason or demand" and where, relatedly, "the questions of how one's relations to others are to be regulated, both in the context of society and more privately, are not detached from questions about the kind of life it is worth living."[6]

The same could be said of the Ancients in early China. For me, it is an article of faith that ongoing conversations with imaginative people, living and dead, are likely to wean me from the usual assumptions. They compel a salutary awareness not only of how peculiar contemporary life is, but also of the verdict that we "have never been modern," as Bruno Latour says.[7] So while this is no more a book about morality than it is a self-help book, the entire subject of the book concerns the good life and techniques to attain it, the earliest and most insistent topic of the moral question. For these early Chinese thinkers studied ethics in its original sense — not as academics, but as skilled questioners looking to spur deeper reflections on the sort of ethos in which to abide.[8]

"With modern science we learn how the world is but not how to live in it," being "overwhelmed by the intangible detritus of twenty-first-century life: unreturned emails; unprinted family photos; the ceaseless ticker of other people's lives on Facebook; the heightened demands of parenting; and the suspicion that we'll be checking our phones every 15 minutes, forever."[9] The writings surveyed in this book attend to the potentially therapeutic: how to create more meaning in our lives while enhancing our vitality, instead of wasting it. Simply put, the group of early thinkers assembled here addressed what Bertrand Russell identified as the central preoccupation of philosophy: "how to live without paralysis in a world that knows no certainty."[10]

Unlike some theories in the modern West, theories proposed in early classical Chinese about the pleasure calculus and attentional economy do not equate rigid "control of the self" with "care of the self," nor do they presume that "all things fatigue us at last."[11] The Chinese theories did not ever predicate the young child as tabula rasa; the child had inherited proclivities, but the world around her and her own sense of it would modify any earlier tendencies. By the Chinese theories, the whole gamut of the emotions stimulate the sensitive nerve endings in ways that can either soothe and sustain or irritate and drain the energies from the organ systems. Some practices, as a result, support the functions of the body and spirit, while others are liable to endanger it. Yet learning to prefer humaneness to destructive activities that shout "power" is not always easy. It requires discipline, great teachers, and repeated insights.[12]

This book tries to "make sense" of all aspects of living in a constellation of ways deploying a range of methodologies. Inevitably, this book has been designed to trace the classical writers' search for meaning. Where this book succeeds, it may, as W. S. Merwin said of his own work, "lead you to speculation about the parentage of beauty itself, to which you will repeatedly return."[13] When all is said and done, it is the unpredictable contours of humanity, past and present, that intrigue, even enthrall, me as well as how people have long justified their disparate views of humanity. And while caveats will be duly be offered — most significantly, for early China, we have little evidence, save that left by a few elite men, most often at court, who are in no way representative of the population at large — these texts evince subtle thinking on a human scale, eluding a number of modern pitfalls. I believe they can tell us things.

Acknowledgments

A book that has taken nearly two decades to produce is the result of innumerable kindnesses. Friends tell you about a book; friends give you a place to write; friends cheer you up when you are in despair. I will not be able to thank all my friends, and these Acknowledgments, however heartfelt, but skim the surface of the debts that I contracted and am delighted to acknowledge.

A group of friends must be thanked first, because they read multiple drafts of one or more chapters with unflagging patience. Of those friends Naomi Richard, to whom this book is dedicated, must come first, because Naomi read and edited all but one chapter, always with her exquisite insights and delicate inquiries about muddle. Frederick Tibbetts read the entire manuscript and is responsible for the felicity of phrasing at many points, especially with the "matching" poems in Chapter 7, which we translated from my cloddish Singlish. Christian de Pee and Alfreda Murck read Chapter 7, repeatedly, and they shared relevant secondary sources with me. Their practical guidance, as well as encouragement, propped me up when I thought I would never get anything right. While I was in Paris, Stéphane Feuillas did me the honor of commenting on Chapter 7 as well, and his expertise on Su Shi also proved invaluable. Andrew Plaks and Li Wai-yee were careful readers (none better) of the material that eventually made its way into Chapter 6. Thomas H. Hahn, when I was too tired to think straight, kindly read the Introduction and Chapters 1 and 6. Ramona Naddaff, of Zone and the Berkeley Rhetoric Department, was a marvelously sensitive reader; she has been a dream to work with throughout this process. Dirk Meyer was a careful reader of Chapter 6, on manuscript culture, as were Marty Verhoeven and Koichi Shinohara, reading Chapter 5. Eleanor Lipsey, a fine musicologist studying at the School of Oriental and African Studies, University of London (SOAS), also read Chapter 2 and gave me astute comments thereon. My teachers and friends Michael Loewe, Geoffrey Lloyd, Nathan Sivin, and Henry Rosemont, Jr., afforded me the opportunity of contributing to their Festschrifts, and some of the arguments that figure in this book I first formulated for them and for compilations edited by Joachim Gentz, Paul Kroll, and David Knechtges, who urged me to publish an earlier, expanded version of Chapter 6 as a monograph; each gave me excellent advice on that version. Alice Cheang, an independent scholar with a real passion for Su Shi, generously gave of her time to read my penultimate draft of Chapter 7, though we knew each other only through our mutual friendship with Cynthia Brokaw.

Next, I must highlight my thanks to Hans Sluga and Robert Ashmore, at UC-Berkeley, whose friendship, organizational hints, and outright corrections mean more to me than I can say. Hans even lent me his copy of Schopenhauer, and Robert slogged through every damn Chinese quotation in the book, comparing my translations to the Chinese text! Meanwhile, the two anonymous readers commissioned by Zone saved me from falls from grace and a few outright whoppers. (For any that remain I take full responsibility, since they warned me.) And Haun Saussy sent notes of encouragement and information about American conspiracy theories regarding music. Among my graduate students, comments by Nicholas Constantino pushed several arguments further, as did those by Trenton Wilson, who is repeatedly cited in the notes. Heng Yin gave me my very own copy of Wang Shumin's compilation of the *Zhuangzi*, which I treasure as a sign of our mutual esteem. Lucia Q. Tang worked to rectify the Works Cited, a thankless task, if ever there was one, along with Sydney Morrow (University of Hawai'i). I owe Scott Davis a great deal, because he painstakingly reviewed the secondary literature in English on Zhuangzi to find the tropes most favored there. Teaching at UC-Berkeley with Carlos Norena has taught me an awful lot about how to present material in a user-friendly way, not to mention Roman history.

Bernard Fuehrer (SOAS) gave me the opportunity to talk about A. C. Graham's translation of the *Zhuangzi*; Lisa Indraccolo and Wolfgang Behr (University of Zurich), about antique notions of friendship; Richard King and David Machek (University of Bern), about the role of the negative emotions in early China; and Roger

Ames (then of the University of Hawai'i), about the *Analects* and *Zhuangzi* — with all these engagements permitting me to try my ideas out on sympathetic listeners. Stefaan Cuypers (KU Leuven, Philosophy Department) steered me in the right direction on the subject of hedonism. Cynthia Brokaw (Brown University, History Department) recommended some writings on Zhu Xi; she has proven to be an insightful and probing listener. Lee Yearley's talk of bewilderment in the *Zhuangzi* was extremely enlightening, as was desultory talk about musical appreciation with Herbert Fingarette (in person and in reading). Andrew F. Jones (UC-Berkeley, East Asian Cultures) was the first reader of Chapter 2; he brought, as usual, his unfailing genius to a very rough draft. My dear friends, Robert Litz and Gerry Boswell, were more patient than I can say with my years of ramblings on the twinned subjects of pleasure and delight.

Institutional support for this volume has come from many places, directly from the University of California at Berkeley and from the Institute for Advanced Studies in Paris, which supported my leaves as I was polishing the book. The fellows at the institute, as well as the staff, were wonderfully encouraging, as were Bruno Belhoste, Françoise Bottéro, Karine Chemla, Anne Cheng, Béatrice L'Haridon, Valerie Lavoix, Suzanne Said, and Olivier Venture, not to mention Kate Overton and Pierre Nicholas, Duncan and Martine Gallie, the last four my genial partners in crime. Support came indirectly from the Guggenheim Foundation, because no one could work eight hours per day on my official Guggenheim project, a reconstruction of the Han-era *Documents* classic, and so idle time gave way to writing draft paragraphs that appear in this text. At Princeton, Gonul Yurdakul, Martin Heijdra, and Soren Edgren offered help. Yang Zhiyi, then a graduate student in East Asian Studies, read Su Shi's poems with me, a rank beginner. Back in Berkeley, I have had unceasing support from Daniel Boyarin (Near Eastern Studies), several colleagues in the History Department, and Katharina Kaiser. In the East Asian Library, I must thank many, in particular Peter Zhou, Jianye He, Deborah Rudolph, Bruce Williams, Toshie Marra, and Susan Xue. Diane Sprouse did me the enormous favor of producing call-outs for the website production, while Susan Stone generated the index with ease and efficiency.

At Zone, no managing editor could have been nicer than Meighan Gale; she is patient, careful, and wise. Ramona Naddaff's editorial comments were predictably nuanced, and the comments by Bud Bynack, Zone's copyeditor, were fun to read and instructive to boot, a form of activist editing that Yang Xiong himself would have applauded. Julie Fry did wonders with the cover (based on the Nelson-Atkins original), and my friends Chen Pao-chen, Chiang Hsiang-wen, Jan Stuart, and Carma Hinton (at the Taipei Palace Museum, Freer Gallery, and George Mason University, respectively), also Ms. Yukwah Cheng, Ms. Biliang Chang, and Ms. Shuhui Chang, helped with securing the best illustrations, as did Marianne Grey, Karen McLean, Adrian Gordon, and Alfreda Murck.

Ami de Grazia let me stay with her in Naxos in the summer of 2017 for two weeks to complete the book. Her generosity was extraordinary, because she had her own writing to do. She fed me with food and with conversation. I owe her warm hugs and many enlightening moments.

Introduction

This book traces the evolution of pleasure theories in early China over the course of a millennium and a half, from the fourth century BCE to the eleventh century CE. To signify acts of pleasure-seeking, pleasure-taking, and imparting pleasure, a wide range of thinkers during that time deployed the single graph, *le* 樂, freely borrowing from one another, sometimes to differing ends, but often with the same goal of arriving at the most versatile model of the human condition.[1] Undergirding their rhetoric was always the dual presumption that pleasure matters a great deal to most people, and how people seek, take, and give pleasure is the truest test of their character.

Why take pleasure as my chosen subject? At first, it was simply because Sinologists for so long sidestepped the topic,[2] and more recently, because serious consideration of pleasure in academia has reentered the realm of ethical and aesthetic theory, also the histories of early modern Europe.[3] Chiefly, however, it is because the steady contemplation of pleasure — not short-term delight or kindred concepts — invites attention to distinctive aspects of Chinese culture, as well as to notions common to Chinese and non-Chinese traditions. Consider the cultural relativity of our division and conceptualization of what we deem the inner states. In German, for example, there are many words approaching "pleasure" (*die Freude, die Lust, das Vergnügen, das Behagen*, and so on),[4] yet none capture the valences of the classical Chinese term *le*. Equally curious and no less significant is the Chinese opposition of pleasure to insecurity, rather than to pain, its classic antonym within mainstream Western traditions.

The verbal use of *le* ("to take or derive pleasure in") in the classical literature in Chinese takes but a very few objects, almost always those that promise deeper satisfactions in return for steady, long-term commitments.⁵ You can take pleasure in intimate friends (*le you* 樂友), in music (*le yue* 樂樂), in a vocation and legacy (*le ye* 樂業), in sharing (*le yu* 樂與),⁶ in being alive and vital (*le sheng* 樂生), in doing your duty (*le yi* 樂義), in learning and emulating (*le xue* 樂學) others of the requisite worth (*le ren* 樂人), in Heaven or the cosmic operations (*le tian* 樂天), and in your true home (*le jia* 樂家). With those thoroughly relational pleasures in mind, this book offers seven chapters to conjure antique scenes for taking and giving pleasure.

Chapter 1, "Coming Attractions," has but two aims: first, to sketch in a preliminary way the key vocabulary items and concept clusters at work in the early pleasure talk in order to prepare the ground for the chapters that follow; second, to distinguish the pleasure theories in China from their much better-known counterparts in classical Greece and Rome and in modern philosophy. I will have recourse to the relevant medical and cosmological theories of the early empires positing the physiology of pleasure in order to elucidate the rationales underlying the Chinese claims.

Chapter 2, "Good Vibrations: The Allied Pleasures of Music and Friendship in the Masters," discusses the metaphors employed for both music and intimate friendship in light of early resonance theories, for all early passages on music and friendship presuppose the existence of unseen sympathies weaving the cosmic and social worlds together — sympathies capable of greatly and indelibly stamping the characters, attitudes, actions, and even the life spans of well-tempered people. Ideally, both music and close friendships illustrate the inherent value of the process of relating, rather than any predetermined end or goal, because what passes for perfection in music and friendship depends upon continual readjustments or "attunements" of temper. To situate these principles better, the chapter contrasts mere sociability with intimate friendships. Then, turning to the enormous "leave-taking" literature in classical Chinese, the chapter reviews the typical pretexts and approved methods for partings.

Chapter 3, "Mencius 孟子 on Our Common Share of Pleasure," focuses on one of the best-known Confucians, the fourth-century BCE persuader whose teachings have too often been reduced to the slogan "Human nature is good." By placing this slogan within the larger context provided by the first chapter of the writings ascribed to Master Meng, whose single focus is, precisely, pleasure, we find the main thread of Mencius's argument to various princes: that all goods in life, material and psychological, are most likely to gravitate to the ruler who shares his pleasures with others, imagining them to be very like himself, with the same desires for pleasure and need for security.[7] Not surprisingly, the chapter ends with a series of contrasts between Mencius and Xunzi, the later Confucian master whom I will review next, partly to give each thinker his due and partly to suggest the degree to which Mencius relies on exciting his leader's pleasurable intuitions.

Chapter 4, "Xunzi 荀子 on Patterns of Brilliance," claims that Xunzi's application of the pleasure calculus to his theories about human nature and court administration — the body and the body politic — are so sophisticated that they became the touchstone for pleasure rhetoric throughout the early imperial period, up to the fourth century CE and beyond. But Xunzi is still more compelling, perhaps, for his daring to pose the question, "What happens if you would teach and advise courts, offering them the best possible advice and remonstrance, but they do not wish to hear you?" Xunzi responds by describing the exquisite pleasures to be had from crafting artful lives of integrity, even when one is ignored by the powers that be. He thereby contributes to the creation of the aesthetic potential for living an unofficial life of beauty, dignity, and worth.

Chapter 5, "Vital Matters: The Pleasures of Clear Vision in the Zhuangzi 莊子," seeks to explain why the much-beloved Zhuangzi compilation resists all post-facto attempts to cast it either as a set of injunctions to model oneself on the spontaneously generating cosmos or as an extended exhortation to "embrace freedom" (aside from freedom from disquietude) or "go with the flow." Starting from the Zhuangzi's "Supreme Pleasure" chapter, Chapter 5 shows Zhuangzi urging readers to conserve their energies so that they may more fully live out the days they have been allotted, rather than vainly wishing for a yet more perfect life. Three of the chief strategic insights fostering Zhuangzi's "fully present" way of life are knowing that no creature, great or small, can attain sufficient understanding of the unfolding situation, so no one can be sure that he or she is ever right (which realization constitutes the only "clear vision" of the world that is available), and recognizing that death must be confronted in small, homeopathic doses if lives are not to be eaten up by needless worry.

Chapter 6, "Yang Xiong 揚雄 on the Allure of Words Well Chosen," considers the monumental figure of Yang (53 BCE – 18 CE), the

Han philosopher and court poet. Yang delighted in Mencius, Xunzi, and Zhuangzi, but for Yang, as for the poet Callimachus (d. ca. 240 BCE) halfway round the world, the deepest and most inviolate forms of pleasure come from immersing oneself in the great writers of remote antiquity, not from sharing with others, say, or from crafting an artful life, or from being fully present in the moment. "Books are as alluring as women," he opined. Since Yang playfully presented his own work through autocommentaries, new genres and styles of writing, not to mention carefully wrought defenses of his writings, there are excellent reasons to deem him one of the first fully self-conscious authors in Chinese history. This chapter surveys his remarkable output once he had gained entrance to the imperial library, including his etymological dictionary, conceived as an entryway into the archaic period.

Chapter 7: "Semidetached Lodgings: The Pleasures of Returning Home in Tao Yuanming 陶淵明 and Su Shi 蘇軾," turns to two of the most famous poets in Chinese history. Tao Yuanming (365?–427) is admired for his poem cycles celebrating his return home and his concomitant refusal to serve another day in office. Su Shi (1037–1101), the polymath who relished court service as his second home, nonetheless used Tao's poetry as a psychic refuge during three increasingly uncongenial exiles from the capital during the last decades of his life. In the end, Su "matched" all but four of Tao's poems through a rigorous use of identical end graphs in couplets, creating a revisionist Tao in his own image. While Su's portrait of Tao is wildly anachronistic, Su clearly thought long immersion in Tao's poetry might help him to resign himself to the reclusion that Tao reveled in. The larger question broached in this chapter is this, then: Do the consolations of emulating an earlier author ever suffice to make up for personal isolation? Judging from Su Shi's case, we would have to say "no."

When I was a graduate student, Wolfgang Bauer came out with his lengthy tome entitled *China and the Search for Happiness: Recurring Themes*.[8] My own search, though in some ways indebted to his, has followed the traces of *le* wherever they led. My analysis rests on retranslations of a wide array of relevant sources, received and excavated, in a studied refusal to play it safe. My reading of the *Zhuangzi*, for example, moves outside the familiar territory of the so-called "Inner Chapters" to take stock of the entire thirty-three chapter compilation, and my research into Su Shi's matching poems complicates the

INTRODUCTION

romantic portrait of a transcendent Su favored in most secondary literature today.

To give readers a brief taste of some early pleasure theories this book chooses for its subject, let me quote passages from two long speeches ascribed to the supremely effective prime minister Zichan 子產 of Zheng (act. 542–522 BCE), which touch upon the main themes of many discussions in early China about pleasure-taking. In the first, Zichan explains to an envoy from the more powerful state of Jin how diplomacy was better managed in the glorious days of yore:

> Nowadays, our humble domain is small and placed among great domains that make insatiable demands on no set schedule. For this reason, we [in Zheng's leadership] dare not take our ease, but instead must try to muster all our meager resources for meetings and court visits.... I have heard, when Lord Wen of Jin was covenant chief, his palaces were small and low, devoid of terraces and towers affording fine prospects.... Hosts and guests shared their cares and pleasures. When something came up, Lord Wen attended to it, instructing guests in those matters of which they were ignorant and taking care of whatever they lacked. The guests, upon arrival, felt as if they were coming home. How could there be any troubles or calamities?... Now your Tongti Palace extends for several miles, while even princes are lodged in abodes fit only for servants.[9]

In the second, Zichan expands upon the theme: "It is very hard, in truth, to be entirely without desires. Let all get what they desire, so they can focus on their assigned tasks and concentrate on completing them."[10]

As Zichan's view gradually unfolds, it becomes clear that his policies of governance are predicated upon calculations of pleasure. All people are bundles of desires, he says, though the type and force of the desires that drive a particular individual are functions of that person's character and inclinations. To thwart people's desires does no one any good at all. Far better, then, to learn how best to accommodate each person's desires in such a way that, at a minimum, they are productive members of society doing the least harm to others in the community. In that way, a widespread sense of satisfaction will unobtrusively reinforce communal ties. Meanwhile, inculcating a desire to emulate worthy models can alter undirected and unbridled impulses and produce more constructive dispositions and inclinations.[11] Zichan's speeches make it clear that refining the desires of power holders and commoners alike, far from being an inconsequential matter best left to the discretion of individual ministers and rulers, is their fundamental business.

Every student of early and middle-period China recalls similar passages detailing pleasure's efficacy. The excavated and transmitted literature — whether the standard histories, medical treatises, "philosophical texts," or bawdy poems — abounds in talk about pleasure, relating the physiological processes entailed in pleasure-giving and pleasure-taking to patterns in the larger realm or cosmos. To the modern reader, the sheer pervasiveness of the pleasure discourse in early writings is startling on first reading, for Zichan is but the tip of the iceberg. Further reflection leads us to see the prevalence of this discourse as great good sense, for nearly every piece of extant writing reflects the preoccupations of the governing elite, who saw that "if power is pleasure, then the way pleasure is managed has direct consequences on the nature of power itself."[12] This book represents a first attempt to build upon recent scholarly insights regarding pleasure, vitality, commemoration, cultivation, insight, and spectacle in order to open new avenues for research. To aid in that exploration, I provide a historical narrative, part of it necessarily speculative, proposing reasons why the pleasure theme arose in classical Chinese at a specific time to address a particular set of problems.

Historical Background

Already in the Zhanguo era (475–222 BCE), treatises by would-be advisors to thrones advocated several ways to increase one's security in pleasure-taking, none of which was particularly easy to follow. The main directives for instilling a greater single-mindedness of purpose were: reduce both the number of one's desires and one's degree of dependence upon others for their satisfaction[13] and thereby decrease the chance of being harmed by a profusion of seductions and allurements;[14] refine and so redirect one's desires to the "higher" (and fewer) sorts of pleasures derived from connoisseurship, even if such refinement does not automatically preclude dependence upon others to achieve one's heart's desire, as in career advancement; and secure one's pleasures by sharing them with others in the belief that pleasures taken in common mitigate envy and resentment.[15] Simply by sharing pleasure with their underlings, those in power might forge stronger bonds within their communities, allowing power holders to savor their pleasures in far greater security. Tighter bonds, in turn, might prompt still more community members to conceive and confer pleasures on behalf of the group—through cooperative ventures or the provision of communal festivals, performances, and spectacles, for instance. This last rationale, whether expressed or tacit, underpinned a great many political calculations.

Why the apparently sudden emphasis on careful or delayed pleasure-taking during the centuries before the common era? I suspect that the vast scale and unprecedented scope of sociopolitical and economic changes during that time elicited two questions: What form of equitable distribution of resources best guarantees a state's stability, and what methods of rule best enable the expanding states to integrate newly conquered populations?

On the question of equitable distribution, admittedly, the available sources consist mainly of recorded pieties. Yet the early texts would have readers believe that before the decline of the Zhou political order in the eighth century BCE, the sumptuous sacrifices offered to the royal ancestors, followed by the division of the sacrificial meats among the descendants, had distributed goods and prerogatives in ways that were generally conceded at the time to be equitable—at least by members of the governing elite. With each member of that elite partaking of the numinous life force contained within the sacrificial meats, each sacrifice served as an outward sign of the inner commitments binding the partakers to their same clan or body politic, notwithstanding their potentially disparate interests.[16] But in the wake of the demise or usurpation of many noble houses, the associated sacrificial orders no longer sufficed to confirm the basic laws of hierarchy, reciprocity, and equitable exchange so vital to any cohesive community. Political elites had to devise entirely new, reliable modes of fair distribution if they hoped to attract the necessary men and materiel to their service. Gradually, the local communal feast, offering a different model of sociability and rewards, came to rival or transcend in importance the blood sacrifices made by ruling lines during ancestor worship, the feast having the signal advantage of being far less likely to entail huge losses of life, including human as well as animal sacrifice.[17] Meanwhile, loyalty to a single superior or group of superiors within the noble lineage (promoted through ancestor worship) yielded to ideals requiring service to a larger community or even to the known world "under Heaven."

Would-be and actual unifiers wanted to instill allegiance and enforce control within vast new populations not persuaded by hereditary ties or local custom to uphold the relatively small kin, surrogate-kin, and cult groups associated with the court. The formal *ming* 命, charge or writ, recorded on Western Zhou bronzes had once certified the obligations due the ruling house by a few allied families.[18] Later, the Chunqiu blood covenants (*meng* 盟) bound far greater numbers of aristocrats and their dependents in temporary agreements. But by the dawn of the fifth century BCE, in Zhanguo times, any state determined upon conquest had to sponsor and direct much larger (even overlapping) networks of loyalty in the social, political, military, and economic spheres, networks that would be capable of mobilizing assorted talents to devote their best efforts and those of their men to the conquerors. The conquerors' usual rationale for this was disarmingly simple: the stability enjoyed by the principal unifier best guaranteed the stability of all other social units, public or domestic, in his realm and nearby. For without stability, no pleasures could ever be secure for any resident, high or low.

In the end, the ruler's force majeure could go only so far in stabilizing the state. The anxieties, insecurities, and disaffection experienced in the sociopolitical order could be laid to rest only if the antidotes spoke to the strong desires felt by all members of the ruling elite, at every level, to preserve and maximize their prerogatives, perquisites, and pleasures while addressing the precarious living conditions of some of their subjects. As one celebrated master put it, "in peaceful times, one may not overlook dangers, nor in secure times forget perils."[19] Effective persuasion pieces had to allay their fears and insecurities in an era of rapid social change and ideally assuage the deep sense of unease that beset many of the most successful in middle age.[20] The idea of converting the consuming pleasures into sustaining ones succeeded so brilliantly at courts all across the central states of the North China Plain, I wager, because it precisely suited the ruling elites' own experiences and hopes.

Throughout their disquisitions on pleasure, the court persuaders made much of a seemingly self-evident truth that contained more than a bit of paradox: most of our present delights merely taunt us by their brevity; even as we indulge ourselves, we suffer from the anticipation of their loss. Additionally, the craving for pleasurable stimuli inevitably generates competition, and the sense of unceasing competition grows more arduous and more frustrating with age. For those who have worked hardest to build constructive orders commonly expect to bask in the results of their achievements, but sadly, as the self-aware ruefully observe, continually fending off all rivals in the bloody contests over territory or material goods requires the stamina of youth. Adding insult to injury, no one can possibly attain as much as they desire before they die.[21] Only those who have managed somehow over the years to reduce or refine or stabilize their pleasures can hope to retain a measure of self-regard. For good reasons, then, the cautious use of pleasure as a way to preclude disorder in the realm and in the individual promised to offset the strange melancholy that pervades the prime of life, especially in those who possess a surfeit of material goods or power. In that respect, the court persuaders' disquisitions on pleasure did far more than provide simplistic answers to two of the first questions most frequently asked: "What pleasure is there in being a prince, unless one can say whatever one chooses, with no one daring to disagree?" and "How can a person in power not only be happy, but happier than other people?"[22]

In this setting at court, talk about pleasure was liable to be pragmatic, geared to the here and now.[23] Only a few assertions about human nature and motivation commonly preceded a persuader's advice to a patron or student about the wisdom of present or intended social engineering policies or the benefits of present and ongoing personal and social cultivation. Nearly all extant accounts omitted systematic treatment of, say, the origins or qualities of pleasure, though our persuaders most probably considered these, if only to equip themselves with defenses against their opponents' rhetorical jabs. Doubtless, those with the luxury of time on their hands asked whether the sensation of pleasure inheres more in an object, in the activity itself, or in the capacity for pleasure developed by the owner or actor.[24] Some surely noticed that the experience of pleasure, no less than of physical pain, is fundamentally inexpressible and therefore potentially isolating. But persuaders were obliged to forgo overspecificity in outlining propositions about pleasure, lest bored or inattentive court power holders start to quibble over minor points or lose the thread of an argument. Accordingly, the surviving passages on pleasure seek to nudge powerful members of the governing elite toward improving their policies and personal modes of behavior, thereby promoting greater contentment and insight among the governed, which in turn would likely preserve their persons and properties from harm at the hands of a disaffected populace.

In proposing the best possible course of action, most of the persuasions ascribed to the late Zhanguo, Han, and immediately post-Han thinkers kept well within a few accepted "talking points." It sufficed for the persuaders to allege that the probable sources of pleasure are easily recognized; the preponderance of human activity consists of the pursuit of pleasure; each person hopes to maximize his or her own opportunities for pleasure-taking, although their objectives may differ; so when humans act "recklessly," it is usually because, having mistaken the nature and hence the effect of their own actions, they have miscalculated the odds that their actions will conduce to pleasure.[25] By such reasoning, they concluded that if the body and the body politic are to continue to flourish, it is crucial for those in charge to distinguish sustaining from consuming pleasures. Naturally, they emphasized that long-term, "sustaining" relations nearly always yield appreciably more satisfaction over the years than impulsive, immediate consumption. So although in their writings no abstract theories assign a fixed, quasi-numerical value to each type of enjoyment experienced by every person, the pleasure discourse figures significantly in the priorities the persuaders allotted to different policies and commitments. In this regard, they took into account the duration and intensity of specific pleasures, as well as the different pleasures aroused by anticipation, by experience, or by memory.

By postulating a generalized "human nature" or "human condition" shared by the ruler and his subjects, persuaders could posit methods whereby the elite might induce a sense of community among their social equals and inferiors in hopes of forestalling all manner of destructive behavior. Analyzing human nature, in other words, was but the prelude to determining effective motivation. "For just as the body had its interests, so did the realm have what spurred it on, and discernment in these things meant just estimates about relative importance."[26] Since the crudest of those at the apex of power were often at a loss to figure out what ratio of carrots to sticks would best motivate underlings and mobilize resources, the Classics and masterworks dealt extensively with that question, at least since the time of Mozi 墨子 (ca. 470–ca. 390 BCE) and his followers. Certainly, the later Mohists and a host of Zhanguo masters (*zhuzi* 諸子) saw the

INTRODUCTION

focus on pleasure and desire as fundamental to every human being, even if few in society had the luxury to choose freely among rival "goods" and courses of action. For those masters, "to act on account of something is to take into account all one knows and judge that something [as in a scale] by one's desires."[27]

In the classical political rhetoric of pleasure-taking, power holders planning to unify the realm or to maintain their standing at court were to supplement the old aristocratic forms of excellence (prowess in warfare, filial conduct toward the ancestors, and practical shrewdness) with the new virtues of fair dealing and self-restraint. One basis for good rule lay in assessing and dispensing the proper shares of access to pleasure through ritual according to the contributions made to the general welfare. The ruler's own model of self-restraint validated the allocation of favors, so the wise ruler would not let inborn inclinations to delight, pleasure, worry, and sorrow move him (*dong zhi* 動之), except "by rule" (*yi ze* 以則). Such rituals tended to confirm the status quo among the elites, whether hereditary or not. But no abstract ideas in support of the ruler's or ruling elite's control could have been imposed on "those below" unless they struck them as admirably suiting prevailing conditions. As I see it, throughout the classical period, in order to feel secure, people at every level of society sought as best they could to place themselves firmly within webs of mutual obligations (*do ut des*), signified and cemented by regular formalized exchanges in the forms of gift, tribute, and sacrifice.[28] Such exchanges, rendered highly visible at intervals by specific changes in the form of the rituals, demonstrated to potential friends and allies, no less than to oneself, the reliable nature of the protection afforded those within a web of obligation.[29] For at the very same time that the person of high status reaffirmed his protected status through public or semipublic acts, those outside the web were put on notice that it would be foolhardy to harm anyone who could call upon the collective strength of the communal network. Obviously, no small store of wealth was needed for the frequent outlays in ritual, but status and safety were secured less through force or sheer spending than through the periodic public manifestations of loyalty by family members, allies, and subordinates. Hence the continual reiteration in the treatises of the period that other people constitute one's own chief security. At the same time, when honor and glory (*rong* 榮) are in the gift of the people before whom they are paraded, then honor and glory can be withdrawn by those same people swiftly and absolutely, as many biographies in Sima Qian's 司馬遷 *Shiji* 史記 and the "Xici" 繫辭 tradition to the *Changes* (*Yijing*) classic so poignantly attest.[30]

It would be hard to overestimate the pervasiveness of the webs of trust formed within and beyond polite society through public display, webs that bound the living and the dead and also — what is infinitely more difficult — people of quite different status. From the court on down, provision was made for nearly all levels of society to experience, directly or indirectly, some of the pleasures of public exchange and display. The extant writings of the period show the royal courts' quite intentional deployment of visual display to render palpable the protective bonds ordering society. And the archaeological record leaves no doubt about the sumptuousness of court extravaganzas, orchestral performances, royal progresses, massive building projects, and spectacles. In illustration, I cite just one anecdote dating from the first years of the Han empire: shortly before 200 BCE, the new chancellor, Xiao He 蕭何, set about building palaces, arsenals, storehouses, and gate towers on a lavish scale while war still raged outside the capital. Xiao defended his priorities on the grounds that "if the true Son of Heaven does not dwell in magnificent quarters, he will have no way to display his authority or establish a base for his heirs to build upon."[31] Perhaps the most striking change documented by classical archaeology, then, is the shift from highly circumscribed rituals conducted for very limited audiences to increasingly splendid displays intended for ever larger groups of onlookers,[32] wherein suasive authority was said to be lodged in four ceremonial aspects: the insignia (badges, seals, tablets, and weapons); the dress (clothing, caps, and coiffure); the demeanor (gestures indicating the degree of poise); and the rhetoric (forms of address and discourse).

Until about 140 CE or so, the capacity of this social display culture to conflate the rewards for public service with the pursuit of domestic pleasures and personal interest seemed one of the best means of insuring stable dynastic rule.[33] But somehow, by the mid to late Eastern Han (roughly by 140 CE), through social processes still not fully understood by historians of the period, elites turned inward. No longer trusting to the expansive webs of relations to protect them and their families from harm, they took to rearming themselves. The signs of collapse were everywhere. Professional and private estate armies replaced conscript armies in the interior.[34] In the absence of adequate cadastral surveys by the imperial court, taxes were no longer gathered fairly. As peasants fled wars and the tax men, they sought protection from local strongmen, becoming tenant farmers on large estates and swelling the ranks of their armies. With the growth of large, virtually independent estates, the court, protected at one point by a mere four thousand guards, lost secure access to the best men in service at court, because they competed with the estates. Less supervision over local governors and their administrations and breakdowns in the court recommendation procedures ensued, so much so that at one point, the court complained that the provinces were sending men who could not read or write to stock the ranks of its bureaucracy.[35] The ordinary business of the court foundered, with failures of water control a symptom of the court's dual inability to organize labor details and to extract the necessary funds from the provinces. Invasions on the borders multiplied as soon as local warlords ignored a central command. Even before 184 CE, when a massive peasant rebellion rocked the empire, the inadequacies of the court to meet the challenges became abundantly clear to all. As Étienne Balazs pointed out decades ago in his essay "Political Philosophy and Social Crisis at the End of the Han," the hundred years from 150 to 250 CE "exerted on China's future development an influence no less important than that of the third century BC."[36] Both periods saw disorder, injustice, and monstrous disparities of wealth, an increasing articulation of the sense of crisis, and a great diversity of proposals for rectifying the defects and reversing the degeneration of state and society. But whereas thinkers in the Zhanguo period had anticipated unification and planned well for the new *pax Sinica*, by the end of the Eastern Han, the most acute thinkers, having lost their serene trust in the ability of civil mechanisms to order society, turned their attention to devising tighter controls.[37]

In hindsight, we see that the practical problems associated with the application of the pleasure theories and display culture had made a mockery of their theoretical elegance. For one thing, the spiraling costs of competitive displays, each more dazzling than the last, led elites to extract ever greater sums from their social inferiors, in this way undermining any potential the display mechanisms had to unify different segments of society by offering many levels of satisfaction. The gap between "names" (titles, assignments, reputations, social roles) and "actualities" widened, in consequence, whence the Eastern

Han complaint that "it is difficult to get names and actualities to correspond" (*ming shi nan fu* 名實難副).[38] The extravagant rhetoric fashioning the emperor as supreme model for all the virtues seemed ludicrous when countered by a string of underage, incompetent, or disinterested rulers. And from the standpoint of the ruling house, it was regrettable that regional powers and local cults, in a refeudalizing era, could so easily appropriate the display mechanisms once emanating principally from the center, to the detriment of the singular authority of king and capital. Meanwhile, the vast estates of the Eastern Han magnates, enclaves of mock-courtly life in a sea of worsening impoverishment, exacerbated perceptions of sociopolitical injustice and fomented rebellions on an empire-wide scale. The formation — not coincidentally — at that point of Daoist religious organizations, which stepped in to fill the vacuum left by the court's withdrawal from the provinces, further divided loyalties and complicated the entwined notions of security and pleasure, even as they made for unprecedented modes of local spectacular display.[39] The widespread interest in Buddhism evinced after the fall of the North China Plain in 316 CE accelerated such trends by calling into question the very reality and consequences of pleasure-taking.

But it was not a case of "Après ça, le déluge." In lives once again nasty, brutish, and short, pleasures seemed more fragile, more fraught, and more difficult to secure, but all the more valuable, for some. Thus the post–Eastern Han Periods of Disunion, with their succession of dynasties that looked to the Western and Eastern Han for models and institutions, hardly prevented those enjoying high cultural literacy from waxing eloquent on the subject of pleasure they found so arrestingly laid out in the early Classics, masterworks, and histories. The survival and application of the pleasure rhetoric of the Zhanguo, Qin, and Han masters were then assured in latter-day situations.

So despite some shifts in the basic presumptions of the cultural elites from the Zhanguo era through the Northern Song, the main threads of sympathetic response theories did not fray badly for well over a thousand years. Readers continued to imagine official and unofficial, public and domestic exchanges, including encounters between fine writer and ardent reader, as embodied fields of relations contributing to the continual process of construction of the cultivated person[40] while confirming the beauty of the densely patterned cosmic fabric within which people lived their lives.[41]

INTRODUCTION

But nearly everything about the discourse of pleasure would have to change once Zhu Xi 朱熹 (1130–1200) and his True Way Learning came to dominate Chinese tradition, for Zhu felt compelled to condemn or erase any older traditions that might impede his ascendancy.[42] So whereas earlier thinkers and writers looked to empathetic reading of the right sort to afford glimpses of the unfolding, unremitting transformations, Zhu Xi and his disciples deplored fond musings on wondrous particulars. (That may explain why Zhu favored his Four Books, to divert attention from the historical *Annals* and *Documents* classics supplying abundant evidence for the wide range of human propensities to messy worldly engagement.)[43] Asserting that the perfectly resonant nature of a sage like himself, pure in mind and body, allowed immediate, unmediated, and comprehensive knowledge of the entire universe, past and present, Zhu demanded that his followers direct their gaze as much as possible to the ineffable grand totality, rather than luxuriating in manifold specificities. As Zhu wrote in the late 1160s, "Let me propose that all under Heaven is just Heaven's Pivot giving life to things (*tianji huowu* 天機活物).... Now how could there be a particular, within time and space, that can be distinctly named," apart from and outside this single flow?[44] Reverence for abstract and unseen principles was to replace pleasure in the near to hand and palpable; by Zhu's theory, pleasure was too subversive, anyway, for lesser men left to their own devices to cultivate.[45] The search for pleasure went underground, in consequence, becoming more transgressive in the process. And so my story, which will, in essence, bear witness to the sustained and sustaining benefits secured through constructing pleasurable relations with people and things in all their distinctive charms, ends necessarily before Zhu and his adherents came to dominate mainstream thinking in China.

Larger Implications

A tired series of dichotomies has occupied far too much of the Sinological community's attention, including inner versus outer, subjective versus objective, pragmatic versus ethical, truth versus rhetoric, nature versus culture, emotion versus reason, and mind versus body. What blessedly seems to have run its course among astute readers is the simplistic impact-response model or a variation thereon, the sender-medium / percept-receiver model. (Unfortunately, Orientalists and self-Orientalists still cling to the gratifying notion of the Western impact on a passive, receptive China, one instance of Jack Goody's "theft of history.")[46] If this book breaks new ground, it charts unfamiliar terrain via more nuanced translations of both familiar and seldom-read texts that imply the deep interpenetration of fact and value, objectivity and affect. The wisdom of consulting the Ancients should be evident in our more distracted contemporary age of anxiety.[47] But I do not wish to contend for "relevance." As a historian, I aspire mainly to acquire fuller evidence, in the firm belief that the historian's task is to "reveal the unpredictable contours of this polygon" that we call "human experience" and "to restore their original silhouettes" to events and ideas that have been "concealed under borrowed garments."[48] If some portion of the litanies celebrating categorical alterity and the "clash of civilizations" is jettisoned, so much the better.

That said, the payoffs from attending to the early Chinese sources seem huge. For example, the stipulation of the precise circumstances for pleasure-giving and pleasure-taking neatly obviates the knotty Anglo-American philosophical problem of how to get from "is" to "ought,"[49] for the sources are at once highly contextualized and praxis-guiding, commonsensical (designed to mirror the world that is), and regulative of human practice. And insofar as they did not hazard a host of unprovable assertions about social units or the cosmos, the early advice comes to today's readers without theoretical superfluity and entanglements. As some have argued, the Ancients were in appreciably better shape than we moderns, if only because they did not cordon off moral from practical considerations when deliberating. So it seems high time, past time really, to recall the unique potentials invested in the word "pleasure" itself.[50]

CHAPTER ONE

Coming Attractions

"Pleasure," wrote Oscar Wilde, the nineteenth-century English aesthete, "is the only thing worth having a theory about." More recently, Andre Malraux's *The Temptation of the West* poses the question, "Of all his ideas, is there any one more revealing of a man's sensibilities than his concept of pleasure?"[1] Either formulation could be plausibly ascribed to the most important classical masters in the region we now call China, since they deemed pleasure one of the most effective rhetorical tools to motivate right action, as each defined it, as well as to discern a person's character. Nearly all of the thinkers from the Zhanguo period (475–222 BCE) through the eleventh century addressed the problem of converting the *consuming* pleasures — those that expend vast wealth, time, and physical energy — into *sustaining* pleasures that could support, rather than deplete or corrupt the polity, the family, or the body.[2] The vast majority held that the sociopolitical realm must accommodate or transform the innate desires for pleasure into desires to do good and — no less pressingly — encourage people to delay certain types of gratification for the sake of future, more secure and lasting pleasures. For the ruler and his ministers intent upon promoting such pleasurable orders through institutions and edifying behavior, the chief task was to devise appropriate policies, because a particular pleasure's pursuit could increase or diminish the health of the body and the body politic and bind subjects more closely to the throne or risk dangerous disaffection. And since health tended to imply stability and duration, more often than not, the relative value accorded a specific pleasure was loosely gauged by its potential for prolongation. In articulating modes of appropriate pleasure-taking, thinkers wanted not only to devise methods to prolong a given pleasure, but also to augment the security of the ruling house and its leaders.[3]

The aims of this chapter are but two: first, to explain the vocabulary considered before adopting the translation of "pleasure" so as to highlight the particularities of the pleasure rhetoric against the larger backdrop of related concepts in classical Chinese, and second, to discuss the physiology of pleasure as sketched in the relevant medical and cosmological theories of the period. Because the English words I necessarily employ have themselves long and convoluted histories, my analysis at points will seek to distinguish the Chinese pleasure theories from the much better-known theories of classical Greece and Rome and early modern philosophy. All this by way of preparing for the six later chapters, each of which examines proposals by different thinkers concerning preferred ways of giving and receiving pleasure.

The Vocabulary of Pleasure

The word I translate as "pleasure" is *le* 樂, denoting an action, pleasure-seeking or pleasure-taking, rather than a state such as "happiness." (More on this below.) To describe the emotions, drives, and sensations, the classical thinkers used a host of terms running the gamut from "desire" to "delight," from "gratification" to "fondness."[4] Many, if not all of the classical masters rely on particular terms to distinguish logically the pleasures that bring deep, rich, and enduring satisfactions (*le*) from the fleeting emotions or those potentially destructive drives deemed "excessive" and "indulgent."

Because their persuasions seldom highlight the gap between the phenomenon that imparts pleasure and the pleasure-taking itself,[5] they help situate the pleasurable sensations in the act of making contact and finding it good. (For that reason, they conceive no pleasure to be purely sensory or purely mental, without the engagement of the evaluating heart and mind.)[6] On the relatively direct experiential sensations, thinkers superimposed a category of relational pleasures, said to require more continuous and more insightful perceptions by the connoisseur of pleasure. Such relational pleasures included, for example, inducing good men to serve in office through suitable politicking; cultivating the fine arts of social intercourse; taking pleasure in virtuous conduct or in music; taking pleasure in one's profession or in family traditions; and taking pleasure in conforming to moral imperatives or cosmic operations.[7] (The expectation that one association of *le* would be romantic or sexual love betrays a modern sensibility.)⁸ Far more than mere sensual gratification or pride of possession, the relational pleasures presupposed an ability to discern the long-term utility and value of things, conditions, and people; seeing beyond the immediate, and the distractions and impulses it engenders. They took a substantial share of curiosity and imagination, astuteness and self-restraint, not to mention commitment and grit. For in not a few cases, to procure a relational pleasure, strong impulses toward quick gratification must be delayed or checked. Gratifying one's taste for exquisite foods and wines may in no way impinge upon pride of ownership in a fine stallion, but routine gastronomic overindulgence likely slackens one's dedication to prudential action.

While the classical writings are not entirely consistent in their use of *le*, over time, master persuaders came more and more to pair the verb *le* with noun objects of consequence, such as *you* 友 (intimate friends), *tian* 天 (Heaven), *de* 德 (edifying, graceful, and charismatic acts), or *ye* 業 (one's family profession or heritage).⁹ Significantly, the word most often paired with *le* is *an* 安, "to secure X" or "to feel secure in X," where X may be a situation or action, and the words most often used as antonyms for *le* are *you* 憂 ("anxious, worried, or concerned"), *wei* 危 ("in danger" or "apprehending danger"), and *ai* 哀 ("grieved by a loss").¹⁰ (See below.) These concept clusters reveal that the classical theories in China were more apt to contrast pleasure with insecurity than with pain, unlike their Western counterparts. (Certainly, an equivalent for physical pain or psychic misery in classical Chinese exists, *tong* 痛 / 恿, but pleasure theories do not foreground it, because some pain is unavoidable.) Essentially, if the conventional objects of desire are to be *fully* enjoyed, many thinkers argue, the subject must know that they do not threaten his person, his livelihood, or his community, now or in the foreseeable future. For that reason, precarious or fleeting pleasures do not warrant the term *le*.

Early in my work, I considered three possible translations for the word *le*: "pleasure," "happiness," and "joy." For four negative reasons, I chose not to translate *le* as "happiness," to the consternation of multiple colleagues. The ubiquity of the anodyne Happy Face in America has ruined the word for some Americans, and worse, confirmed a link between happiness and psychosocial clinical normality. Also, for much of Euro-American history, "happiness" meant "favored by fortune" (a connotation grossly inappropriate in the Chinese context, as will become clear). Moreover, by 1725, happiness was associated (in the work of Francis Hutcheson and later writers) with the utilitarian concept of the "greatest happiness of the greatest number," hardly a theory found in early China, even with Mozi.[11] And finally and most importantly, "happiness" refers to a state of being, whereas the Chinese graph for "pleasure" implies actions: pleasure-seeking, pleasure-giving, and pleasure-taking.

But more positive reasons call for the word "pleasure." Li Zehou, the premier philosopher of aesthetics in contemporary China, has made the case — to my mind, superbly — that the early Chinese men and women of letters came of age in a culture uniquely "alive to pleasure" (*legan wenhua* 樂感文化).[12] The English word "pleasure" encompasses many feelings: those of enjoyment, gratification, satisfaction, sensual or sexual contentment or stimulus, even elation, among others.[13] But pleasure is always a response to something or someone present or in the mind's eye; it cannot be experienced in the mind or body in true isolation. The early thinkers contended that the keenest sense of pleasure comes from long-term relational pleasures, presupposing steady practice and the patient accumulation of "embodied" knowledge about other people and surroundings. "Immersed" 潛 is the Chinese word, the antithesis of half-attending. As Mihalyi Csikszentmihalyi's discussions of "flow" and the poet Donald Hall's *Lifework* suggest, full enjoyment in an activity often, if not invariably, entails total absorption in it.[14] Importantly, too, "pleasure" is the only English word capacious enough to allow for the complex bodily processes registered in the senses and emotions and in the sensitive heart and evaluating mind.

The complexities of those processes becomes evident when one sees the early thinkers in their pleasure calculus generally taking into account the different time frames for pleasure (pleasure in anticipation, in experience, and in retrospect),[15] the duration and intensity of the pleasure, and the person's changing sense of belonging and identity vis-à-vis the communities in which he or she participates. The pleasure calculus recognizes no arbitrary distinction between "sensory pleasure" (percepts, experienced feelings or sensations) and "attitudinal pleasure"[16] (that is, the consciousness of taking pleasure in something or someone). Instead, Chinese theories of the body's physiological processes, construed as one locus among countless sympathetic resonances reverberating throughout the cosmos, tend to entangle sensory experiences and conscious attitudes, autonomic movements and intentional awareness, motivations and actions, in the social world, as well as in the imagination (Figure 1.1).

Such theories deeply embed each person within a series of relations whose members ideally sustain and certainly transform one another. So by the end of this book, I hope to persuade readers that my true subject is pleasure and related propositions about the best conceivable lives, with "best" meaning "most sustained and sustaining" and "most fulfilled."[17]

In English, however, if we accept the central hedonist contention that the "good life" is a life full of pleasure, that contention can convey at least five different ideas: that "good" is identical with "morally good" (as if *that* concept were clear!);[18] that "good" means "causally good" (having good effects); that "good" means aesthetically pleasing or beautiful, albeit with elements of tragedy or pain; that the "good" is embodied in exemplary people; or that "good" simply means "good for the one living that life" (no matter how much selfishness or depravity a person requires for "personal welfare" or "flourishing"). Add on the potential conflicts when local, global, actual, and hypothetical preferences are factored in.[19] So how are we to assess the competing claims of such seemingly absolute goods as virtue, pleasure, knowledge, flourishing, and justice, or, for that matter, the general good against an agent's own "pleasant life"?[20] To understand what follows requires probing the vocabulary that Euro-American academia still adopts reflexively. For that reason, I briefly rehearse some of that vocabulary below.

Pleasure, Not Happiness or Joy
As one writer insisted, "Don't mistake pleasures for happiness. They're a different breed of dog."[21] Certainly, upon first reading, "pleasure" may sound disreputable,[22] whereas "happiness" sounds morally acceptable, thanks in part to the famous, if ill-understood phrase in the Declaration of Independence guaranteeing the citizen's right to "life, liberty, and the pursuit of happiness."[23] Perhaps because of the Declaration's lofty aspirations, few have troubled to think through the connotations attached to the words "happy" and "happiness."[24] True, there is an idea behind Bhutan's Gross National Happiness index, that money can't buy enough contentment, just

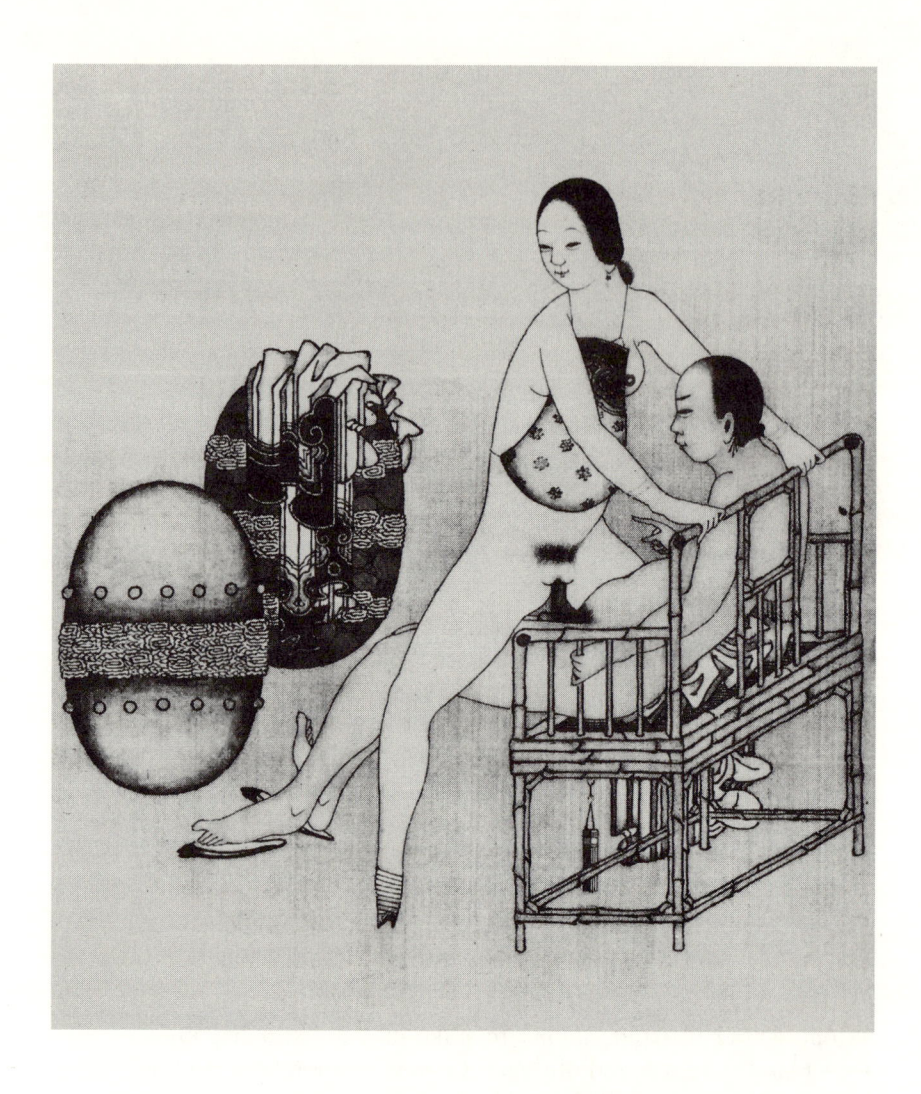

Figure 1.1. Erotic Chinese wedding manual, after Phillip Rawson, *Erotic Art of the East* (1968), plate 189. Unidentified, except for "Chinese album painting, nineteenth century."

This erotic scene reminds us that every sexual act, like every social interchange in China, as a comingling of yin and yang *qi*, has a cosmic dimension, here indicated by the cosmic eggs incongruously placed in the bedroom, with one egg sporting drapery suggesting mountains and clouds, with mountains generating the "clouds and rain" that signify sexual fluids.

as ideas spur UC-Berkeley's "Science of Happiness" project and the academic revival of interest in phenomenology. But evidently, the modern preoccupation with happiness is a symptom of the industrializing and industrialized world. To today's smart thinkers who equate "happiness" with "being satisfied with one's life as a whole" (a state of being, rather than a response to a given stimulus),[25] the early thinkers who are my subject would have replied that seeking such an ambitious and unattainable goal would likely end in dissatisfaction and greater unhappiness. Indeed, sheer overreach may have brought about the current Euro-American paradox; ergo talk of happiness is most prevalent in the very populations heavily reliant upon antidepressants and opiates.[26] It is moreover hard to ignore the disturbing racist and culturalist overtones of recent "Happiness" projects, lodged in the universalist presumption that all cultures everywhere have replicated the same set of emotions and emotional triggers as US citizens today. As readers will discover, the vocabulary for several American virtues relating to happiness (the virtue of "cheerfulness," for example) do not seem to exist in the classical writings in China, though an absence of literary evidence does not insure that cheerfulness was absent from daily life.[27]

The Ancients with clear-eyed, even brutal frankness noted the manifold ills to which all people are prey: sickness, decrepitude, death, natural catastrophe, and slander among them. Fully cognizant of the level of destruction that ill luck, bad timing, or vengeful powers can wreak upon the innocent, the thinkers discussed in these pages advised followers to devise and adhere to programs and practices that promised a fair chance of shielding people, as much as humanly possible, from the worst slings and arrows of outrageous fortune. Each person, they said, can learn to give and take pleasure, despite the calamities that beset ordinary lives. In addition, those already favored by fortune may learn how to magnify their blessings.[28] The closest counterpart to the modern tropes linking autonomy to pleasure is the continual profession by Chinese elites of their determination to avoid enslavement to other people and things. This admirable clarity about life's constraints—so greatly at odds with American positive thinking—precludes mindless optimism.[29] In consequence, no writings in classical Chinese, so far as I know, denote or connote "happiness," either in its older Western sense of "favored by fortune" or in its modern connotation of "a free state of blissful autonomy." The thinkers reviewed in this book would mock the fond hope that one can "stumble upon happiness."[30]

"Joy" is no more suitable as a translation for *le*, since "joy" too often implies the "disembodied," "detached," and "elevated" in Western languages, given its old associations with religious ecstasy and more recently with "sublime" or "transcendent" feelings. (Of course, the phrase "joie de vivre" fits the Chinese rhetoric very well.) Use of the word "joy" is apt to privilege the interior and the mind, making it quite unsuited to the classical theories under examination. Not surprisingly, "joy" as a verb, "taking joy in," cannot take mundane objects such as "rich food" or a "quiet smile," whereas the objects of *le* are ordinary and all around us.

Pleasure, Not the Converse of Pain
No matter how sustained and sustaining, fulfilled and fulfilling a particular life, each life knows a heap of trouble: a sciatic nerve acts up during a convivial dinner; one's favorite colleagues, relatives, or friends sicken and die; the cat is run over; the newspaper is replete with tales of poverty, destruction, and hate crimes. For a person who considers herself part of the world and sensitive to the plights of others, such events inevitably usher in feelings of misery, bleakness, and depression, apart from bodily pain. Entirely mindful of this, the early Chinese theorists did not advocate avoidance of pain, although medical practices sought to alleviate it. This is one crucial difference between Euro-American ideas and those in China discussed here.

A second: current philosophers in the West, intent as they are at describing universals, have responded predictably to the vagaries of modern life by using "pain" as an umbrella term, defining it as any

> unpleasant experience or feeling: ache, agitation, agony, angst, anguish, annoyance, anxiety, apprehensiveness, boredom, chagrin, dejection, depression, desolation, despair, desperation, despondency, discomfort, discombobulation, discontentment, disgruntlement, disgust, dislike, dismay, disorientation, dissatisfaction, distress, dread, enmity, ennui, fear, gloominess, grief, guilt, hatred, horror, hurting, irritation, loathing, melancholia, nausea, queasiness, remorse, resentment, sadness, shame, sorrow, suffering, sullenness, throb, terror, unease, vexation, and so on.[31]

As noted previously, the early thinkers consistently opposed pleasure not to pain, but to anxiety or insecurity. "Anxiety," "cares," or "worries" (*you* 憂 in classical Chinese) are equated with being "uneasy" or "insecure" or "unsettled" (all *bu'an* 不安). As one early text puts it, "human nature is incapable of feeling pleasure in an insecure situation, yet one cannot gain any benefit from what one does not take pleasure in."[32] Thus, the classical Chinese texts name a perceived lack of security as the chief barrier to experiential pleasure and "security" and "ease" (*an* 安) as the chief preconditions for maximal well-being. Not uncommonly, the excavated and received texts leap quickly from *bu'an* ("insecure" or "unstable"), to *bule* ("not pleasurable") feelings, to the person who "lacks charisma" (*wude* 無德), as if such ties were commonplaces.[33] "Pleasure" is the right word to translate *le*, not only because it so clearly contrasts with negative feelings ranging from "aggravation" to "despair" to "distaste for the graceless," but also because its attractions are not necessarily only of the moment and entail the senses without excluding heart or mind.[34] Keeping in mind this heady refusal to understand pleasure as simply the absence or surcease of cares or pains affords a potential for pleasure to be much more active, much less self-preoccupied and more participatory, while much more selective.

Pleasure, Not Hedonism
Many students of China feel downright uncomfortable with the word "pleasure," perhaps because of the strong strain of Protestant moral purism threading through the Anglo-American tradition: pleasures of the flesh are purportedly the chief road to perdition. The English and Americans, in particular, fret about "disgraceful" and "base" pleasures.[35] But references to "wine, women, and song" or to carpe diem hardly convey the most sophisticated versions of hedonist theory in Anglo-American philosophy, let alone in the classical Chinese sources.[36] In the Western tradition, thinkers since Epicurus have coupled "pain in the body" with a "disturbance in the soul."[37] A group of famous thinkers — Epicurus, Bentham, Mill, among others — share the view that essentially "every person by nature ultimately seeks only pleasure and shuns only pain; therefore pleasure is the only intrinsic good and pain is the only intrinsic evil."[38] Thus it behooves serious students of pleasure to linger, for the moment, over the past and current arguments for hedonism.

The pleasure calculus in classical Chinese parts company with Western hedonist theories mainly because of its total disinterest in the pleasure-pain dichotomy, as previously noted. But hedonism in Western philosophy today, having lost its original moral thrust, currently embraces such a broad spectrum of diffuse good feelings as to be philosophically incoherent and thus a stumbling block to clear analysis. (The *Stanford Encyclopedia of Philosophy* entry mentions contentment, delight, ecstasy, elation, enjoyment, euphoria, exhilaration, exultation, gladness, gratification, gratitude, joy, liking, love, relief, satisfaction, Schadenfreude, and tranquility.) By contrast, in the majority of cases,[39] the Chinese word translated as "pleasure" (*le* 樂) denotes, far more narrowly, "long-term relational pleasures," the continued investment in which may augment one's sense of gratification and vitality. No less significantly, *le* in classical Chinese cannot refer to any antisocial or destructive impulses as Schadenfreude, selfishness, and sadomasochism, regardless of the intense sensations they may engender in a suitably twisted individual.[40] And since immediate urges can easily override any prudent desire for ultimate security, the Chinese proposals often portray the steady pursuit of disinterested and other-regarding pleasures while opposing self-indulgence as the best way to sustain and increase one's own personal stock of pleasure.

What Pleasures?

If we compare *le* to its nearest synonym, *xi* 喜 ("delight"), we begin to apprehend what distinctive work *le* does in the classical Chinese rhetoric. Whereas almost every instance of *le* points to relational pleasures that accrue through the long-term beneficial practices and associations that promote deep satisfactions, the uses of *xi* 喜 are a mixed bag, at best. For *xi* connotes short-term delight that is not necessarily wrong in itself, but that may conduce to eventual harm, insofar as it consumes bodily resources without replenishing them.

Xi, "short-term delight," is a neutral state, in theory, whose impact can be assessed only after calculating how often the person expends his or her resources through "movements" toward specific sorts of objects in specific sorts of activities. *Xi* is revealed readily in the face (as when a person erupts in laughter); it often simply describes "enjoyment in doing X," where X can be anything: swordplay, dancing, composing a poem, resting after hard work, stumping others in debate, or even going to war. *Xi* is the feeling one has in response to good food and good wine at banquets, indeed, to whatever activity goes smoothly. *Xi* is coupled with getting possessions or rewards, also with pride of possession or the sense of novelty to be had from luxuries, curiosities, and baubles. *Xi* can convey self-congratulatory moods and airs.[41] By the masters' theories, *xi* need not bespeak measured responses to external stimuli; *xi* activities can roil the senses without refining one's capacity for appreciation, heightening one's taste for reciprocity and communion, or inducing a willingness to delay immediate gratification for good reasons.

Nothing is inherently wrong with either delight or brevity, of course, and that explains why it is best to regard *xi* as "ethically neutral" at the start. I can take innocent delight in many things and many activities, even if such small delights pale beside the weighty and significant pleasures associated with *le*. As the Chinese phrase goes, "*xi* is reserved for small gains of what one likes, and *le* for the greater" (小得其所好則喜, 大得其所好則樂).[42] But since it's equally possible to delight in fine weather, in a narrow escape from just punishment, or in backstabbing an enemy, the term "delight" does not entail either profound transformation or moral commitment. Delight and anger tend to be paired in the sources, because the one emotion can turn on a dime so easily into the other.[43] Thus, *xi* often describes "having a good time" or "having one's way," whether or not that conduces to one's edification or maturity. In fact, a decidedly unreflective air hovers over the entire set of inclinations covered by the word "delight" — and hence *xi*'s associations with casual informality (often misplaced), lack of deliberation, negligence, and even moral benightedness. Whence the early leitmotif repeated in so many sources: "the noble man is not delighted when good fortune comes" [in contrast to his less astute peers]."[44] For only a fool takes illogical delight in a stroke of good luck, thinking it due recompense for her own greatness; by contrast, the wise distrust pieces of luck, lest they render them arrogant and inattentive.[45] Manipulating other people, judging other people capriciously, or treating dependents as pets or commodities to be collected — these, too, can reek of short-term delight. Still worse, some take acute delight in another's sorrow, downfall, or degradation, so long as these promise personal advantage. Finally, *xi* is the word used in connection with an underhanded or fraudulent scheme to curry favor with a higher-up — actions likely to call down calamity upon oneself or another, sooner or later. In other words, *le* is not apt to call forth a negative object, but *xi* can, regularly.

The sheer number of ordinary experiences occasioning delight suggests an allied problem: one cannot easily anchor so ephemeral a feeling firmly to a set of principles or to the performance of discrete social roles. For that reason alone, what the ordinary person delights in is often at variance with the preferences evinced by a more refined or prudent person. It is meanwhile a matter of no little consequence that the "movements" out from the body to bring the object of delight in contact with the senses are so many and various that all of them cannot be reliably compensated. Damage ensues when a voracious appetite for short-term delights deadens a person's sense of the problems that may arise in the course of their pursuit or afterward. Voracity proves especially problematic in the case of the ruler, for he is to provide a good example, lest his court and his subjects imitate (and so exponentially multiply) his worst habits out of fear or cravenness.

But *le* and *xi* hardly exhaust the rich vocabulary in classical Chinese for feeling in relation to thinking and acting. An enormous list of vocabulary items appears in the Chinese pleasure discourses, beginning with *yu* 欲 ("desires" or "to desire"); *ai* 愛 ("to love or care for");[46] *dan* 酖 or *ni* 泥 ("to indulge in or be addicted to");[47] *yin* 淫 ("to go to excess"); *yue* 悅 ("to think well of," "to be gratified by"); *hao* 好 ("to be fond of or to prefer"); *shu* 舒, signifying the physical ease, comfort, or release of a superior;[48] *wan* 玩 ("to play" or "to toy with"); *xin* 欣 ("to appreciate");[49] *yu* 娛 ("to amuse or be amused" or "to enjoy"); *wan* 婉 ("to take delight in"); *yu* 愉 ("at ease," "unstrained"); *kai* 愷 ("to be stirred or thrilled");[50] *kang* 康 or *yi* 怡 ("unruffled," "unflappable," or "calm" and "at ease"); *yi* 斁 ("sated" or "satisfied"); *yan* 晏/燕 ("satisfied," "contented"); *yin jin* 殷勤 ("on good terms"); and *xiang* 嚮 ("to enjoy" or "to savor"), originally used of the relish the ancestors felt in receiving offerings of the finest sacrificial meats.[51] Taken together, such terms attest a wide spectrum of human responses in early China; separately, each term suggests a discrete response to a precise set of circumstances.

In this lengthy catalogue of vocabulary items, what is striking is how often, when the graphs are deployed as verbs, they take on different ethical valences, depending on their direct objects; also how numerous are the terms implying specific durations for the pleasurable experiences.[52] Needless to say, pleasure comes in three temporal aspects: confident expectation of the future pleasure, the sensory experience of it in the present, and the pleasant memory of it in retrospect. Memories, even false memories, can be the most powerful of the three, there being ample time after the sensation to dwell upon it and incorporate a well-considered appreciation of the act's social construction. The classical persuaders were probably right in regarding the communal reaction to an action or text as a major determinant of the person's memory of it, making for a gratified smile or a grimace. Since man is a social being, a memory of pleasure can gain added force from social approbation.

The Physiology of Pleasure

The classical rationales for constructive pleasure spurred the elaboration of two new major theories, the first being that of *ganying* 感應 ("sympathetic resonance," rather than "stimulus-response")[53] and the second that of human nature. Because debates on human nature will be taken up later, in Chapters 3 and 4, this section focuses on resonance theory in relation to the physiology of pleasure.

Resonance theories in early China posit continual, complex forms of interaction more akin to quantum entanglement or the butterfly effect than to any simple, mechanistic cause-and-effect model drawing a straight line from impact to impacted. In resonance theories, all the myriad things are part of one inseparable whole (a macrocosm), and movements in any part of the whole affect the other parts in ways hidden or visible. Human beings, as a result, have no "essential self," but are always undergoing transformation.[54] One piece of evidence often adduced for the "truth" of resonance theory is the audible reverberation of stringed instruments at some distance after only one string on a different instrument has been plucked.[55]

This observation prompted other, far less testable ideas: chiefly, that contact with external things, people, situations, or events stimulates and moves (*gan* 感) the senses, whereupon the organ system identified with the heart, or *xin* 心, rapidly evaluates these externalities and the dispositions they produce; also that such exchanges—dubbed "motions" or "feelings" (that is, emotions) (*dong* 動) or "responses" (*ying* 應), whether visible or not, leave nothing quite as it was before. For either as part of the first wave of these motions or as a secondary ripple effect, the body registers the values attached to the stimulus along a scale ranging from pleasure to the converse, values that may be more or less conscious, depending on how palpable the motions are. I see apples in my room, and I like to eat apples, but I am not hungry, so the lion's share of my attention lands elsewhere. Or I see Thomas with Jamie, and I like being with them, but Jamie seems ill. "Oh, no!" I focus hard on how she looks. Pleasure in prospect or in memory happens through somewhat more mysterious processes triggered by old associations, presumably rooted in earlier contact traces stored in the *xin* and accessible throughout the body.

On firmer ground, most thinkers in early China viewed pleasure "as a basic necessity of life, like food and air,"[56] assuming that the distinctive forms of pleasure-taking and pleasure-giving, along with other marvelous human creations such as symbol systems and communal conventions, are what distinguish humans and raise them above the brutes.[57] Two important observations follow: first, that pleasure was never viewed as a luxury, let alone an evil or temptation to evil, and second, that the capacity to experience complex pleasures defines the highest achievement, that of being fully human, whole, and at ease.

Probing further into the pleasure rhetoric's tying of the desires to experiential pleasures and pleasurable outcomes, we notice the central role assigned the *xin* in these discussions. For by the Chinese theories, humans are born with five sensory organs designed to produce sensory percepts: the eyes see, the nose smells, the ears hear, surfaces of the body touch, and the *xin* 心 organ system simultaneously identifies and evaluates the perceptual information that it receives, either directly or through the other four organs, after which it is again the *xin* that decides the person's commitments. In a person's life, in consequence, the most important thing is to govern the *xin*, which is liable to be excited by any touch (that is, perceptual awareness of contact).[58]

As soon as it has left the womb, if not before, a person naturally comes into recurrent contact with external things, other people, and situations through the organ systems. At contact, the appropriate organ receptor registers first the perceptible qualities inherent in the object encountered (in the simplest examples, "black or white," "sour or salty"), after which that organ, working in tandem with the powers of discrimination located in the evaluating heart and mind, assesses the value of the object or person contacted through the senses (for example, "beautiful or ugly," "shrill or sonorous," "tasty or foul"). Once external phenomena stimulate the senses, moving (*gan* 感) and quickening them, the inclinations and dispositions (*qing* 情) arise (*qi* 起 or *xing* 興) in their wake. Depending on the strength and quality of these percepts, the percepts sooner or later generate within the body more defined inclinations and dispositions toward objects, people, and events external to the person.[59] In the classical language of the time, the heart (*xin* 心) produces evaluations (*si* 思), which lead to the formulation of commitments (*zhi* 志). As in impressionable children, in adults, these inclinations and commitments are inevitably shaped by memories of previous experiences, even if a specific contact and form of resonance is new. "He looks intelligent and kind, something like Gerry." Impulses (*ying* 應) that are simple and unreflective are shared with the birds and beasts, but commitments mean directing one's attention and activities in a sustained fashion toward specific aims beyond mere satisfaction of the first-order desires for food, sex, and survival. Second-order desires crucially include the mimetic desires to have what others have and to emulate impressive people. Such mimetic desires, arising within the context of the basic human drive for safety and security in all social settings, can be turned to either destructive or constructive ends; thus, they acquire an ethical valence only because they lead to personal, familial, and social consequences. Whereas first-order desires are common to all living beings, the collocation of second-order desires define the *xin* and indeed each person.

That distinct experiential world unique to each person inevitably exerts a cumulative effect on the configuration of his or her habitual ways of responding, because experience, reinforced by habit, by memory, and by schooling, has the *xin* putting additional labels on an ever-increasing number of stimuli.[60] Over time, fewer and fewer experiences come to a person entirely free of labels, and some labeling activities become so firmly entrenched as to constitute a virtual "second nature" (*xing* 性) barely noticed by the agent. So failures of recognition or lapses in judgment occur easily, as much as through occluding habits as through perceptual flaws or competing, but incommensurate desires. Mencius phrased the problem nicely: since "the sensory organs of hearing and sight do not [themselves] analyze, they can be deluded by things."[61] Yet human misery is the inevitable result of wrong assessments beclouding the mind ("delusions by things"). Besides, any interaction between the object, the senses, and the heart requires successive acts of coordination, correlation, categorization, and prioritization, making the correct identification of suitable ways to gratify the whole person a hugely complicated business, even were there no distractions, misconceptions, and delusions.

Fortunately, the heart and mind can and do think, under prompting by the sensory organs and associations, but they will be predisposed to evaluate things, persons, and events correctly only on three conditions: if the organ system is trained and well practiced in making categorical connections, if material conditions conduce to calm thinking, and if the will to think hard is present. It takes deliberate, sometimes arduous physical and mental effort to overwrite experiential labels, alter fixed inclinations, and transform one's outlook on life, to allow a better alignment between one's actions and one's goals. Thus, men and women of probity take their first task to be to muster sufficient will and fixity of purpose (*zhi* 志) to fashion a more constructive and sustained set of inclinations, in the hopes of removing the most obvious barriers to superior penetration (*tong* 痛 or *ming* 明) by the sympathies. By these theories, the initial requirement for self-discipline and accommodation to others will become less burdensome once the rewards for a more intense and pleasurable mutuality (*xiang le* 相樂) manifest themselves at each step along the way.

The medium facilitating these forms of resonant interaction and sensitive exchange is *qi* 氣, often translated as "spirit" or "vital energy." Since all things, supposedly, are composed of *qi*, all phenomena, including human beings, are sufficiently alike to be in continual sympathetic communication with one another, even if certain connections are too subtle for ordinary people to apprehend with the senses. The movement back and forth in this sympathetic communication is *dong* 動; significantly, in classical Chinese, as in English, the words for "motive" and "emotion" are cognate with "moving" and "being moved." A secondary analogy links the internal drives propelling these contacts with "driving [a carriage]" (*yu* 御); along much the same lines, one famous text likens the *qi* in motion with steeds racing.[62] (In English we are "driven" by an idea or a compulsion; erudite readers may recall Plato's image of a chariot drawn by two horses.) All sensory perceptions stem from resonant exchanges of *qi*, the *qi* of one thing or one person venturing out from its accustomed site to meet the *qi* of another that attracts it before traveling back, ineluctably changed, to the original site. To establish contact between two persons, for example, the *qi* of one party must be exuded and penetrate the skin and vital organs of the other before it returns, carrying an impression. And although the theories seldom specify whether each type of *qi* (for example, *qi* colored by "delight" versus "anger") is prone, right from the start, to operate in a particular way on these journeys, they insist that it is only refined *qi*, usually called *jing* 精 or *shen* 神, that permits the most exquisitely sensitive (and hence gratifying and enlivening) exchanges to occur. (Note the Chinese "failure" to assign the tasks of identification and evaluation to different parts of the body and thus to sever fact from value.)

In describing such patterns of *qi* circulation, the early texts caution against the undue leakage of *qi* associated with inappropriate desires and ill-conceived acts. For the early medical treatises identify the leakage of *jing* 精, the most potent *qi* 氣, as a major cause of illness, debility, and death, second only to attacks by vengeful ghosts and spirits.[63] After all, *qi* continually flows, even floods sites inside and outside the body, and repeated outflows from the body's core organ systems inevitably threaten to deplete the body's fairly closed system of blood and *qi* (blood being but one form of *qi*). Each leakage lessens the finite amount of the vital spirit allotted to a person at birth. Once spent, that allotment of vital spirit is difficult, if not impossible to restore. In extreme cases, our texts warn, the leakage may irreparably damage the true self (*shi zhen* 失真), preventing "wholeness." (By analogy to leakages from the body, ruinous outlays and expenditures are said to weaken the realm, right along with divulging court secrets.)[64]

Discharges were known to accompany sexual arousal and a variety of other conditions. "Excessive" (*yin* 淫) and "overindulgent" (*chi* 侈) contacts exponentially increase both the number and volume of the leaks until they reach a pitch of exhaustion from which the body can never recover. Too many palpitations of the heart or too much outward flow make it very hard to retain enough *qi* at the vital center, the *xin* organ system. Hence the proverbial warning, "profligate of ear and eye ... this is what the sages forbid."[65] One satirical poet saw it this way:

To give free rein to the desires prompted by ear and eye, indulging in the physical comforts — these harm the equilibrium required by the blood's flow in its channels. Furthermore, going out by chariot and coming in by sedan chair are rightly deemed "portents of rheumatism and paralysis." Cavernous halls and cool palaces are called "inducers of chills and fevers." Gleaming teeth and the moth eyebrows of fair maidens have been dubbed "axes that chop away at the vital nature." And the sweet and the savory, the fat and the rich at table — these are known as "drugs that rot the innards."[66]

Clearly, some forms of flow are necessary to the body's survival or equilibrium. For instance, the regular circulatory flows within the body are crucial to bodily health. Even movements outward can be beneficial, for example when seeking food to prevent starvation or sex to quell the body's craving for it.[67] Curiously, the animal passions were not usually thought to be a bad thing per se for humanity. After all, the same passions and desires that are so necessary to survival drive people's engagement with the outside world, creating the possibility for the constructive exchanges that define highly cultivated human beings. Desires for food and sex, in point of fact, provide a good model for all vital engagement, for if both eating and sexual stimulation *can* be solitary, both afford endless occasions for convivial and memorable exchange. Then, too, eating alone is usually less enjoyable than eating in company, and a wide range of satisfying sexual activities call for a partner. Whence the notion that the pleasures that tend to consume a person, by definition excessive or inherently unsatisfying, tend to be either solitary pleasures on which no natural social checks are exerted or commodified exchanges that divert time and energy from long-term social relations. For that reason, indulgence in food and wine had better take place within the commensal contexts of choreographed banquet rituals.[68]

According to the early thinkers, too-frequent arousals and outward flows of *qi* to nonessential ends — the desire to have an object "right now," a yen for a concubine or a dinner of bear paws — principally endanger a person when satisfying those impulses engenders few or no compensatory gains in the form of sustaining social relations. Participation in art and ritual are liable to create more secure and satisfying communities, but even these activities cannot guarantee ipso facto that an act of striving for participation will not cause an excessive outflow of *qi*.[69] Only by exercising discrimination when moderating and refining one's drives can one hope to offset outflows of *qi* with inflows, thereby staving off physical harm. This way of thinking is very old, and it can be found in all sorts of texts, including the *Analects* ascribed to Confucius, which says, "The person of superior cultivation guards against three things: when young, the blood and *qi* are not yet settled, so he guards against lust; when mature, the blood and *qi* are vigorous, so he guards against combativeness; when old, the blood and *qi* are declining, so he guards against covetousness."⁷⁰ In other words, as the body alters, the wise person, to sustain well-being, continually adjusts his approaches to the outside world.

The Politics of Pleasure
In statecraft, theorists were quick to invoke the pleasure calculus when outlining techniques for the acquisition and maintenance of political power. In their analyses of diplomatic relations and the workings of the court at center, the terms for "insecurity" (*you* 憂 and *bu'an* 不安) loom large. So numerous are the ruler's opportunities for pleasure-taking that the wise ruler has to assess his position continually vis-à-vis the shifting threats to his security and sanity. Does he fully deserve to enjoy that form and amount of pleasure? If the answer is "no," he risks having the righteous or the merely envious deprive him of the very pleasures he wishes to enjoy. Thus, it was vitally important that those in high positions "not err in [their own] likes and dislikes" of things and people, because their subordinates and inferiors may consciously and unconsciously copy those same likes and dislikes, for better or for worse. The sixth section of the Guodian "Black Robe" chapter (ca. 300 BCE) is typical in asserting, "If those above show humaneness, then those below will fight to take the lead in showing humaneness." It continues in section 8, "In serving their superiors, subordinates do not follow their [overt] commands; rather they follow what they do. If those above like something, those below will invariably find it even better." A third outlines the symbiotic relation between ruler and subject: "When the people (sometimes meaning "the king's people," the leading men of the court) take their ruler as their own heart, the ruler takes the people as his own body.... Therefore the heart can be wasted by reference to the body, and the ruler ruined on account of the people."⁷¹ Far-sighted administrators perforce studied how to apply their insights into

pleasure to the quotidian tasks of ruling. Xunzi, for example, devised a famous paradox that, once understood, convinced men in service to the court who "aimed to live a life of ease" of the advantages to be had from "risking death on the battlefield" for their ruler's sake, not the least, the advancement of kith and kin.[72]

The genius of such early prescriptions lies in their profound recognition that the inclination to pursue pleasures can never be fully suppressed nor the desires eliminated. Enforced asceticism, because unnatural, is counterproductive, particularly so because the inclinations and the desires, if properly directed, could advance the good. Fully satisfying the desires for pleasures can benefit society, so long as rulers or their advisors are smart and determined enough to turn both the nature of their desires and the forms of their gratifications to civilizing effect. Let us watch Xunzi or a later disciple describing the sage-king's invention of music:

> Now people by nature have blood and *qi*, and conscious evaluative faculties (*xin*), but they lack any [innate] constants with respect to their inclinations towards sorrow and pleasure, delight and anger.... The sage-kings, mightily ashamed of their chaotic tendencies, instituted the elevated sounds of the refined odes and hymns to guide men's natures. These were rooted in human nature. One could contemplate the sage-kings' regular expressions, and so there were instituted the ritual ceremonies. The sages' music joined with the harmony of the *qi* in life and led [the people's] conduct via the Five Constants [the main social relations]. It caused them to incline to either yang or yin and not be scattered or brittle or unduly hamstrung. The firm and resolute did not become angry, and the weak did not display fear, with the result that the four full expressions [of the emotions?] cohered in [each person's] central core and yet could be expressed outside. Each person then rested in his or her rightful place, without detracting from or harming others. This was sufficient to move the hearts of human beings toward the good and to prevent the misaligned *qi* from gaining a hold on them and their inclinations.[73]

So although the most consuming pleasures were seen as "vermin" eating away at the vitals, there were ample ways to remedy the disastrous consequences of "giving free rein to the desires."[74] One optimistic theory claimed, "It is true, of course, that the Five Tones make people deaf, and the Five Tastes dull the palate.... But if a person [does not allow his desires to rule his life and so] retains the *qi* he was born with, he will be able to prolong his lifespan, surely!"[75]

And by Xunzi's influential theories, the skillful resort to sumptuary regulations can achieve three entirely laudatory aims at once: rewarding the good, advertising to lesser people the signal benefits of emulating good models, and insuring a fairer division of wealth. Still, the prospect of future loss was daunting, given that pleasures could be fully enjoyed only to the degree that they were reckoned fully secure. So what additional devices beyond ritual and music could help to secure a power holder's position? Spectacles in moderation, employment of the worthy, careful allocation of rewards and punishments — these three strong incentives, together with sumptuary regulations, could inspire "those below" to invest in constructive sociopolitical behaviors. With these in place, even subjects lacking good parents and teachers could be schooled to build upon their mimetic desires so as to take justifiable pride in their own upright conduct and derive pleasure from it.[76] Social cohesion ideally meant that everyone had achieved and was secure in his own proper place in society (*ge de qi suo* 各得其所). But this was a lofty ideal, and even to realize it to a minimal degree nearly always necessitated ways-and-means calculations,[77] as when the ruling family's desire for more discretionary income contradicted its compelling need to procure unwavering support from subject populations through lenient tax rates and generous boons.

The deeper one delves into the early sources, the more obvious becomes a warning threading through the pleasure theories: those in power had better not be enslaved by their desires. A typical early text in classical Chinese, bringing to mind rhetoric from other classical civilizations more reliant upon slave economies, puts it this way: if the direction of our wills depends overmuch upon "external things," "pleasures," and "practices," the inclinations (*qing* 情) will be pulled this way and that by "external things seizing upon them."[78] Thus, to call a person "witless" (*yu* 愚) was tantamount to saying that he had "no master" (that is, he lacked self-mastery), while to say that a man lacked sufficient will or commitment (*zhi* 志) was to allege that he lacked the very quality that spelled true nobility, an abundance of courage.[79] It followed that to be "seduced by others" (*you yu ren* 誘於人) is a condition to be avoided at all costs, for "plainly, for a person to rely on others is less good than relying on oneself, and, equally plainly, what others may do for one is never as good as what one may do for oneself."[80]

In essence, to be enslaved by desires is to forget that the true utility of things and relationships lies in their potential to enhance occasions for serene pleasure-taking of the sort that will not undermine position, person, or family, now or in the foreseeable future. Conflicts between insufficiently deliberate responses and carefully considered inclinations are to be quelled, lest they harm the body and eventually give rise to increased dissatisfaction with the world's abundant pleasures. After all, there existed proven ways to master the body's natural impulses to respond indiscriminately to countless phenomena generating innumerable desires. Granted, since the pleasure theories were addressed to the upper ranks, no early text goes so far as to presume unconditional equality or autonomy in their discussions (unlike contemporary misbegotten theories). Still, calls for reciprocity and *relative* autonomy in giving and taking pleasure figured prominently in the early thinking.

It would be absurd and false to imply unanimity of opinions among thinkers competing for court favor over the course of the early empires. Brief mention can be made of the disputes between the advocates of No Desires and those promoting Refined Desires.[81] Some thinkers deemed honor, rectitude, or dignity a prerequisite for maximum pleasure. Some didn't. Logical objections, needless to say, could be hurled against any and all parts of the basic hedonistic justification for acting constructively, such that persuaders scrambled to evade the grim implications of Yang Zhu's insights into the self-defeating search for worldly glory.[82] But most of the theorists under review here were not so reckless as to hazard absolute claims, anyway.[83] They stated only that *on average*, appropriate conduct undertaken after careful calculations about personal, social, and cosmic consequences is more likely than destructive conduct to conduce to a life of maximal pleasure and minimal distress. Once the wisdom of that assertion was conceded, persuaders could easily get their intended audiences to accept the allied point: that wise administrators could encourage others to choose right actions by a combination of techniques, not least of which was the power holders' self-fashioning as exemplary models. Fancying the charismatic potential of one's own splendid example could be too gratifying to resist, and if a power holder's deliberate model was propagated for wider consumption, the multiplier effect of his actions would resound on a far grander scale. Thus would "love of one's own fleshly person and love of the societal become one and the same."[84]

In consequence, many of the early texts make an implicit or explicit promise: that a single person *in society* (never construed as an autonomous "individual"), with the help of society's teachers, living or dead, has the potential to retain integrity and to manifest an estimable unity of purpose, despite the "pulls" and "leakages" occasioned by the excessive or inappropriate desires that begin or end in internal conflicts, two-facedness, debility, and worse. But by definition, the "whole and intact" (*quan* 全) person (a worthy or sage by another name) had achieved that ideal state where an appropriate level of self-regard and self-reliance, even of *sprezzatura*, went hand in hand with the keenest receptivity to the unfolding situation. A readiness to engage productively with others had the ideal person's feelings (*gan* 感) reliably comporting with the realities of his experiential world. In the most common metaphor, the worthy man is a skilled archer whose arrows always hit the target,[85] but other metaphors suggesting superb reliability were common, as well. Dubbed gracious and "self-propelling" (*ziran* 自然), his actions come reflexively, with virtuosity, as he modifies his models and commitments as necessary. The consummate artfulness of the refined life, by these theories, lies in circumspection in both its senses: "looking at things from all sides" and "prudential acts." Yet these extraordinary capabilities instilling order in the body and body politic are predicated on an often long and difficult period of preparation.

As everyone knew, prudence, however much trouble it may entail at one point, could in the long run secure two of the most profound pleasures, security and unsullied honor. For the person no longer "in two minds" meets fewer obstacles to fluid action and good judgment, and moral courage, when combined with sublime skill in negotiating difficulties, can win the admiration of peers. Thus, worthy advisors were to encourage those in power to ask themselves such questions as, "How much pleasure of what sort under what circumstances sustains or consumes?" If the ruler and his ministers could be persuaded of the value of learning to direct and elevate their tastes, they might then find that acting on their "desires and addictions"—far from being a severe drain on scarce resources—had become a secure way to "gratify hearts and minds" (*yue xin* 悅心), their own and those of others. Not surprisingly, then, the great social engineers of the Zhanguo, Qin, Western and Eastern Han, and the post-Han eras repeatedly combined theories of pleasure, detailing the human urges and motivations, and of public display into a single, if highly versatile package, a package designed to introduce a measure of reciprocity and mutuality into the hierarchies they intended to buttress, a package that disseminated meritocratic slogans alongside modified aristocratic models.

To "Warm Up the Old"

I have tried to suggest the ubiquity of the pleasure rhetoric in theoretical discussions and in public applications during the classical period and beyond.[86] But by no means did the pleasure thinkers under consideration in this book emphasize the same "goods" pleasing to the same human faculties. Naturally enough, the moralists laid constant stress on the need to "do good" (and hence to "act constructively") for the community, the payoff for long practice being the attainment of an enviable state of acknowledged expertise in forging effective policies and reliable allies in the social sphere. Some of the thinkers under review (Mencius being a prime example) evidently believed that human beings, by their very natures, come equipped with a capacity for moral intuition that can be developed with relatively little difficulty.[87] Many scoffed at such claims, however, proclaiming improved habits the keys to success in ordering the competing claims on one's attention, lest the person be doomed to live a life that is "frantic and fragmented" and so less than vital (see Chapters 4 and 5, on Xunzi and Zhuangzi). Those who wondered whether gods exist, and if so, what their interactions with people are typically like, as well as those more willing to entertain life's mysteries (for example, the legendary Zhuangzi and historical Yang Xiong, the protagonists of Chapters 5 and 6) rejected the notion that a person can always know *why* something is the case or even if something *is* the case. They stressed that a person's apprehension of the larger context is inevitably partial and flawed, however persistent the search for greater certainty or awesome erudition. (Modern neuroscience confirms that insight, as it happens, even as most modern Western philosophers blithely continue to name "full awareness" and "freedom" as preconditions for real pleasure.)[88] What the early Chinese thinkers agreed upon is merely this: that life itself is good and valuable, so a person should try to maximize life and vitality, not only by "increasing the lifespan" but also by "enhancing the pleasure taken during life." Obviously, such a consensus entails distinguishing the pleasures likely to shorten or reduce pleasure from those likely to extend or enhance it. Hence the incessant cautions against the "vices of disproportion" (egoism, selfishness, one-sided views, cowardice, and cynicism)[89] and the theories that separate consuming from sustaining pleasures, which form the subject of succeeding chapters.

As the Preface to this book suggests, all of the antique pleasure theories detailed are worth serious consideration, insofar as the human condition, despite huge technological advances, still confronts each human being with the age-old dilemmas about how to best utilize a limited time on earth so as to glean meaning from it. Admittedly, as Sheldon Pollock, the distinguished Indologist, has noted, the "meaning" of any classic work is never simple to ascertain, for every authoritative text has no fewer than three logically distinct "meanings": the meaning it had for its author(s) or compiler(s) within the textual community that generated it and to which it was addressed; the meaning or meanings it acquired over time, down through tradition(s) (usually plural); and the meaning it holds for people today, which often differs hugely from the first two meanings.

> The triad, historicism, traditionism, and presentism can be seen . . . at work in an analysis of a[ny] text . . . for example, [one may speak of] what Homer really (or, "really") meant at first; what the Hellenistic and later commentaries up to almost-now have taken him to mean; what Homer means to me here and now. But the point of the tripartition is the same. There is no singular truth about words. None of the three . . . "enjoys a closer relation to reality" than any of the others. "The meaning" of an idea or text can only be the sum total of meanings that have been attributed to it . . . the point being to attend to all three, hold them in balance, and understand that all are real and consequential, that they don't cancel each other out, and that we present each as a truth with reference to the kind of work we want to get done, the form of life we want to understand.[90]

Doubtless, there will be people who disagree with my readings of the texts I deploy in successive chapters. But the underlying motive for this pleasure book is to adopt the Confucian injunction to "warm up the old," reinvigorating long-standing traditions (plural) from China so that they can be brought to redirect and illuminate the ordinary lives that we construct through daily practices and hourly lapses.

Figure 2.1. "Friends and Music," detail from anonymous album leaf (ascribed to Liu Songnian) entitled "Listening to the *Qin* (Zither)," ca. 1150–ca. 1225, ink and color on silk, 23.8 x 24.6 cm, with mat 33.3 x 40.5 cm, The Cleveland Museum of Art, Leonard C. Hanna Jr. Fund, 1983.85.

This album captures the refinements of character and taste that gave musical performance (here, in a domestic space, as indicated by the solitary *qin*) an important role in forming intimate friendships.

CHAPTER TWO

Good Vibrations: The Allied Pleasures of Music and Friendship in the Masters

The Way of Humanity is social relations (*jiao* 交) — Yang Xiong

One has music to gladden close friends 樂以嘉友 — *Zhaoshi Yilin*

Music is used to harmonize spirits; it is the mark of true humanity
樂以和神, 仁之表也 — *Hanshu*

Pleasant dissipation on idle days may not instruct the people,
but what harm does it do? — Song Yu

The Ancients did not merely perceive the world differently; rather, they saw, heard, and touched different things.[1] That includes music, and the palpable experience of intimate friends who are seen, heard, and touched. This chapter treats music and friendship together for several allied reasons. The same resonance theory underlies early thinkers' understanding of both friendship and music. Because of this, the two were often seen as mutually reinforcing and paired in practice. Also, musical performance and appreciation often served as the most forceful metaphors for friendship, as well as the best preparation for and clearest illustration of the subtle forces of attraction that facilitate friendship (Figure 2.1). As exquisite relational pleasures, music and intimate friendship ideally promoted artful enhancements of the persons trained in them. Trusting to the extraordinary potential value that music and friendship made to personal development, thoughtful people sought to cultivate these arts and, happily, they left behind abundant testimony describing their early experiential worlds. These efforts, in turn, made performances of music and friendship the best possible advertisements for the admirable characters such people evinced. In the end, music and friendship share a chapter because a wide range of early extant sources speaking about pleasure (beginning with Ode 1 in the archaic *Odes* classic) so continually couple the two that separate discussions would impede the comprehension of either and both.[2]

By the antique laws of attraction, the world was less often viewed in terms of mechanical cause and effect than in terms of sympathies, affinities, and resonances both subtle and moving (see Chapter 1). Musical performances provided the basis for elaborate theories relating aesthetics and harmonics to human development through friendship as far back as the *Odes*.[3] For example, all early *Xunzi* passages on music and friendship presupposed the existence of unseen sympathies weaving the cosmic and social worlds together—sympathies capable of greatly and indelibly stamping the characters, attitudes, actions, and even the life spans of well-tempered people.[4] Already by 300 BCE, in the first datable excavated manuscripts, one mysterious effect relating to music struck early thinkers forcefully: a string always vibrates, producing the same note in response to a sound made by a tuning fork or stringed instrument at some distance.[5] (Exploration of the properties of the magnet no later than Han times seemed only to confirm the strength of such unseen ties.)[6] Because people were perceived as microcosms of the sympathetic powers operating across time and space in the wider macrocosm, the early thinkers began to speculate about the animating impulse inherent in things and people that propels like to seek like, but dissonance to repel—the same impulse, supposedly, that led culture heroes to locate the best sources of personal and social expression within constructive social relations aided by the arts.[7]

Music epitomized mutual attractions, since many songs—temple hymns no less than folk songs—were designed to be sung antiphonally.[8] And insofar as music renders perceptible the innermost feelings (constituting the "mark" of humanity), it was held to facilitate communications between those with the requisite sensibilities.[9] Not coincidentally, pleasing musical intervals indicated by lengths of strings in simple numerical ratios such as 1:2, 2:3, or 3:4 were dubbed close "neighbors" in early China.[10] Connections between music and friendship could not have been formulated, however, if the attractions of music and friendship were not predicated on the marvelous interplay of diverse elements, summed up in the four-character phrase "in harmony, yet not identical" (*he er butong* 和而不同)[11] or in lines about the "ocarina and flute."[12] Emphatically, neither music nor friendship was merely a metaphor for the other. And when music and friendship happened to overlap (as when old friends met to make music or to listen to it together), the waves of sympathetic response could crest, sending out resonances far beyond the participants' circle into the world at large.

Friendships, intimate or merely sociable, built upon the "base" of familial reverence, extending a person's social connections well beyond the family. Music, as the premier art, was to be learned after a solid ethical "groundwork" had been duly laid through training in ritual.[13] Surviving records assert the political utility of music (sometimes cast as a subset of the rites, but just as often as complementary or superior to them). More than one masterwork, for example, employs zither music as an analogy for adept governing. Others portrayed friendship as the "glue" binding the court and realm together.[14] "Private" or "solitary" pleasures were apt to be labeled "ugly" (*e* 惡), inferior, and verging on the antisocial,[15] but both music and friendship miraculously evaded that danger, surmounting an important divide between official duties and the domestic sphere. Habits and tastes formed in one space directly informed actions played out in another, with discipline, empathy, and cultivation the best possible preparation for service on behalf of the common good. Entertaining friends and guests, for instance, could be "public" in the sense that wide circles of people might be meant to learn of the proceedings and find them edifying, even when the actual guest list for the party excluded all but a select few. At the same time, throughout the early empires, music and poetry (initially the tunes and later the lyrics) gave voice to aspects of life outside court, because these ostensibly more "private" forms of music were repeatedly recited and recirculated, achieving the widest possible impact in society.[16] Then, as now, shared interests in music and poetry brought together those whose official affiliations were ordinarily at odds.[17] Music and friendship helped temper sharp distinctions between people, inducing a greater sense of community and common purpose among otherwise disparate groups, "from the Son of Heaven on down to the common people."[18] Music and friendship even facilitated the crossing of gender boundaries (more porous in early and middle-period China).¹⁹ Singsong girls openly dallying with drunken patrons, loving couples engaging in a range of heterosexual and homosexual acts (see Figure 1.1), and homosocial bonds tying members of the governing elite — all these and many more types of relationships are to be found in the literary and visual sources relating to music and friendship.²⁰ Aficionados of these two relational pleasures valued this process of relating, rather than any predetermined end or goal, acknowledging that what passed for perfection in music and friendship depended upon continual readjustments or "attunements" informed by cultural conventions.

Theorists expatiated upon further links between music and friendship. For example, both music and friendship occasion in many instances a need for touch, at once spontaneous and almost ritualistic. Friends "held each other by the hands, taking turns singing all day long until evening" and thus, without doing violence to themselves or others, "restored a sense of harmony and balance," drawing deep from a reservoir of feelings dubbed the "spirit font."²¹ In like fashion, musical instruments require sensitive touch if they are to transmit the most haunting intimations of the musician's mood.²² Profound appreciation of a friend or a type of music constituted a form of rapt contemplation ascribing beauty and excellence (*mei* 美) to trusted relations of ineffable coherence.²³ Two hearts "where the *qi* and intentions beat as one" were dubbed "soundless music" — their strength supposedly "pervading the Four Quarters" of the known world, breaking down or transcending the usual barriers.²⁴ All the while, music and friendships took many improvisatory forms in disparate contexts, making for wide variations, with the result that at their best, interludes of music and friendships offered a "primordial freshness" that theories and abstractions could almost never claim. As rich reservoirs of vital feelings, both music and intimate friendship ideally insured that mutual interactions would lead to that truly enviable state of security known as "self-possession" (*zide* 自得), the composed state of self-awareness in which people "realized their best selves."²⁵ In that state, people could derive the highest level of personal satisfaction from acting in concert with others while ridding themselves of unworthy desires to inflict harm.²⁶

However casually experienced, an initial attraction to a particular piece of music, instrument, or companion invariably would reflect and reinforce certain habits of the heart,[27] while mediocre music or lackluster friends could sap a person's devotion to cultivation and his inner circle.[28] Since the status into which one was born generally mandated the choice of one's marriage partner and profession, it was commitments to particular friends and types of music that signaled personal priorities and tastes. Thus, the selectivity shown in music and close friendships constituted ready gauges for the tenor of a person's character, without resorting to the crude language of choice.[29] Not coincidentally, performances of friendship and music (often in the same public or semipublic settings) provided one of the few socially sanctioned venues in which people could advertise their preferences and accomplishments. By general consensus, a person who lacked a good name could blame his friends and his own taste in friends.[30] Learning among the elite required but three tests: loyalty to superiors (described as actual or potential friends), faithfulness to close friends (generally imagined as peers), and cultivated social behavior, with the last presupposing reciting, singing, and knowing the musical *Odes* as well as connoisseurship in playing and hearing an instrument.[31] But the salutary effects of good music and good friends were lost on those without the requisite discernment — so much so that even sages such as Kongzi were hard put to introduce their planned improvements into social and political life.[32]

The precise pleasures to be had from music and friendship may strike some readers as so universally shared as to demand no specific acts of conscious translation from another time and place to our own.[33] Indeed, a multitude of pleasing resonances (pun intended) exist between ancient and modern times as regards music and friendship. To begin with, an ancient bell set that once belonged to Marquis Yi of Zeng (d. 433 BCE) shows that two of the ancient twelve tones in the standard pitch system precisely correspond to the standard Western notes of middle C and A.[34] (See Figure 2.3.)

And that a single Chinese graph 樂 — now differently pronounced, to be sure — signified both "music" and "pleasure" recalls the fortuitous correspondences between the two English senses of "glee."[35] Such coincidences notwithstanding, historians of early and middle-period China are right to stress the enormous conceptual gulf that distinguishes the Central States empires from those of the modern nation-state.[36] To give one example: the role of music in politics loomed far larger in early China than moderns might suspect, for today, it beggars belief that power holders would credit music with so dramatic an impact upon state and society (*pace* Tipper Gore) or that legendary concert masters would act as policy advisors on serious matters domestic and foreign.[37] And since the role of friendship constructed in early China likewise bears but passing resemblance to that found in today's Euro-America, the roles of music and friendship in pleasure-taking ideally could unpack the kindred early notions of "virtuosic mastery" premised on sublime "transformations," if a project of that complexity did not require its own lines of inquiry.[38] That said, the precise timbre of music or friendship involves a concatenation of evanescent pleasures, with both as reliant for their effects upon silences as on tones, movements, or gestures (see below). Small wonder that the available writings never outline a method by which to link "the hazy outlines of music history to social details."[39]

The topic of friendship alone warrants nearly thirty comments in the "Confucian" *Analects*, beginning with the opening lines exalting the pleasures of "friends coming from afar" and including the famous passage that has the disciple Dian recounting his fervent wish to participate in the spring lustration rites with friends, accompanied by troops of young boys who "would take the air at the Rain Dance altars and go home singing."[40] But early authors and compilers rarely needed to explain themselves after the fact,[41] since they addressed small textual communities. Hence the resort in many early sources to a kind of shorthand whereby friends in conversation simply are said to "hit the mark," to a "musical effect."[42] Or the standard portrayals of "knee-to-knee talk" among friends enjoying themselves while exchanging banter or sharing the same song, the same mat, the same goblet, or the same peach,[43] as in Cao Pi's 曹丕 (Emperor Wen of Wei, 187–226) elegiac letter to a friend: "In bygone days, when traveling we drove our carriages side by side, and when we halted we placed our mats together. When were we ever apart? Whenever the wine was passed round, accompanied by the music of pipes and strings, our ears reddening in intoxication, we would lift up our eyes and chant lyrics, beyond even awareness of our own pleasures."[44]

The seemingly bland passage gains depth when read alongside the *Odes* and *Changes* classics, where the white rush mat symbolizes authenticity, simplicity, and humility.[45] Such four-character phrases as "setting out a couch for the guest to stay" encapsulate a delicious aura of special intimacies.[46] Needless to say, to teach another a beloved melody was to bestow a signal favor upon that very dear friend.[47] Delving more deeply into the sources, it seems evident that the most minor of acts could impart vivid color to friendship and music long afterward.[48] In gathering material, the modern historian need but consider how far to stretch the allied topics of "music" and "friendship."

But how early Chinese music really sounded, what sorts of musical arrangements conveyed the keenest pleasure to connoisseurs of the time, how performances paired music with singing and dancing and pantomimes, even what scales or tropes were most frequently employed — doubtless the fine texture of musical experience will continue to elude us, no matter how many pieces of literary and archaeological evidence accumulate.[49] The same opacity clouds our understanding of antique friendships. How friendships operated, endured, or faltered in early China and the specific social constructions investing intimate friendship with far greater powers to bind the perceptible world together — these remain somewhat closed to us.[50] For even memorable descriptive texts cannot map ephemeral social realities exactly, and our extant texts constitute but a small fraction of those that once existed.[51]

All we can safely surmise at this remove is the main thrust of the available evidence, yielding rough contours of the past. Many early texts, for example, rate music more highly than ritual when it comes to powerful civilizing effects. Supposedly, the legendary sage-kings of antiquity ruled entirely through music and dance, while their successors during the less ideal Three Dynasties made do with rituals.[52] The early writings concern themselves with music performed at court banquets or at community gatherings, with the instruments and songs finely gauged to the degree of formality of the occasion, or with solemn offerings of bells, drums, and voices lifted to the ancestors at temples and shrines.[53] Over time, the sources at our disposal evince greater interest in small-group performances of music (typically on solo instruments such as the *qin* or *se*) strummed at or near home for a small circle of intimates or even the unseen powers.[54] The discourses of friendship witnessed changes as well, with the first lyrical bursts celebrating the particular joys of intimate "friendships" (*you* 友) eventually outnumbered by angry polemics berating the petty man for his failure to prioritize correctly a wide array of less-than-intimate associations forged with colleagues, acquaintances, allies (*pengyou* 朋友, *liaoyou* 僚友, and so on), and contacts (*jiao* 交).⁵⁵ Because not only intimates, but also quasi-intimates exerted influences for good or ill through bonds that lingered in memory and endured well beyond the grave,⁵⁶ extreme caution was required to avoid becoming mired in webs of destructive relations.

In our texts we can trace the increasing theoretical sophistication of treatments of music and friendship over the centuries⁵⁷ and a sharpened sense of the role of music in creating social and political relations.⁵⁸ As changing conceptions of the locus and effects of music's power became more widely disseminated,⁵⁹ a disinclination grew among the early empires' social engineers to credit court entertainments with being the single most powerful force for good in court and society, offset by an increasing concentration on the transformative effect of music on cultivated persons in more intimate settings.⁶⁰ The result was the association, among elites, at least, of the values promoted by music with the values—personal, social, and political—of friendship, based on the consonance of the myriad things. After exploring theses topics, what follows examines the nature of friendship itself in ancient China—how friendships were made, the nature of intimate friendships, the benefits said to flow from them for the friends themselves and the social order, and the way in which friendships served as a model for social relations in general and as a reflection of the cosmic order of things.⁶¹

This chapter offers but preliminary treatments of these literary trends in the assurance that more finely grained research will follow. My arguments nonetheless challenge the platitudes alleging that the "horizontal" bonds of friendship were less important than the hierarchical ties between ruler-minister, husband-wife, parent-child, and older-younger siblings;⁶² also that music took a back seat relative to ritual in early and middle-period China. The volume of rhetoric and pictorial imagery devoted to music and friendship makes such commonplaces untenable.

Music: Theoretical Pleasures and Practical Utility

The earliest depictions of musical performance in China, inlaid decorations on bronzes dating to the fifth century BCE, show four types of musicians grouped alongside rows of dancers: players of drums, players of sets of tuned bronze bells, players of sets of tuned stone chimes, and players of wind instruments (Figure 2.2).⁶³

The early theoretical literature often includes a still broader range of musical topics, adding piping, pitches, sounds, and animal calls — essentially, any rhythmic movement of *qi*, the configured breath or energetic stuff thought to animate all life forms, visible and invisible, human and divine.[64] Less expansive passages in the literature posit two main divisions within instrumental music: the bells and drums featured in ritualized musical performances at court, at feasts for the living, and at sacrifices for the ancestors, versus the winds and strings as instruments enjoyed in more informal surroundings, on the road or at home. At the same time, the sources list no fewer than eight materials for instruments (metal, stone, earthenware, leather, silk, wood, gourd, and bamboo), each producing a different quality of sound or timbre.[65] Those Eight Sounds in turn, corresponded to the Eight Winds — the entire universe, in effect — with one sound and instrument allotted to each of the four cardinal points and their midpoints, though later these eight, propelled by Five Phases categories, would devolve into five (the four directions plus the center), said equally to account for the whole of time and space.[66] Because music and sound were at once produced and carried by the vibrations of those winds, music figured largely as a subtle, yet intelligible medium whose regular effects were knowable to experts equipped with extraordinary insight.[67] Thanks to broad interest in the operations of *qi*, discussions of music remained relevant to a host of moral and sociopolitical issues for millennia.

For music, unseen and intangible, was held to work directly, swiftly, and powerfully upon the imagination and emotions, in contrast to the other forms of pleasurable experience, which were believed to stimulate parts of the body composed of grosser materials.[68] As one proverbial line put it, "If a person hears the [refined] sounds...then, as a result, his ambitions and commitments are enhanced" and suitably improved.[69] Ideally, music functioned to "extend the person," unlike the rites, whose emphasis lay in measured requital.[70] For when the exemplary person "hears the tones, he does not merely hear the sound of tinkling jade"; the music supposedly "translates" in him into a longing for sublime harmony in all his dealings.[71] Associations with motion and moving — already embedded in the rhetoric of pleasure-seeking and pleasure-taking (see Chapter 1) — are especially prominent in discussions of music, because music not only "moves" people, producing strong emotions,

Figure 2.2. Jannings *hu*, late Zhanguo period (475–222 BCE). Height 31.8 cm, diameter (at the mouth) 11 cm. Originally part of the Werner Jannings Collection, now in the National Palace Museum, Beijing.

This bronze wine vessel, often viewed as an early milestone in the history of Chinese portrait painting, is covered entirely with pictorial scenes arranged in three main registers that are each subdivided, with the central register occupying the upper half of the vessel's body. On it are featured three scenes: one of hunting birds with stringed arrows in a natural environment suggested by the fish below, one of people in a roofed enclosure preparing offerings in vessels, with other people dancing under the eaves of the building, and the scene shown here, of musicians beneath a large set of chimes and bells that hang from a stand supported by bird-shaped legs.

but also literally sets them in motion,[72] providing the pulse that drives many human activities forward. Hearse pullers, loggers, and boat pullers working in gangs coordinated their pushes and pulls through the singing of work songs. Invocators and shamans and other religious experts danced and sang to solicit the aid of the gods, sometimes interpreting birdcalls as revelations of the spirits' will.[73] And the best charioteers used signal bells and jingle bells to impel the four horses of a quadriga to run in unison.[74] Given this powerful effect, "Music is what the sages take pleasure in" (*yuezhe, sheng ren zhi suo le* 樂者, 聖人之所樂也).[75]

A charming story cycle sketches the sunny continuum from music to harmony to friendship, centering on the legendary figures of Bo Ya and Zhong Ziqi, who lived sometime prior to 221 BCE. The basic story goes something like this:

> In olden times, the zither player Bo Ya 伯牙 strummed his zither and Zhong Ziqi 鍾子期 knew what was on his mind. If he was thinking of a man, Ziqi knew it, and if he had his mind on a river, Ziqi knew it. Zhong Ziqi always grasped whatever came into Bo Ya's heart and mind. Once Bo Ya was roaming on the north side of Mount Tai when he was caught in a sudden rainstorm. He took shelter under a cliff, and feeling somewhat pensive, he took up his zither and strummed it. First he composed an air about the persistent rain, then he improvised the sound of avalanches. But whatever melody he played, Zhong Ziqi never missed the direction of his thoughts. And so Bo Ya put away his zither and remarked with an admiring sigh: "Good! Good! How well you listen! What you imagine is just what is in my heart and mind. Is there nowhere for my notes to flee to?" Later, when Zhong Ziqi died unexpectedly, Bo Ya broke his zither, because he knew that no matter how well he might play, he would never again in his lifetime experience such a good listener as Zhong Ziqi.[76]

Bo Ya knew that his friend's way of listening immediately intuited his own heart. Zhong Ziqi did far more than merely recognize a melody. For him, the timbre and phrasing revealed Bo Ya's motivation and mood, as did Bo Ya's occasional retreat into silence.

Thanks to the strong sympathies binding Bo Ya and Zhong Ziqi through life and death, for over two millennia, the single binomial phrase "knowing the tone" (*zhi yin* 知音) — a set phrase occasionally associated with the homonymic *zhi yin* 至音, the "ultimate sound" whose nuances elude the ordinary run of men — has evoked the unique satisfactions of good music and close friendship.[77] Patterned tones being inseparable from human expression and from human apprehension of the social, "to know the tone" of another person was to find a soul mate. And so the brilliant Boli Xi 百里奚 reunited with his long-lost wife after each recognized the other's tone in music and song,[78] their voices and emotions being absolutely distinct. Because an "appreciation of music resides not in individual tones, but in the relationship between one tone and another,"[79] it suggests friends' and lovers' mutual responsiveness and empathy. The historian Sima Qian pushes this notion quite a bit further in his treatise on music when he states categorically of the same phrase, "knowing the tone,"

> To know a sound, but not recognize the tone [equally of a voice or a musical piece, with both expressions welling up from deep within the human heart] is to be less than human — no more than a bird or a beast. To recognize a tone while failing to understand music is to belong to the common run of mortals. Only the exemplary person is able truly to appreciate music.[80]

By Sima Qian's account, people, as the most strongly mimetic of living things, will derive real pleasure in training themselves to be more exemplary figures once they have thrilled to the distinctive quality of the "tone" that charismatic people acquire.[81]

From the dawn of theorizing, then, a talent for friendship and an appreciation for music bespoke eagerness for the immersion into something or somebody else capable of sparking and quickening one's own sensitivities. To understand the power of this claim, one need only consider the famous anecdote where the Supreme Sage Confucius learned through practicing a single musical composition to identify and emulate the spirit of his hero, King Wen, who lived five hundred years earlier.[82]

> Kongzi was studying the zither with Music Master Xiang, but he had made no progress for ten days.[83] Music Master Xiang said, "You had better improve a bit." Kongzi complained, "I have learned the tune, all right, but I have not got the rhythm, certainly." After some time passed, Music Master Xiang said, "You have got the rhythms down, so you can go on now." Kongzi protested, "But I have not yet got at the intention of the music!" After some time, Master Xiang said, "You have got to the core of its intention, so you can go on now."
>
> After still more time spent in practice, there came across the countenance of Confucius a solemn and profound longing, followed by a joyful look upward, as if he were making a far-reaching commitment. Kongzi said, "Now I know the composer! He is dark and tall. [Note that Kongzi himself was tall and

reputedly the scion of the Black Lord, so he was no less "dark and tall.'] He is infinitely great, with great pools of eyes that seem to command the kingdoms in all four directions. Only King Wen could have composed this!" The Music Master Xiang rose from his seat, and bowed to Kongzi, "Ah, yes. My teacher once told me that this is "King Wen's Zither Tune" and that is the very piece you've been playing."[84]

In the end, music was the medium that made King Wen's presence fully "available" to the ritual master Kongzi, which encounter inspired the latter-day sage in his committed pursuit of the Way.[85] For powerful strains have the potential to induce a dual awareness comingling the listener's *felt* emotion induced by the music with the *perceived* emotion that the listener imputes to the composer of an artful composition.[86] Thus, listening to and playing music, whether at court or outside it, could heighten the imaginative powers until heart and mind came to perceive the ineffable.[87] After all, in some powerfully affecting cases, music places "practically nothing but tonal structures before us: no scene, no object, no fact."[88] As one early text puts it:

> In general, what we call sound is born in a person's heart and mind. What is felt there then moves to the aural. When the aural takes form outside, it transforms the inside. That explains why, when you hear a particular sound, you know the local customs and airs [literally, its "wind"], and when you investigate the local customs and airs, then you understand the will and aspirations [of those residing there]. Observing their wills, you know their characters. Be they successful or declining, worthy or unworthy, noble or petty, all takes form in music; nothing can be hidden. Hence the saying, "Music as spectacle is profound...." The music being harmonious, it induces perfect accord [in the focused listener]. The music being harmonious, the people are attracted from everywhere to it.[89]

Small wonder, then, that the early stories portray the sage Kongzi (that is Confucius) as coming, through long musical practice, to apprehend the sublime qualities of King Wen of Zhou's character.[90] Thus, music's transformative processes in the cosmos and society work in such a way as to make people simultaneously the agents, subjects, and objects of beneficent change. Indeed, insofar as music operates through resonance, rather than mechanical cause and effect, it illustrates a process of mutual transformation ill served by the modern dichotomies of subject/object or agent/acted upon. True, in a clever essay entitled "On the Absence of Emotions in the Music" (聲無哀樂論), Ji Kang 嵇康 (223–262 CE) had an unnamed interlocutor try to refute the idea that it is possible to "know a person's proclivities and feelings (*qing* 情) by examining his music."[91] Still, Ji's dialogue inadvertently reveals how very prevalent the notion of music's boundless expressive, affective, and communicative powers had become.

This theory of music had many practical implications and ramifications across the early North China Plain societies. For one thing, because of its powers, music was by no means secondary to the rites; rather, music was taken to be "the highest" or "ultimate" experience and the supreme form of social cultivation, as the term *zhi yin* 至音 indicates and the rites master Kongzi came to learn. Its cultural significance therefore was unparalleled, insofar as music conduces to mimesis and cohesion, for in music "similar sounds mutually respond to one another" and "spirits conjoin."[92] Somewhere near the wellspring or "bedrock" of human feelings,[93] music served as solid foundation for the rites, which are but social constructions, however rooted they may be in the basic human feelings of love and respect.[94] For "only music cannot be treated as artifice or faked."[95] And because music by a graphic identity *was* pleasure, music became the supremely "human way" to refine oneself and others, the conscious cultivation of a taste for music being one of the most effective ways to cultivate personal character[96] as prelude to the all-important arts of social cultivation.[97]

Long-standing traditions link two more aspects of music closely with friendship: because only humans can learn to appreciate music, development of this capacity demarcates not just human society from that of lesser beings, but also noble spirits from the less mature, and curiously, only music and friendship can each compensate in hard times for the lack of the other. Ordinary people may thrill to a piece of music, but only the most worthy can fully experience the special impacts of an instrument or a human voice in song. Then, too, "nothing compares with music in its ability [to forge relations] for one who dwells in poverty and seclusion,"[98] and nothing compares with friendship in its ability for turning simple social settings to the confluence of hearts and minds. Thus, music and friendship could each serve as antidote for the lack or loss of the other. Evidently, when despairing at the loss of true friends, only music can console:

Xie An 謝安 [320–385] once said to Wang Xizhi 王羲之 [d. 361], "In middle age, I suffer from grief and pleasure. Whenever I take leave of a relative or friend, strong feelings plague me for several days." Wang replied, "When a person is advanced in years, this is natural. For that very reason, I rely on flutes and strings to transform these feelings, affording me a sense of catharsis."[99]

Never does one regret having had a true friend who was later lost to distance or death: one can always find a casual companion to enjoy a drink with, if musical instruments or entertainments are not near to hand, but the experience or memory of a true friend imparts a unique blessing down to one's dying day, despite the "weariness of longing" felt for the absent companion.[100] Bereft of friends near to hand, the cultivated person could still gain that satisfying sense of comingled release and connection by using his body as his instrument and whistling, as we see from Chenggong Sui's 成公綏 (fl. Western Jin 265–315) charming "Rhapsody on Whistling":

> To make this sound one needs no instrument,
> To effect it requires no other thing.
> One takes what is near at hand in his own body,
> And makes it, using his heart and controlling his breath.
> He simply purses his lips and out comes a tune.
> He opens his mouth and creates a sound.
> Moved by whatever he encounters, he
> Responds in kind and sings forth accordingly.
> The sound is loud, but never boisterous,
> The sound, howsoever faint, is never inaudible.
> In clarity and intensity, it matches syrinx and mouth organ.
> In richness and smoothness, it equals zithers, great and small.
> Its mysterious wonder, enough for communing
>   with gods and wakened spirits.
> Its refined subtlety suffices to explore the hidden and fathom the deep.[101]

Although conceptions of human nature, cultivation, and music's status in social relations altered in ways both subtle and significant over time, for early readers of Chinese, this tight bond between music and the communion of the cultivated person with others naturally elicited an awareness of music's political utility and its role in creating social relations.[102] Shortly before unification in 221 BCE, Xunzi remarked, "Music is the most effective way to govern men."[103] Sima

Qian explained why through a play on words: "Whoever devises or takes part in a musical performance (*yue*) aims at modulating and moderating pleasure (*le*)" (凡作樂者, 所以節樂);[104] after all, the chief business of policy making (if not always music practice) was to instill in the people a sense of moderation and the rhythms of pleasure-taking and pleasure-giving. Music prompted a desire for self-regulation in preparation for being "in sync" and in concert with others. "If each and every bird and beast is moved by music," Sima Qian writes, "how much more is it true of [the exemplary people who] cherish the Five Constant Social Relations in their hearts!"[105] And since music managed to bring quite ordinary people together in constructive ways, it was bound to be even more effective in forging links with the exquisitely sensitive gods and spirits, some proportion of whom had once walked the earth as human beings.[106] So important, then, was music to good governance that while certain aspects of music could be left to functionaries (for example, the stage management and performance of grand musical spectacles at court and temple), superintending "the heart of music"—that is, its core civilizing power—was a responsibility assigned to none other than sage-rulers and sage-ministers.[107]

The *Hanshu*—composed circa 100 CE, roughly two centuries after the *Shiji*—takes a different tack, one suggesting a gradual downplaying of court music's significance relative to that of court rites.[108] It displays the more self-conscious classicism of the late Western and Eastern Han courts and contains no separate treatise devoted to music.[109] A single treatise considers music and rites there,[110] with rites its chief focus, perhaps because several new arts, including elegant prose and elegant calligraphy, challenged music as chief emblem of personal and social cultivation at around the same time.[111]

Though the *Hanshu* lifts several passages wholesale from the *Shiji*,[112] it introduces subtle departures from the *Shiji*'s language. Whereas the *Shiji* alleges that neither music nor the emotions it evokes are inherent in basic human nature (unlike the innate desires for food, sex, and companionship), they being instead responses that well up when percepts meet external stimuli, the *Hanshu* repositions those emotions *within* basic human nature. By that logic, performance of the sages' music regulates the physical person and facilitates communications with the divine aspects of the cosmos, all through the subtle animating medium or "quintessential *qi*" (*jing* 精) that allows for "divine insight" (*shen ming* 神明).¹¹³ In consequence, music becomes the single best tool by which to align with the unseen processes, in contrast to the rites, which are designed to adorn visible acts.¹¹⁴ Reinterpreted in the later text as an intuitive feel for order, music propels the emotions and sets off the vibrations that follow, although its expressiveness cannot replace more deliberate actions. As the text says, "A delight in harmony and intimate feeling is hard to form or shape, but then [finally] one lets it out in lyrics, songs, and chants, in bells, stone chimes, flutes, and strings."¹¹⁵ Needless to say, this pleasurable feeling consists of being at once securely centered and more open and in tune with what's outside.¹¹⁶

Notably, according to the status-conscious *Hanshu*, professional musicians (with rare exceptions) benefit little from music's amazing properties. Whereas Sima Qian, like the pre-Qin masters, had taken for granted the ability of professionals to civilize themselves and others, the *Hanshu* writings on music focuses less on the type of music played than on the status of the player or connoisseur-listener, with status affecting music's ability to transform human beings suitably.¹¹⁷ By the *Hanshu* logic, mere professionals, such as those employed at the imperial Music Bureau (*Yuefu* 樂府), might "know how to sing or play the instruments, but few of them really understood the theory and underlying significance of music."¹¹⁸ Only highly cultivated members of the governing elite had the requisite innate refinement and prolonged training to understand the theoretical framework of the musical arts, whether they took up an instrument or not, so only they possessed the hearts and minds with the capacity to be transformed profoundly by music.¹¹⁹ (Lady Catherine de Bourgh would concur!)¹²⁰ So by Yan Zhitui's 颜之推 era (531–591), men of high standing were specifically advising their sons not to play "too well," lest "they gain a reputation in that art and be forced to entertain the high and mighty" like men and women available for hire.¹²¹

Changing Conceptions of Music's Power
A complete history of music or musical appreciation lies well beyond the scope of a book whose principal subject is early pleasure theories. However, a short sketch of musical appreciation and performance in the pre-Song and Song periods with regard to the role of music in producing social bonds can help to prepare the ground for what comes next — the consideration of theories about intimate friendship.

Figure 2.3. Zeng hou yi 曾侯乙 tomb bell set, assembled before 433 BCE and excavated in Leigudun, Hubei Province, 1977–1978.

This bell set, weighing some 2,500 kg of bronze and wood, was housed in the central chamber of four in the tomb of Marquis Yi, ruler of the small state of Zeng. Its lavish gold inscriptions provide two sets of pitch standards, attesting the close ties between the small state of Zeng and its powerful neighbor Chu.

For most of human history, until the invention of record players, transistor radios, and MP3 players, music typically was heard when humans came together, in ritual settings or at leisure times, for a common experience, which in turn often became the basis for trust in a common project. Generally speaking, "music" in early China tended to refer to orchestral pieces performed by musicians in concert with dancers and singers, so when the masterworks debated the proper uses and kinds of music, they usually confined their remarks to musical performances at court. Nearly all of Mozi's 墨子 (ca. 470 – ca. 390) famous harangues against "music" expressed dismay at the great expense of putting on spectacles for the ruler, given that so many subjects lived at or near the margin of subsistence.[122] Were musical performances to be curtailed, Mozi reasoned, the savings could go to assist the truly needy. Besides, the mere presence of seductive male and female entertainers distracting power holders from their responsibilities made highly orchestrated musical performances of dubious moral utility.

To see what Mozi criticized, we need only look at the ensemble of sixty-five bells found in the central chamber of the tomb of Marquis Yi of Zeng (d. 233) (Figure 2.3). At a minimum, this bell set totaling some 2,500 kilograms of bronze and wood with ornamental inscriptions in gold would have required more than twenty professional performers at a time.[123] There were also forty-one stone chimes for grand performances after death. In this way did elites plan to deploy musical instruments to attest their status in the afterlife and to adorn their persons.

Xunzi countered with a reasoned defense of musical productions, emphasizing the harmony and balance (both *he* 和) that could be instilled when members of a community watched a single performance and harnessed their moods and movements to its rhythms and melodies. Any expense was warranted, Xunzi thought, so long as people could derive a satisfying sense of participation in the greater community of the living and the dead.

> When music is performed in the ancestral temple ... there are none not filled with a spirit of harmony and reverence. When it is performed within the household, fathers and sons, elder and younger siblings, listening together, are filled with a spirit of harmony and closeness. And when it is performed in the community, with young and old listening to it together, all are filled with

a spirit of harmony and accommodation. Thus music brings about complete unity and induces harmony.[124]

The value of true harmony, in Xunzi's view, far exceeds that of mere conformity, and hence Mozi is illogical and wrong when he excoriates musical performances at court due to their cost. For Xunzi, the word "harmony" suggested a finer mix resulting from myriad adjustments designed to "add to whatever was deficient and trim the excessive."[125] Moreover, the desirable unity of purpose that was the final goal of communal interactions could never be attained without a balance between moderation or frugality and lavishness, form and substance, striving and attainment.[126]

Xunzi had far more to say on the subject: that music quickens the conscious desire to join with others, which drives humane action; that music provides a necessary outlet for the emotions and the senses; that taking pleasure in musical performances reinforces the profound truth that hierarchy and order, humility and restraint, are as fundamental to successful social interactions in civilized orders as an aptitude for empathy; that music can heighten and extend a person's emotions, imagination, and intellect; that training in music and dance teach martial vigor and imbue performer and patron alike with awesome majesty; and that rhythmic movements can teach a joyous awareness of time's passing (which otherwise would tend to inspire fear). "Take but degree away / Untune that string / And hark what discord follows."[127] Finally, just as music's sound floats out to destinations unknown, music signifies and strengthens the profound and lingering influence radiating out from the man of cultivation, be he player, connoisseur, or commissioner of the piece. (That music was at the same time an excellent tool for prognostication was assumed long before Xunzi,[128] and Xunzi alluded to this belief, though he felt no need to elaborate it.) For all these reasons, Xunzi would have the music's presence "prove that it is possible to savor pleasure in harmony without any loose behavior."[129]

At least two of Xunzi's ideas warrant serious consideration: first, the strong ties he posits between musical airs and the invisible energetic "airs" sustaining not only the life of the person, but also the potential for civilized and humane conduct, and second, Xunzi's belief in music as microcosm or distillation of the larger forces of attraction at work in the universe. Good music, in Xunzi's view, prompts good airs or humors—an environment that makes ears and eyes keener, soothes dispositions, and tempers the person's predilections.[130] In theory, therefore, music allows each participant to edge somewhat closer to the goals of beauty and excellence (*mei* 美) (see Chapter 4), leaving a strong sense of satisfaction in performers and auditors, however conscious they may be of other constraints in life.

> The drum represents a vast pervasiveness; the bells, fullness. The sounding chimes represent restrained order; the mouth organs, austere harmony. The flutes represent a spirited order, the ocarina and whistle, a range of tone. The zither with more strings represents gentleness, and the zither with fewer, a soft grace. Songs represent purity and fulfillment, and dances, the spirit conjoined with Heaven's Way. The drum is surely the lord of music, its sounds recalling those of the sky. The bells resemble earth; the sounding chimes, water; the mouth organs and lutes, the sun; and the scrapers, the myriad things of creation.[131]

Thus does music become, in Xunzi's telling, "the great arbiter of the world, the key to fitting harmonies, and a precondition for [properly aligned] human feelings"[132]—an ineffable series of sounds that successfully engenders and animates an entire resonant order. Such themes would be taken up by many later writers.

Xunzi thought it sufficed to say that music has the capacity to civilize those in power and to bring subjects into willing allegiance. Two pieces of Qin and Western Han writing associated with Xunzi were even more insistent on elaborating the cosmic dimensions of music—the *Lüshi chunqiu*'s eight essays on the relation of sound and music to pleasure (compiled in Xunzi's time, around 239 BCE) and the later "Record on Music" from the *Record on Rites* (*Liji*) usually ascribed to Xunzi's disciples and their disciples. Regarding music's cosmic dimensions, the *Liji*'s "Record" makes the following claims, which would be impossible if music did not simultaneously constitute "perpetual flux" and "underlying order":

> In clarity and brightness, it is the image of heaven; in breadth and scope, the image of earth. Moving in time, it is the image of the four seasons. Just as the five colors, fitted into orderly patterns, do not clash, so, too, the eight winds, channeled through the pitch pipes, create no discord.... Altogether, the short and long pipes bring [the melody] to completion, in each piece engendering beginnings and endings. Melody and harmony, high tones and low, alternate

to weave the fabric [of the listening experience]. Therefore, when music goes forth:

> Orderly relations are clarified,
> The ears and eyes grow bright and sharp,
> Blood is put in harmony; the *qi* is balanced.
> Transforming ways and refining customs,
> Music brings peace and security to all the world.[133]

And yet, not all accounts of the role of music in the sociopolitical order stressed its ability to harmonize both the governors and the governed on the basis of cosmic order. Some found the major role that music played in the operations of the state downright debilitating. Responding to the heated debates between the followers of Mozi and Xunzi, the *Lüshi chunqiu* is particularly interested in exploring the paradox that for some ruling elites "music (*yue*) is no pleasure (*le*)" (*qi yue bu le* 其樂不樂). By the *Lüshi chunqiu* account, the decline of the major states in history came about when their rulers emptied their coffers to buy impressive sets of musical instruments, in large part because they committed the all too common error of "mistaking *bigness* for *greatness*." The resentment felt by the common people was augmented by the music these instruments produced: too booming, too exciting, too striking. Such assaults on the emotions diminished the vitality of ruler and subject alike, upsetting the customary balance between cold and heat, work and play, hunger and satiation. Imbalances weakened the body's natural defenses, making it all the harder to control the proliferation of addictions and desires,[134] particularly because the music offered no compensations for the negative "pulls."

Worse, the resentments, febrile excitement, and preoccupation with cares prevented the ears from fully attending to the right sorts of music. For these reasons, the ideal sage-rulers and their music masters had to pay attention always to due proportion—the size of the instruments, the type of melodic lines and orchestration, the size of the expenditures on instruments, and so on. They moreover carefully adjusted their music to the season and occasion to correct the tenor and temperament of the time.[135] From this it follows that maximum pleasure and the ultimate music (*zhile/yue* 至樂) can be secured only through orderly reigns, when natural sympathies link people to society and cosmos and dissolve the worrisome anomalies and disasters that cut short life spans and dynastic lines.[136]

And as time went on, music itself changed, as appetites for popular lyrics and foreign melodies triumphed in the imagination over solemn musical performances conceived as paradigmatic of the cosmic and social harmonies, except in the sacrificial rites. From the early empires on, some scholars had argued that the "high" or "elegant" (*ya* 雅) music thought to evoke the eternal verities came, around the late third century CE, somehow to be conflated with the "low" or "popular" (*su* 俗) music and lyrics, though supposedly the educated had once found it easy to distinguish "high" from "low" registers in terms of tone, lyrics, and moral import.[137] As one modern styles it, by Han times, "light" music no longer elicited the salutary censure or contempt, because members of the court unwittingly accepted it increasingly as "an unnamed category of high music."[138] Such an anachronistic narrative rests upon a firm belief in the accuracy of polemical accounts prone to distortion and a selective reliance on a few systematizing statements by biased parties.[139] In all probability, few court advisors or professional musicians ever cared much about any theoretical distinctions between "high" and "low,"[140] old and new.

The casual impulse by the Western Han founder, Gaozu, seems typical of the way pieces of music entered the court repertoire during the Zhanguo and Han periods: when Gaozu witnessed a performance of martial dances whose accompanying songs pleased him by their patriotic sentiments, he duly ordered his court musicians to learn the songs by heart, without posing a single question about their origin, "barbarian" or not.[141] And when the influential Han thinker Yang Xiong 揚雄 (53 BCE – 18 CE) belatedly queried the songs' origins, his usually admiring disciple Huan Tan 桓譚 (43 BCE – 28 CE) dismissively remarked that it was really too bad that Yang understood so little about music.[142] By and large, as the dynastic histories make plain, the ritual music of the Qin and Han courts happily reframed and adapted melodies and lyrics that were already circulating widely,[143] producing works that were at once "original" and yet simple rehashes. Similarly, in the immediate post-Han centuries, the proponents of "light" music were always present at the imperial and royal courts, ready to rename reworkings of older performance pieces, as needed.[144]

Writings about Wudi's Music Bureau (Yuefu), or the later Music Bureaus of Western and Eastern Jin (265–420), provide much the same picture. No *Music* classic was ever "restored" during the two Han dynasties or the successor state of Wei, despite the unwavering belief that such a "lost" classic had once existed. Evidently, contemporary music was good enough to gratify elite audiences.[145] So although the *Hanshu* "Treatise on Rites and Music" describes several abortive attempts to "restore [ancient] music," it duly records the abject failures, as well. Sadly, there came no triumphal restoration of the "old" music that could lend the dynasty a decided air of respectability and inviolability.[146] Likewise, with the Jin court, we read of the technical writers consulting contemporary musicians for advice, despite the enduring belief that dynastic legitimacy depended upon forging stronger ties with antiquity.[147] Chance finds of genuinely ancient instruments[148] and the customary and conventional celebrations of "legitimate" dynastic successions spurred debates over the best use of music to promote good rule, but the debates were never resolved.

Nor did the term "old," when applied to music, invariably denote "Western Zhou," let alone times before that, a fact that seems to be lost on most modern historians.[149] Certainly, as late as Southern Song (1127–1260), Zhu Xi and his followers decided to employ twelve melodies of Tang or Song for the twelve odes to be sung during the newly invented community drinking rituals whose origins they traced back to the preunification era (see below).[150] Reform-minded idealists handily foisted makeshift, fanciful, and even downright bogus "returns to antiquity" (*fugu* 復古) on the unwary and cognoscenti alike. To help verify the old harmonies, the standard measurements for the pitch pipes were continually reset and the bells recast, even as the court orchestra and its repertoires were adjusted through the addition or deletion of various instruments.[151]

As prospects for recovering the antique music dimmed, a related long-term trend surfaces: the gradual "privatization" of the musical pleasures and the valorization of intimate musical experiences, accompanying the downgrading of the status of professional musicians (see above). Judging from the extant sources, court performances of music and dance by trained professionals served less and less to mark superior sophistication and erudition, once increasing numbers of aristocrats and *nouveau riche* could purchase for their own enjoyment everything needed for spectacular displays to rival the court's. More and more, private (that is, unofficial) renditions by elite amateurs tended to be touted as the true insignia of culture, especially when zithers, great and small, were involved, instead of the clamorous bells and chimestones played by experts. The more selective the gathering, the better. This marginalization of showy court spectacles among high-status amateurs, coupled with status concerns about entertainers for hire, seems to have fostered the governing elite's propensities to cast "soundless music" as the most elevated sort of music, apprehensible only to rarified characters with exquisite tastes.[152] (An undoubted advantage: with soundless music, no gentleman need bother practicing before claiming a reputation as master.)

The general trend favoring intimate music was surely facilitated by a very slow shift in types of musical assessment, from judging melodies to judging lyrics, a shift likely encouraged by the belated realization that no amount of erudition would ever permit recovery of the original sounds of the antique music. This requires a brief consideration of the frequent denunciations of "the music of Zheng and Wei." In Sima Qian's time, the phrase unquestionably referred to the melodies (qu 曲) associated with those two small kingdoms, believed to be corrupting because they were plaintively sounded on wind instruments, rather than on chimes or bells and frequently lapsed into (possibly foreign-inspired) minor keys.[153] That initial focus on the musical sounds (yin 音) shifted gradually to an emphasis on the content of the lyrics (wen 文). In the end, it was Zhu Xi 朱熹 (1130–1200) who finally condemned certain lyrics as "depraved," over stiff protests by those who deemed Zhu to be misreading, possibly willfully, the earlier sources on music.[154] Zheng Qiao 鄭樵 (1104–62), for example, argued forcefully that "the main point of the *Odes* was [originally] their musical phrasing [*yuezhang* 樂章], not the meaning of their lyrics [*wen yi* 文意]."[155] But assent to Zheng's complete argument entailed a virtual admission that probably no music from the original *Odes* classic had survived the fall of the northern capitals to "barbarians" by 316 CE,[156] making many shy away from a principled embrace of Zheng's position, particularly when Zhu Xi gave such an elegant account suggesting that his practice of disciplined recitation could "make up for" the loss of those antique musical effects.[157] Rather than consign the study of the transmitted *Odes* classic to irrelevance, many classicists, like Zhu, welcomed unfounded, even preposterous theories.

The preceding—helpful to Sinologists confused about the destructive properties ascribed to the music of Zheng and Wei—may well strike ordinary readers as too much information, but the rhetorical shift away from a fascination with music's power to a focus on the damage incurred by listening to sensual words is symptomatic of a constellation of changes that Zhu Xi fostered, in acknowledging irretrievable cultural losses from the antique sage-kings' eras.

Music and Friendship: Consonance

Music was far more than a metaphor or template for friendship. Music offers a unique pleasure to humans, a stimulus to motion and emotion.[158] In early and middle-period rhetoric, where social cultivation — not self-cultivation — seemed the highest goal, achievable through steady practice, music, and intimate music in particular, played a seminal role in elevating the quality of intimate and even merely affable personal exchanges.[159] What did such music practice confer that little else could, according to the prevailing theories? Performances could embrace and convey complexities that logical thought could not and be all the better for it, serving as compelling "emblems" for the achievements of the sages.[160] So also with the intimate gestures of close friends, which eluded facile explanation. Yet music was held to "ornament" (*shi* 飾) the feelings, releasing a regulated and equable state.[161] The common wisdom made "music the flower [the glorious product] of virtue."[162] No less germane, a love of music habituated a person to a special kind of listening, a hearkening to subtleties of the situation beyond the obvious, a focus on the unseen, even, and a concomitant willingness to give oneself over to the larger schemes unfolding as the music proceeded. (Careful listening and a kindred willingness to hear the person out are, of course, fundamental prerequisites for intimate friendships.)

Herbert Fingarette has outlined a multistage process that takes place when a powerful musical performance enthralls listeners: at first, before settling down, a receptive audience may experience the opening bars as separate notes in a sequence, but eventually, there comes a more concentrated attentiveness to the structural patterns of the music, including its pauses, prompting a kind of curiosity as to how the music will roll out and a reach or lift, an anticipation and a yearning as each listener abandons herself to the pace and symmetries of the music. Finally, there is an overwhelming communication of beauty and order whose impact is likely all the greater when shared with fellow lovers of music.

Listening here, as in Barthes's formulation, is a psychological act, in contrast to physiological hearing.[163] It both recapitulates the past and predicts a future, based on stored knowledge. True, the spell cast by the music, that sense of being transported together with the music, its composer, and the audience, may be broken as soon as the listener would put a name to what has happened. For one, "It reminds me of my mother singing," for another, it is "Ella in Carnegie Hall." But a poignant awareness or residue of the communal immersion often lingers after the post-facto attempts to register personal reflections,[164] imparting a sense of coherence and satisfaction.[165]

The most famous bell set we have, the set that once belonged to Marquis Yi of Zeng, alerts us to yet another feature of music. The pitch standards inscribed on gold on the bronze bells always describe a tone to be sounded in terms of its related pitch standard; then two sets of standards, one from Zeng and one from neighboring Chu, are correlated with one another, with all the pitches fashioned as reflections of the others. What is going on? The two sets of pitch standards, geared to different intervals, attest the close relations between two neighboring states, suggesting the bells figured in a diplomatic exchange between trusted allies; the stipulation that the pitch standards be understood as mutually generative of recurring relations with variations that created pleasing consonances, rather than as static and unrelated points, swiftly conjures two words describing music and friendship equally: harmony (*he* 和) and the "binding" nodes (*jie* 節) that punctuate the rhythmical patterns. For just as *harmonia* is tied to *nodus* in classical Greek and in Latin, which means "joining," "tying," or "binding," so, too, in classical Chinese, certain configurations of aural movements at intervals of perfect proportion, though the product of tension, could lend grace to time's unbroken and amorphous flow. Insofar as the binding force inherent in rhythm and harmony nicely balanced music's improvisational and momentary character, it could govern, even restructure quivering sensibilities, tensions, and emotions, nudging people and whole communities to embrace shared actions and values. This kinetic energy stemmed from the sheer dynamism inherent in the music's primal force as "the originary phenomenon of the human spirit and the cosmos itself."[166] (Note the presumed deep interpenetration of fact and value, objectivity and affect, regularity and change, as the music resonates within and without.)

Together, rhythm and harmonics offered ideal approaches to consider the perfect consonance attaching microcosm to macrocosm in the natural world of heaven and earth, just like friendship in the web of social relations; such approaches set aside simplistic models positing a single line of effect traveling through sender-medium-receiver or neat divisions between the biological and the cultural. With consonance rooted in the marvelous coincidence of percussive beats, the listening ear became the "central organ of the metaphysics of presence," with doubts reserved for the question whether the pleasures of music derived principally from the tonal concatenations themselves, from the listeners' inclinations, or from the profound communications transpiring between the living or the living and the dead.[167] Regardless, to feel at one with others made for worldly blessings and good feelings, as one early Chinese text puts it.[168] Thus, a musical performance is the paramount, tried-and-true way to "nourish heart and mind,"[169] far surpassing alignment with ritual norms in its ethereal unforcedness. And, bestowing a not inconsiderable boon, music's pleasurable symmetries comfortably implied that "ways of explaining parts of the universe can also explain other parts," since the myriad things of the universe are symmetrical in their parts and in their dimensions, clearing a path for connectivity across the ethical, theological, and physical orders and quelling fears of unpredictable contingency.[170]

So, too, did intimate friendship, with an occasional assist from witty repartee or raillery (*ji jie* 疾節), usher in a welcome coherence, proportion, and unity, reinforcing life's goodness.[171] As it happens, in music, the most pleasing consonances result from the simplest ratios between pitches,[172] and much the same could be said of true friendships: often the less artful the ties, the higher the gratifications. One Western authority reckons, "The human voice seems most pleasing to us because it is the most directly attuned to our souls. By the same token, the voice of a close friend is more agreeable... because of sympathy or antipathy of feelings."[173] Friendship adds savor and spice to life; in the Chinese way of speaking, it "nourishes the heart and mind," rendering it all the more receptive to the "quintessential, marvelous, refined and delicate."[174] The "Yueming" chapter of the *Documents* classic has one close friend saying to another, "Do teach me what should be my intentions. Be to me as the yeast and the malt in making sweet spirits, as the salt and plums in making a well-blended stew."[175] Personal perceptions, inflected with a sense of history, entwined logically separable topics in nearly all the extant stories.[176]

The Nature of Friendship

The "common wisdom" retrojected from late imperial China holds that all important extrafamilial relations appealed to the single paradigm of the family, whether in the typical small household of some five persons or the much larger webs woven by marriage and mourning ties, making the best of friends but quasi-kin.[177] But nearly the opposite was the case. The preponderance of the extant early texts sketch a broad range of friendships presumed to yield ever greater satisfactions as intimacy increased. Intimacy defined one's sensibilities and oriented that fusion of desire, energy, and intellect whose continuous operation "moves among and responds to particular objects of attention, [acting as] the force of magnetism and attraction that joins us to the world, making it a better or a worse world."[178] At play were the various attractions and commitments binding people to those *beyond* the family and how best to express those feelings and obligations. (Admittedly, some few family members—wives, siblings, and even fathers—might also figure as intimate friends, as the early texts readily acknowledge, but family members were not invariably soul mates "to confide in.")[179] Even master-disciple relations could be likened to "mutual friends."[180]

No early Chinese thinker, so far as I know, would have agreed with Seneca's pronouncement that "the wise man is self-sufficient," "living happily even without friends."[181] As far back as we have texts in China, friendship seems to have been construed as vital to the good life replete with sustaining pleasures. Chinese thinkers seem much closer to Aristotle, who says that friendship "is most necessary for our life," insisting that "no one would choose to live without friends, even if he had all the other goods" such as wealth, power, fame, comfort, and beauty.[182]

That said, friendships, like music, were seen as hard to capture in words, being at once preverbal and postverbal (see below), intensely personal, even idiosyncratic, and liable to wide variations.[183] So one often gets the sense, when reading about past friendships, that "you had to be there" to appreciate the coherence and fitness of the people making the scene of friendship.[184] Unhesitatingly, the early and middle-period sources assert that pleasure-taking in the company of worthy and convivial companions is the most proper theater for human enjoyment, which meant that the most satisfying forms of pleasure-taking typically combined friends and music, whereas solitary pleasures were said to be prone to melancholy, excess, and morbidity. While friendships forged inside and outside the court surely differed in circumstance and type, the early descriptions of intimate friendships, regardless of origin,[185] define this sort of friendship in much the same way: as "friends who act to help their friends realize their potentials," what we might dub "their personal bests." It matters not a jot whether the stories told about famous friends are historically accurate, since these legendary figures personified the pleasures of intimate friendship down through the ages.

When examining examples of friendship understood as the perception and cultivation of a friend's true character, the natural place to begin after Bo Ya and Zhong Ziqi (see above) is with Guan Zhong 管仲 and Bao Shuya 鮑叔牙, whose legend appears, with variations, in many story cycles.[186] During the Chunqiu period, Guan Zhong (d. 645) was the most famous prime minister to hold office in the Central States.[187] He served Duke Huan in Qi with great distinction, paving the way for Qi to assume the role of hegemon, that is, effective overlord for all the princes nominally allied with the weak Zhou royal house. So impressive were Guan's achievements that Confucius purportedly opined that had it not been for Guan Zhong, the Central States inhabitants living in the North China Plain would probably have been forced to give up their ceremonial robes for nomadic garb.[188] However, routinely laid at Guan Zhong's door were the ignoble last days of Duke Huan's reign, with the duke sunk in undignified excesses of every sort. So by the time of Duke Huan's death, encoffining, and funeral (the first hastened and the last two grotesquely delayed by mismanagement at the Qi court), many were ready to castigate Guan Zhong (and somewhat less so, his liege lord) for wasting Qi's moral, political, and financial capital, turning it into "the laughingstock of the empire," despite every inherent advantage.[189] Many were equally ready to term Guan's long service to the duke not "great loyalty," but "second-rate [that is, superficial] loyalty," given Guan Zhong's failure to reform the duke's own wretched proclivities and thereby achieve Qi's long-term interest.[190]

If Guan Zhong proved less than a paragon as prime minister, his unblemished reputation as friend to Bao Shuya earned him a secure place in history. During the struggles of succession among various Qi princes that preceded Duke Huan's rise to the throne, Guan, then a staunch partisan of the future duke's chief rival, had tried to kill Duke Huan. But as luck would have it, Guan's arrow was deflected by a metal clasp on the duke's robe. The duke not only survived the assassination attempt; he went on, quite unexpectedly, to vanquish his rivals, including Guan's prince. Once he assumed the throne in Qi, he immediately named Bao Shuya as his chief advisor. Given Guan Zhong's earlier strike at Duke Huan, Bao Shuya, who had known Guan for some time by then, had his work cut out for him when he decided early on in his tenure as advisor to recommend that Guan Zhong replace him in the highest office of the realm.[191] Bao Shuya argued that his friend Guan excelled him in five areas, which both separately and together made him a much better candidate for the position: "In generosity and concern for the people, I, Bao, am not his equal. In loyalty and fidelity, the very qualities that attach him to others,[192] I am not his equal. In managing rituals and mandating laws in the four quarters, I am not his equal. In deciding lawsuits equitably, I am not his equal. In . . . inspiring troops with valor, I am not his equal."[193]

Bao Shuya was adamant in Guan Zhong's defense, and unlike Guan Zhong in later times, Bao had no intention of letting his ruler off the hook simply because of either past deeds or private biases — not when the destiny of his beloved Qi realm was so clearly at stake. So Guan Zhong was eventually named prime minister, and he set Qi, at least during the first part of Duke Huan's reign, on the path toward political greatness, thanks in large part to his careful attention to diplomacy and finances. That Bao Shuya was so persistent a backer for Guan was somewhat surprising, for the story goes that Guan, time and time again, had acted so dishonorably that a lesser man judging by conventional standards would have severed all contact.

Five incidents are named in one of the longest cycle of stories: When Bao and Guan were in business together, and Guan took more than his share of the profits, Bao let him do so, since his friend was poor. When Guan Zhong made plans for Bao Shuya, Bao never blamed him when the plans went awry; instead, Bao decided that the times were not yet ripe to fulfill their plans. When Guan was dismissed three times from low-level offices, Bao did not charge him with incompetence; instead, Bao realized that the job had yet to be offered that suited Guan's unusual talents. When Guan fled the battlefield three times, Bao Shuya never thought him a coward, knowing that Guan had an aged mother to support. And when Guan Zhong's patron, the senior prince of Qi and rightful heir, died during the succession struggles, Bao Shuya credited Guan Zhong with practical wisdom, since Guan never once considered committing suicide so as to accompany his lord honorably in death. (Bao Shuya knew that Guan would never willingly die before he had made his name.) In every instance, Bao Shuya had discerned the true character of Guan Zhong behind the shameful appearances. Hence, the truth of Guan Zhong's summary judgment late in life: "My father and mother are the ones who bore me, but the one who knew me really well was Bao Shuya."[194] Keen discernment about a friend's character and actual potentials is the one true basis for friendship (and indeed life), as echoed throughout the classical literature.[195]

Accordingly, many acclaimed Bao Shuya a far greater figure than Guan Zhong. In one imaginary dialogue, Kongzi told his disciple Zigong that Bao Shuya was responsible for Qi's ascendancy, since it was Bao Shuya who had forcefully recommended Guan Zhong for the highest post: "To recognize a sage is to be wise. / To advance a sage is to be humane. / To recommend a sage to the ruler is to do one's duty. / Who, then, is greater than the man possessing these three virtues?"[196]

Among the "three great difficulties and two bars to excellence" at a court, Huan Tan listed a distant and distrustful relationship between ruler and minister. Predictably enough, Huan Tan exemplified the ideal relationship as that among Guan Zhong, Bao Shuya, and Duke Huan of Qi, saying, "Unless they are willing to tear out their hearts for one another, unless their actions are above suspicion and doubt . . . it will always be difficult [for those in power] to carry out their proposals and bring their ideas to fruition."[197] In contrast to Bao Shuya, "depraved, disobedient, and rebellious ministers" are greedy men, locking horns over trifles, eager to sabotage "eminent gentlemen of extraordinary talent and ability."[198] In the "official" friendship recorded of Guan Zhong and Bao Shuya, Bao proved to be astoundingly perceptive in his assessment of his friend, while Guan was the lucky recipient of Bao's insight and magnanimity.

One may contrast the stories told of three other legendary sets of friends, as those whose friendships were unofficial and more reciprocal: Bo Ya and Zhong Ziqi (mentioned above); Zhuangzi 莊子 and Hui Shi 惠施, and the intimate friendship groups portrayed in the *Zhuangzi*'s Inner Chapters. Zhuangzi and Hui Shi were certainly an odd couple. Hui Shi went down in history as a lawmaker and sophist par excellence,[199] a "wisdom bag" who traveled with five cartloads of manuscripts he had assembled during his travels to the courts of the various rulers. He was moreover a man who relished puzzles about how the attributes of an object such as "hard" and "white" relate to the object itself, and a creator of paradoxes such as "I set off for Yue today and arrived yesterday," "Linked rings can be separated," and "The thing newly born is the thing dying." By such means, presumably, Hui Shi intended to persuade his fellow rhetoricians to a less limited view of the world, since logic could be so easily turned on its head. His fellow rhetoricians obliged with such sayings as "An egg has feathers," "Wheels never touch the ground," and "Fire is not hot," but it never became clear to what insights such clever exercises led them. Zhuangzi, the master logician, on the other hand, wielded superb logic chopping in order to ridicule logic chopping, in the belief that it seldom clarifies life's mysteries or finally convinces anyone of anything. With acuity, Zhuangzi saw that Hui Shi never managed to find any serenity or pleasure in his strenuous efforts; he flitted around more like a gadfly than a seeker after practical wisdom. Zhuangzi therefore composed a satirical verse in Chinese doggerel, "Though Heaven has made you a shapely sight / You care only about 'hard and white.'"

The book ascribed to this same Zhuangzi supplies other moving portraits of perfect friendship, underscoring the idea that close friends, like the Dao itself, allow each other to evolve naturally in distinctive ways.

> The four Masters Si, Yu, Li, and Lai were talking together. "Whoever can look upon formlessness[200] as his head, on life as his back, and on death as his rump? Who knows that life and death, existence and annihilation, are all but a single body? I will be that person's friend!" The four men looked at each other and smiled, there being no disagreement in their hearts, and so the four became friends.
>
> Sometime later, Master Yu fell ill quite suddenly, and Master Si went to ask how he was faring. "Amazing! The Great Fashioner is making me all crookedy like this! My back sticks up like a hunchback; my vital organs are on top. My chin is buried in my navel. My shoulders are up above my pate, and my pigtail points to the sky. It must be some dislocation of yin and yang!" Yet he seemed calm at heart and unconcerned. Dragging himself by fits and starts to a well, he looked at his own reflection in the water. "My, oh my! So the Fashioner is making me all crookedy like this!"
>
> "Do you resent it?" asked Master Si.

"Why no, why should I? If the process continues, perhaps in time it will transform my left arm into a rooster, in which case I'll keep watch in the night. Or perhaps in time it will transform my right arm into a crossbow pellet, and I'll shoot down an owl for roasting. . . . I received life because the time had come. I am losing it because the order of things is passing on. Be content with this time; dwell in this order; and neither cares nor extraordinary pleasures can touch you. In ancient times, this attitude was called "unloosing the bonds." There are those who cannot free themselves because they are bound by things, but nothing can ever win against Heaven! . . ."

Suddenly Master Lai grew ill. Gasping and sneezing, he lay on the point of death. His wife and children gathered round in a circle and began to cry. Master Li, who had come to ask how his friend was, said to them, "Shoo! Get back! Don't disturb the process of change." Then he leaned against the doorway and talked to Master Lai: "How marvelous the Fashioner is! What will it make out of you next? Where is it going to send you? Will it make you into a rat's liver or a bug's arm?"

Master Lai replied, "A child, obeying his parents, goes wherever he is told. . . . And yin and yang, how much more are they to a man than his parents? Now that they have brought me to the verge of death, were I to refuse to obey them, how perverse that would be! . . . The Great Clod burdens me with form, labors me with life, eases me in old age, and gives me rest in death. So if I think well of my life, I must think well of my death. When a skilled smith is casting metal, if the metal were to leap up and say, 'I insist upon being made into a Moye sword,' it would surely be thought very inauspicious metal indeed. Now, having had the audacity to take on human form this once, were I to say, 'I don't want to be anything else; I would continue as I am,' surely the Fashioner would think me a most inauspicious sort of person. So now I think of heaven and earth as a great furnace, and the Fashioner as a skilled smith. Where could the Fashioner send me that would not be alright? I will go off to sleep peacefully, and then with a start wake up."[201]

And woe betide the man who realizes his own failure to form intimate friendships too late in life to rectify his errors. That lesson is brought home, I think, in the moving "Letter to Ren An," traditionally ascribed to the historian Sima Qian. The letter is typically read, rather simplistically, as a rationale for his decision not to commit suicide (as honor demanded) in order to complete the monumental *Historical Records*, despite the mortifications heaped on him as a castrate, "a mutilated being dwelling in degradation," a person "whose very substance is marred."²⁰² Granted, the letter does provide the expected rationale, following the standard rhetorical devices of the day—whether the letter was composed by Sima Qian or by a later author who framed a literary impersonation to divulge Sima Qian's motivation. But the rationale occupies so small a proportion of the surprisingly lengthy letter that more must be going on. And that "more" is a cautionary tale about the value of establishing and maintaining close friendships.

The letter opens with Sima Qian's observation that Ren An 任安 has cautioned him to be more "careful in my dealings with people" while "recommending men of ability" and "advancing worthy gentlemen" at court. Immediately, then, the letter confronts readers (initially, Ren An alone, but eventually the court of opinion in "later generations" who will judge Sima Qian's caliber) in view of the familiar conundrum faced by most officials: while intimate friendships posed certain risks, not to recommend others to the court could prove equally risky. For "whom will you get to listen to you," and "who will act on your behalf," if you hold yourself aloof from other men at court?

The axiomatic is laid out: those who have benefited from moral training "delight in giving to others" (this being the "origin of humane compassion"); they take it as their duty to engage in "proper giving and taking." That is how men of worth are to make their mark. Reviewing his twenty years at court, Sima Qian or his literary persona admits how few contributions he made to court life, either in military or in civil ventures. As a younger man, disdainful of "village and district" alike, Sima Qian had sought only the favor and love of his ruler, Han Wudi. The words in which the letter writer chooses to express this fact are curious: "Day and night I exhausted my paltry talents, working wholeheartedly to serve [the emperor] in my post, seeking to draw close to and curry favor with my lord, in the manner of a seductive woman."

In the *Shiji*, Sima Qian in memorable phrases insists that ardent lovers will do anything for each other, even die, if need be.²⁰³ Sima Qian unaccountably mistook his relation with the emperor for mutual, unswerving devotion. That is why Sima Qian was so misguided as to think that he could safely protest the injustice being done to Li Ling 李陵, one of the finest generals of the realm, whose willingness to share every hardship with his troops had earned him their undivided loyalty. But the emperor and his court, dismissive of true loyalty of the sort displayed by Li Ling (and also, clearly, by Sima Qian himself), touted Li only so long as he served up victory after victory on the battlefield. As soon as Li failed for lack of the reinforcements he had pleaded for, the fickle court rushed to condemn him, because scapegoating the absent Li served a dual purpose: to obscure the role that the court itself had played in Li's rout and thus to absolve themselves of any guilt in the matter. Hoping to restore more balance to the court discussions, Sima Qian recklessly "took the opportunity to speak of Li Ling's merits," an act that swiftly drew the harshest of reprisals. Not a single friend or ally rose to defend Sima Qian at court, for the obvious reason that he had never tried to make friends among the court officials. Not even his rich relatives bothered to bail him out.

He had, therefore, but two options: either to commit suicide, the honorable course dictated by convention for a man of his standing, or to complete his father's work in hopes of salvaging some of the family's reputation, like other noble failures in antiquity who had met misfortune with "rankling in their hearts." In his present situation, "Though I might possess the value of the fabulous jewels of Sui and He and my conduct be as faultless as that of the legendary recluses Xu You 許由 and Bo Yi 伯夷, in the end I could never achieve any glory [in my lifetime]. To the contrary. I would only succeed in arousing mocking laughter and bringing further shame on myself."

Harboring no illusions any longer about his standing at court, where before his castration he had had the chance to make friends with his peers and colleagues, Sima Qian went to his grave bemoaning his lack of true friends in life and hoping for better treatment from later readers of his masterwork: "Only after the day of my death shall the rights and wrongs of the matter be determined. But when I have truly completed this work, I shall deposit it in the Mountain of Fame. And if it is handed down to men who will appreciate it ... though I should suffer countless mutilations, what regrets will I have?"

The letter, from an outcast to a man condemned to die, is almost a macabre fantasy on "cutting off" relations (see below) on both sides — whence its perennial fascination, no doubt. While Sima Qian may have believed that every close friendship (*qin you* 親友) promised the most sustaining of life's pleasures, intimate friendships tended to end well only when the friends kept each other from excess and harm by schooling each other in maturity.

Hence the cautionary tale of another famous friendship, this time dating to Western Han, between Dou Ying 竇嬰, Lord of Weiqi, and General Guan Fu 灌夫, as recorded in the *Shiji* by eyewitnesses.[204] The story, in brief, goes like this: during the reign of Emperor Jing (d. 141 BCE), Dou Ying (son of a cousin of Jingdi's mother) held a variety of important posts, even though Dou sometime earlier had angered the empress dowager. Jingdi, rather desperate for talent during the Seven Kingdoms Revolt of 154 BCE, had been looking for trustworthy leaders from the Liu and Dou families, at which point he "discovered that no one could match Dou Ying for practical wisdom." Despite Dou Ying's repeated refusals to accept office, Jingdi forced an appointment as general-in-chief upon him, and subsequently he conferred on him a gift of one thousand catties of gold. Dou told his junior officers that the gift money was theirs to use in a military emergency; he would not add the smallest coin of the realm to his own coffers.

Dou Ying marched east to conquer the strategic center of Xingyang and to oversee troop movements farther east, in Qi and Zhao. As a result of the swift military victory over the Seven Kingdoms that rebelled, Dou Ying, along with Zhou Yafu 周亞夫, a second general, helped to determine policy for the rest of Jingdi's reign. But Dou, stubborn and proud to the last, continued to rebuff the favors proffered by both the emperor and the dowager. Before long, Dou acquired a reputation for self-righteousness and for being rather too fond of having his own way. In truth, Dou preferred to concentrate his time and effort upon his closest friend, Guan Fu, a stalwart war hero who had survived multiple wounds on campaigns in the south. Guan, like Dou, was famous for his absolute honesty and generosity toward subordinates. Their friendship was apparently as close as that of father and son. "They never tired of the pleasures they shared, and their only regret was they had come to know each other so late in life."

Emboldened by their friendship, Dou and Guan did not hesitate to express their disdain for the careerist Tian Fen, newly appointed to the post of chancellor, the highest office in the land. Continual ill will and altercations led the emperor finally to refer the friction to his court officials for adjudication. Guan Fu, the least well-connected of the three, was soon made the scapegoat, while Dou Ying was held under house arrest. Dou, aware that he was ultimately responsible for his friend's downfall, felt obliged to intervene on his behalf. He turned up in court with an imperial deathbed decree granting Dou the extraordinary privilege of a personal audience with Jingdi's successor. But when no duplicate copy of the testamentary edict could be found in the imperial archives, Dou was charged with forging an imperial edict, a crime meriting execution in the marketplace. In the year 131/130 BCE, both Dou and Guan died the most excruciating judicial deaths. "Both men tried to help one another, but they only succeeded in calling disaster down upon them both."[205]

The story of Wei Sheng 尾生 unfolds to a similarly fatal conclusion.[206] Wei Sheng made an engagement to meet his girl under a bridge. The girl failed to appear, probably because she was stopped by a downpour, and the water began steadily to rise. To keep his word, Wei resolved to stay; wrapping his arms around the upright beam of the bridge, he soon drowned in the rising waters. History has condemned him, believing him so "ensnared by thoughts of reputation" that he took his own death too lightly.[207] True, "a man of good breeding (a *shi* 士) who is worthy of the name dies for the sake of the person who profoundly understands him."[208] But close friendships were meant to enlighten and enlarge a person's capacities, not to muddle judgment and precipitate death.

Making Friends

Writers in early and middle-period China felt no need to elucidate their presuppositions for the sake of their peers, but the following sections will tease out, as much as is possible from the texts now at our disposal, how friendships (casual or close) were thought to work in the distant past. Let us begin at the beginning with some ordinary forms designed to foster elite sociability, through good manners and mutual deference.[209] Gift giving — an activity conferring honor upon both donor and recipient[210] — was one of the central features of conventional social exchanges in the distant past, as a selection from the *Mencius* demonstrates: "Men presented, among other gifts, jade and silk, animals and birds, for greater or lesser occasions, by this indicating and advertising their status."[211] Wan Zhang 萬章, Mencius's interlocutor, asks Mencius repeatedly to specify whether and in what contexts one may properly decline a gift (ranging from a live fish, to a fief, to the gift of an empire). In an ideal age, Mencius blandly intones, "gentlemen filled baskets with black and yellow silk to bid other gentlemen welcome, and the common people brought baskets of food and bottles of drink to bid other commoners welcome." But Wan Zhang probes further, begging for more guidance in the practical matters relating to the mundane realm of social intercourse (*jiaoji* 交際). Thus pressed, Mencius offers that all social relations must reflect a deeply rooted sense of respect for others and that it demeans the giver to inquire whether the gift was honestly got.[212] Then, to our great surprise, Mencius ends by insisting that an exemplary person may not decline a gift under any circumstances, "even in his heart," if he hopes to engage in dignified and productive social relations. No polite formulae can possibly excuse a distrust of others' generosity or a willful disdain for due ceremony, even if some hold that receipt of an improper gift implicates the recipient in the giver's impropriety.

Mengzi evokes the Confucius of the *Analects*, who always finds clever ways to fulfill his ritual obligations to others while avoiding the outright gift refusals that could sever a social relation. The Confucius of that book is hardly averse to engaging in behavior that Kant would deem ethically suspect, urging his followers to maintain sociable relations with as many people as possible for the good reason that gentlemanly sociability among those with claims to birth and education is likely to facilitate the attainment of Kongzi's twin goals of civilizing the individual and injecting more civility into the larger community.[213] As Mencius puts it, Confucius "wanted to make a beginning." Prescriptions for the proper way to offer gifts or conferrals with due propriety are never lacking, and the wise person who aims to be effective in the realm "dare not presume" to destroy the sense of affable courtesy that such exchanges foster, despite others' misconduct. Difficult it may be in daily life to negotiate the sometimes conflicting demands of magnanimity, ritual decorum, prudence, and basic good will toward others, especially where one is duty-bound to old friends and old ties. Still, it is always best not to advertise oneself as "too pure" to engage in ordinary social relations.[214]

One legend had Confucius assisting an old friend with his wife's burial, and when the old friend sang over the corpse, Kongzi made "as if he did not hear." His disciples, struck by the ritual impropriety of the song, wanted their master to "have done" with his friend, to which Kongzi retorted, "I have heard that close kin do not forget their kinship, nor old friends their time-honored ties."[215] In another instance, when the local strongman Yang Hu 楊虎 aims to force Confucius into a patron-client pseudofriendship relationship through the presentation of a gift, Kongzi famously goes to thank Yang Hu when he believes that Yang is not home, so that he might fulfill the ritual requirement of gratitude while preserving the reputation of one not easily bought. Critics of such archly ambiguous actions typically lambasted the Confucians' propensity to value "horizontal" friendships above the "vertical" relations binding superior and subordinate (and not the reverse), but the literary records attest that within and well beyond Confucian circles, special importance was ascribed to courteous adherence to the social forms, including gift exchanges.[216] As one Eastern Han text says, "Most of the prominent families in the commandery befriended one another through extravagant [gifts and parties]."[217]

Though many high-status settings and practices for sociability in early China could be mentioned here, including "asking after a person" (cun wen 存問), "asking after another's health or activities" (wen ji 問疾, or wen qiju 問起居), and presenting one's calling cards, notes, and gifts at set occasions, it may prove most useful to consider two prime settings for elite sociability in the early empires whose origins lay in the pre-Confucian aristocratic age: the male wine drinking ceremonies[218] described in the earliest prescriptive ritual classics and in the *Zuozhuan* (where many drinking parties go disastrously awry),[219] plus the elaborately choreographed mourning rituals that reaffirmed kinship and friendship ties.[220] While the lion's share of Sinological research has gone to mourning (with the wine banquets largely ignored), the local community banquets and mourning rituals shared an important function as early "status theaters" carefully designed to train the participants (entirely male in the first case) to perform their status successfully, whether drunk or sober, via supremely civilized gestures and speeches. There existed two supreme tests for social cultivation: Will the person, when drinking, continue not only to be amiable, but also trustworthy? Will the person remain no less mindful of his indebtedness to others when they can no longer exact repayment? For this reason, communal drinking for the living and funeral feasts honoring the dead dominate discussions about friendship in Chinese. The trust that ensues among participants who can predict how their peers will act in future also supports a sense of community, a commitment to mutual aid and comfort among like-minded participants and their families.

Settings such as these provided typical backdrops for sketches of intimate friendships in early and middle-period China. As readers will recall, intimacy requires considerably more than mere sociability; that explains why so often scenes of real intimacy are juxtaposed to those of ordinary decorum, with the intimacies revealing themselves in either more or less show. Looked at from outside, the faithful friend may appear to be a naïf, "a moron," or downright "bizarre."[221] For confronted with unusual circumstances, close friends must be ready to defy social conventions or the laws and embrace the possibility of spectacular failure, if necessary, in order to succor an intimate in distress. Attention to duty and appropriate action (*liyi* 禮義), in the end, never suffices to construct intimate friendships. Thus, the reserved quality that many early sources attribute to the relation between true intimates is offset, even undercut, by the many portrayals where intimate friendships, built upon extraordinary levels of mutual commitment, seem to call for reckless acts.

Drinking Ceremonies as Performance Sites for Friendship
Formal "drinking bouts" and classical forms of music are allied topics from the pre-Qin period onward. "The ceremonial usages that come closest to our human feelings are not those of the highest sacrifices" offered to the ancestors, according to one *Rites* classic.[222] Instead, they are events such as the local community banquets, for gracious participation in such rituals signified a willingness to act in concert (*he* 和) with nonkin, presupposing the virtues of humility, mutual respect, and willingness to forego precedence that mark communications between superior men. Patterns of commensality can be quite as revealing as overt symbolic actions when tracing the cultural assumptions that underpin community solidarity.[223] The organization of the communal space, in tandem with the participants' observance of a well-crafted hierarchy, exemplified and reinforced collective values and norms. In most early societies, drinking rituals were the primary sites for training young adult males outside the family, preparing them for eventual service to the realm (recall the classical Greek symposia). The drinking ceremonies (*xiang yan* 鄉宴 / *xiang yinjiu* 鄉飲酒) allowed younger members of the governing elite to interact convivially with older authority figures who could act as "mentors and friends" (*shi you* 師友).[224] These more senior members of the community, well versed in refined sociability, were

gratified to find the same feasts affording ample opportunities to express communal loyalties and thereby strengthen or expand their necessary social networks.[225] We may lament the signal lack of unmediated reports from early and middle-period China of spontaneous performances or casual cross talk (the "beautiful words and flattering laughs"), but much can be teased out of the highly self-conscious scripts, visual and literary, that depict members of the governing elite performing their expected roles, badly or well (Figure 2.4).[226]

Drinking sessions in China, like the Mediterranean symposia, provide the backdrops to many scenes of male friendship, although the two venues at first glance have little in common. Once we get beyond three simple differences—that the Mediterranean elites reclined on couches, whereas the early Chinese sat on mats on the floor, that the Chinese did not always invite *hetaerae* to the entertainment, and that the sexual mores for males in the two societies differed dramatically—we begin to notice the similarities between the two sets of drinking ceremonies. For example, the "key masculine activity" took place in dedicated rooms or buildings set apart from the usual domestic space, presumably because these male drinking sessions had their origins in triumphal endings to campaigns and hunts far from home (as did many features of the ritualized music and dance performed in them).[227] The drinking sessions in these dedicated settings replicated some of the comforts of home, thereby blurring the boundaries between kin and nonkin, promoting the convivial exchanges likely to cement political alliances and allowing distinguished men the satisfaction of seeing and being seen by their peers.[228] The deliberate emphasis at these gatherings was on drinking, though some food was served to prevent the most unruly behavior, since men in their cups were most likely to reveal their true character, having lost most of their inhibitions. The most civilized men were purportedly able to conduct serious business while relaxing (hence the story cycles about Confucius/Kongzi at leisure, as well as about Socrates). And both cultures found ways to associate conventional features of the good life, including satisfying sexual relations, with celebrations of virtue. A strict division between official and domestic duties was hardly germane to men who enjoyed aristocratic privileges, nor would it necessarily help those "who labored with their hearts and minds" to know men better or to foster social cooperation and cohesion.[229] Hence the genius of using formal drinking sessions to bridge the gap between domestic life and office-holding.[230]

Of course, since no pederastic education of the Greek type was practiced in early China, the local drinking ceremonies were largely homosocial, rather than homoerotic: intense, but unconsummated feelings of attachment were allowed to bind males in the social arena, presumably as a way of releasing or circumventing the sort of explicit sexuality between the participants that might distract from larger communal ties.[231] At such gatherings, preening males did bandy about the lyrics employed as diplomatic code by rulers and their men in service, sometimes during reciprocatory toasting bordering on the competitive. One relatively lengthy account describing career officials "always coming, always going, / cultivating contacts in the capital, / forming cliques, cementing factions, / glibly chatting and playfully conversing" gives us a sense of the drinking parties that more typically merit a shorthand phrase or two:

> They hold a feast in the high hall
> To entertain fine guests.
> A bronze jar is in the middle of the mat;
> Mincemeats and fruits are placed on all sides.
> In the cups there is clear wine;
> For sliced fish, they have little "purple scales."
> As winged goblets are raised in "rounds of competitive toasting,"
> Strings and reeds begin to sound.
> A beauty from Ba strums the zither;
> A maiden from Han beats the time.
> They begin with the "Western Melody" in rapid tempo,
> Sing "On the River" clear and resonant.
> Twirling their long sleeves, they dance again and again,
> Lightly wheeling and turning, gracefully flowing.
> Goblets gathered, they press closer together on the mats,
> Drawing full cups and drinking penalty toasts.
> They happily imbibe this night,
> And stay drunk for month piled on month.[232]

A recently "discovered" (that is, not scientifically excavated) Tsinghua manuscript also discusses leisure diversions as prime occasions for cementing alliances, evoking both the *Odes* and the *Documents* in the process,[233] while a received text, the *Zuozhuan*, tells

Figure 2.4. Detail of the "Feast with the Married Couple," from a late Eastern Han (25–220 CE) or early Sanguo (220–280) tomb mural at Zhucun, near Luoyang, Henan Province. Full mural: length 4.76 m, width 3.3 m, distance from floor 1.1 m. From Huang Minglan and Guo Yinqiang (eds.), *Luoyang Han mu bihua* 洛陽漢墓壁畫 (Beijing: Wenwu Press, 1996), p. 190, plate 1. Permission to reproduce this detail granted by Wenwu Press.

This detail shows an idealized scene where the hosts, presumably the tomb occupants or their relatives, receive guests and allies of high status, who have arrived in fine carriages. This scene may show a reception either before a drinking feast or before a grand funeral.

us bluntly: "There are rites that consist of formal entertainments or feasts with wine, the first offering instruction in reverence and restraint, the second, in showing kindness and favors."[234]

In the few delicious accounts of male drinking bouts that we have from the distant past, wine, music, and friendship infuse a set of cultural exchanges that were deemed simultaneously witty, provocative, and highly "educational," for it was no mean feat to maintain the proprieties while drinking. Supposedly, the formal drinking rites began with the local prefect inviting all men "of worth" (that is, men of sufficient status with claims to good breeding) in the community, and once the guests convened, they would be ushered up the steps and into a great hall, where they made three sets of ritual bows to their host. After the bows, the local men settled themselves in their places according to rank, then witnessed the ritual pouring of a libation, before they carefully washed hands and wine vessels. Then, in an edifying display of civility, the guests in unison would raise their cups to partake of a brew like mead or sake whose dark color symbolized simplicity, cleanliness, and impressive substance.

Because repeated toasts were delivered, the sharing of the brew seems to have been as significant to the participants as the ritual distributions of baked meats after sacrifices. Certainly, the ritual classics insist that "guests and host alike shared" in all the rounds. At some point, snacks furnished by the host were brought into the drinking hall from the east room, and once more, "guests and host alike were to take a share." The host was busy, however, for it fell to him to serve the principal guests after he had performed further solemn ablutions. In so acting, the host recalled Heaven's own beneficence and, we are told, "attained the [true] self-possession" enjoined by the antique sages. He also saw to it that whenever the wine was passed, precedence was accorded the old over the young. Ideally, each of the participants took the time to admire the host for his elegant performance of his duties before tasting the snacks and sipping the brew. To express mutual esteem, the guests returned to the host some portion of the meats and sauces that he had just presented to them and positioned themselves on the edge of their mats, as if ready to serve the host at a moment's notice. In the idealized prescriptions drawn from the *Rites* classics, such ceremonial honors were prized far more highly than displays of wealth, motivating those of lower status to emulate their social betters.

Conventions of status and seniority decreed that those aged sixty or over take their seats in the drinking hall, while the younger men remained standing, patiently awaiting their elders' orders. Before those aged sixty, three dishes were placed. Four dishes were placed before the mats of those aged seventy; five before those aged eighty; and six before those aged ninety. "This illustrated the principle that the aged should always be cherished and properly nourished." In addition, we are told, the food stands and dishes were set out "in proportion to the ranks of the assembled guests."[235] By such cues, the communal drinking rendered visible the ritual proprieties and the political hierarchies within which the participants were to play their parts.[236] Shortly before the entertainment interlude, the host made sure to ask each guest about other family members who might be in attendance, whether seated or standing. Once these formalities had been observed, the hired entertainers (male and female) ascended the stairs into the great hall, where they played three pieces, after which the host thanked them and offered each a cup of wine. Next a troupe of flutists played three tunes in the hall before receiving their due rewards. Singers and flutists alternated three more sets each and then performed three additional numbers in concert. With that, the conclusion of the musical interlude was announced, and the professional musicians left the hall quietly, so as not to disturb those still drinking.

Immediately afterward, a person appointed by the host as master of ceremonies took up his goblet, and began its passage round among the guests. This was the signal that pledges could be drunk all around, with the master of ceremonies insuring that no disorderly behavior spoiled the general merriment. Long before the drinking ended, shoes came off, and the younger men also sat down on mats, at which point guests and their attendants were solemnly invited to drink as much as they wished, so long as the drinking would not interfere with the performance of their duties the next day. Tipsy at best, each guest eventually left whenever he wished, but the evening's host made sure to escort each stumbling guest part of the way out, so as to uphold the ritual courtesies to the last.

Ideally, the local drinking ceremonies for elite males adhered to four sorts of principles, principles that the local elites, not to mention the ruling line, had to uphold in order to maintain their places:[237] the adjustment of usages according to hierarchy; the careful avoidance of disorderly conduct and maintenance of harmony; studious deference to elders by the younger; and, finally, during the drinking, the adoption by local elites of an air of easy bonhomie.²³⁸ So long as these governing principles of sociability were observed, the sheer conviviality of the scene would attract exemplary men to participate in the civilized order, the creation of a succession of sage-kings. Hence Kongzi's insistence that it would be very easy to implement the king's Way through proper participation in the local drinking rites. Normative pleasure-giving and pleasure-taking, reinforced in vital ways through the drinking, enacted the signal advantages attached to tempering hierarchy with reciprocity and to ornamenting one's person with affability.²³⁹ That the host received his guests with drinks and snacks suggested that as a rule, "rulers do not receive anything from their subordinates without compensation... lest their subjects suffer want when the ruler imposes taxes or levies. For when harmony and good faith prevail between high and low, neither high or low are dissatisfied with one another."²⁴⁰

By definition, each exemplary person had to be a highly socialized being whose exquisite sympathies propelled him beyond the mere formalities to intimacy and community in which, seeking the best for others, pleasures and benefits would redound to him, as well.²⁴¹ Casual friendships had the potential to shift the center of the world from the self to another and in this salutary "process of unselfing"²⁴² enable individuals to see, respect, and cherish each other. As noted above, these elite male drinking rituals at the same time afforded safe spaces in which "to observe the participants' bearing and likely fates" and so gauge the degree of trust that might be invested safely in others in the group.²⁴³ In such relaxed situations, with tongues loosened by wine, a man's true colors would likely reveal themselves in the way he conducted himself and by the companions who attracted him.²⁴⁴ With men of true worth, the value derived from such leisurely and genteel settings far outweighed any commercial benefits associated with the marketplace.²⁴⁵ The proper "return" for friendship was affectionate services voluntarily tendered, not contractual "debts" or "obligations." (It was an article of faith that " associations contracted merely for gain" [*bi zhou* 比周] deserved condemnation; besides, such associations would likely prove draining, rather than life-affirming on both sides.) Affectionate service did not devolve into "altruism," since any improvement of a friend's lot so clearly spelled an improvement in one's own lot, as well.²⁴⁶ Ideally, then, during a local drinking ceremony, music and the sight of budding friendships, in heady combination, would temper any sharp differences among the participants and instill a sense of common purpose within an otherwise disparate group of people.[247]

Invariably, reconstructions of the old drinking scenes stress the improvisational tone of these exercises in male bonding. Supposedly, the elite men gazed politely enough at the performers, out of respect for their host's arrangements, but the guests, strictly cordoned off from the entertainers during the interlude, really set to drinking in earnest only after the entertainers had left. As perpetual "amateurs," the primary qualification that these men of standing offered was passionate commitment to the code of gentlemanly behavior, rather than any technical training they might have acquired along the way. But their devotion to the larger community and their ingrained cultivation had to be on display in order to reap full value. So what good men exemplified via the ordinary rites of sociability was a cultivated tolerance (*kuan* 寬),[248] an ability to "make allowances for others" (*rong* 容 or *shu* 恕), a refusal to "get things" through coercion or intimidation, and an easy style comporting with any company.[249] They did not "swill or slurp the soup,"[250] neither did they hold themselves aloof from petty men who knew no better. By the time a man of good birth and education reached maturity, the ordinary courtesies would have become second nature, but the man of consummate social skills learned to "live between what is permissible and what is not," aware that the impulse to appear correct at all times could signify a desire "to elevate himself above his contemporaries," an impulse that might ruin his chances for friendships or alliances.[251] Cooperation with people of different temperaments was a skill that had to be honed through continual effort. Even the most hidebound moralists therefore urged members of the ruling elite to use convivial settings as a way to soften their habitual arrogance and heighten their willingness to work together. Making allowances (*duo ke* 多可) for other people's failings was far less shameful and detrimental to one's flourishing than suffering a lack of friends.[252] This understanding of ritual is what any honest reading of the early sources yields.

Texts such as the *Zuozhuan*, the *Guoyu*, and the *Shuoyuan* remind us that these banquets were the contrived remnants of contests and that real-life consequences ensued when inept, immature, or arrogant power holders mixed heavy drinking with cultivating contacts (*yang jiao* 養交). Huge fights (some quite deadly) erupted over hierarchy and rank among elite males of the classical era.[253] Three oft-cited examples place the pious portrayals of respectful bowing and formulaic politeness in their proper context: as thoughtful attempts to mask, counter, and remedy the undeniable fact that few power holders ever learned what education and rituals tried to teach. From the *Zuozhuan* we learn:

> In autumn, in the ninth month, the Prince of Jin entertained Zhao Dun with wine. He had in hiding armored soldiers to attack Zhao. Learning of this, Zhao's aide on the right, Timi Ming, rushed forward and ascended the steps, shouting, "For a subject to exceed three rounds of drinking while waiting upon his ruler at a feast does not accord with ritual propriety." He then helped Zhao Dun down from the dais. At that the prince whistled up his fierce hounds, but Timi Ming wrestled them down and then killed them. Zhao Dun commented, "He deserts good men, preferring to use hounds. They are truly fierce, but to what avail?" Scuffling and struggling all the while, Zhao managed to leave the hall, but Timi Ming died in the defense of Zhao Dun.[254]

One of the musical *Odes*, blandly entitled "The Guests Take their Seats," says tartly:

> It is always the same with drinking brew:
> Some get drunk while some do not.
> So we appoint a master of ceremonies,
> Perhaps supplying him with an aide.
> When the drunks are misbehaving,
> Those still sober feel uncomfortable.[255]

The poem continues, "But if they, when drunk, do not retire / This destroys the power of the gathering [to promote good fellowship]."[256] Then, too, Liu Yiqing's 劉義慶 (403–444) *New Account of Tales of the World* (*Shishuo xinyu*) reminds us that some drunken hosts and guests had sex with one another or with the entertainers on stage:

> At the height of the summer, the emperor went to escape the heat... and he drank and feasted all night long. Palace women between the ages of fourteen and eighteen (all gorgeously got up) peeled off their upper garments, stripped down to their underwear, or went bathing in the nude.... Only when the candles were set before the hall [at dusk on the second day] did the emperor finally come to his senses.[257]

Hence the continual admonitions that pepper the extant literature: "In drinking bouts, one may observe the quality of the participants' majesty and bearing, and so discern [the likely] calamities or blessings [that will ensue]." That is also why the *Odes* intones, "The rhinoceros drinking horns curve up, / The fine wine is soothing. / Neither haughty nor arrogant, / To him myriad blessings come and gather." For any "fine man" to be arrogant is the sure way "to bring disaster on!"[258] "Good men are always alert to dangers . . . / Good men are always seemly."[259]

Funerals as Performance Sites for Friendship
Given the importance placed on mourning rituals among the same leading families of early and middle-period China, funerals likewise figured as premier sites for making, renewing, or breaking off friendships and alliances. As in modern society, the classical-era funerals tended to be formal and protracted. Any minor divergence from the strict performance of the liturgies mapped out for host and guest in successive stages might mean loss of prestige, loss of an official post, or worse.

That made the spectacular swerves from the mourning rituals all the more striking, although eccentric behavior regularly occurred, if the extant records can be trusted. A small selection of the most memorable examples demonstrate this point. The first comes from Liu Yiqing's *New Account*:

> Zhongxuan [aka Wang Can 王粲, one of the Seven Masters of the Jian'an Period] loved donkey brays. After he was interred in the ground, [Cao Pi, the future] Emperor Wen of the Wei, came in person to mourn him. He turned to his companions and said, "Wang loved the braying of donkeys, so I think it would be fitting if each of us gave a bray to send him off!" At that, all the guests duly let out donkey brays.[260]

In life, the prince had acted as Wang's patron, but Wang was a proud descendant of the prominent Wangs of Shanyang. Here the prince's gesture publicly confirmed his intimate friendship with Wang. Cao claimed to know Wang best, for he alone had divined what would delight and calm Wang's spirit in the grave. So although nothing would "offend the ritual" more than a raucous chorus of donkey brays, Cao did not hesitate to offer his friend this peculiar final tribute. Critics might wonder whether Wang Can, as connoisseur of such coarse sounds, was a vulgar aristocrat lacking the requisite refinement to truly "know the tone."²⁶¹ If so, what did that say about his patron's ability to cultivate worthy friends?

But not everyone was so rigid about the rituals. Many people, in fact, did not find it "peculiar" when intimate friends, at times of crisis, gave vent to an irrepressible urge to proclaim the exclusivity and closeness they had invested in their most intimate friendships. A second anecdote, from the *Hou Hanshu*, hints as much:

> When Dai Liang 戴良 was a young man, he was never constrained by propriety. His mother liked the sound of donkey brays, so he constantly imitated them, thinking in this way to delight his mother. When she died, Liang's elder brother lived in the mourning hut, eating only gruel. If something was not strictly according to ritual, he refused to do it. But Liang ate meat and drank wine by himself, and he burst out sobbing whenever he felt sorrowful.... Someone [sarcastically] asked Liang, "Do you really think the way you observe mourning accords with ritual?" Liang replied that he did. "Ritual is the means by which a person directs and controls wayward feelings. If these are not to be dealt with, then why bother discussing ritual at all?"²⁶²

The anecdote remarks that both the seemingly inobservant Dai Liang 戴良 and the scrupulously observant Boluan 伯鸞 "acquired a haggard appearance," a sure sign that both men were exemplary mourners.

Of course, these were not the only unceremonial ways of mourning an intimate. Ying Shao's 應劭 *Fengsu tongyi* (compiled before 203 CE) describes many more elite funerals where the mourners acted in comparably unconventional ways, insisting the outbursts were entirely spontaneous, even when they were carefully rehearsed. (Some of these performances Ying Shao applauds, though he disapproves of others.)²⁶³ One of Ying's entries relates the story of Xu Zhi 徐穉, who had gained a national reputation for steadfastly refusing offers of official employment in the realm. Notwithstanding, Xu Zhi had a habit of traveling long distances to a gravesite to mourn a would-be patron.

> Xu would then go straight to the site of the burial mound [of the would-be patron whose advances he had refused] and soak a piece of cotton in wine to release the odor of the wine. He would set out a platter of rice on a mat of white rushes, with a cooked chicken in front of it. And when the libation of wine had been duly poured, he would leave his formal calling card (*ye* 謁) at the

grave mound. After this he would depart forthwith, without ever attempting to speak with the chief mourners at the funeral.[264]

At one such event, we read, "The crowd [gathered for the funeral] thought his behavior bizarre; none of them understood the reason for his actions." But at this remove, we can see Xu Zhi's motivation plainly enough: only when the would-be recommender could no longer entreat him to take up office did Xu feel it right to exhibit his undying gratitude to the person who had discerned in him a set of sterling qualities. And while we might wonder, along with the typical mourners, how this expression of loyal friendship conduced to the good of the deceased, let us not be hasty in our judgments. By acknowledging his profound psychic debts to his would-be patron, in the absence of societal expectations, threats, or contractual obligations, Xu was acting as a true friend fulfilling the highest duty in mourning: to declare the bond publicly and so confer due honor on the dead man.[265] Xu was meanwhile asking, with great respect, "Where does a man stop?"[266] Xu Zhi's peculiar style of mourning is hardly unique in the early sources.[267]

Though many a curious story of the same type could be added, here it is important simply to behold the import of the texts: that persons bound by genuine ties of respect will usually want, sooner or later, to make clear the freewheeling character of their mutual appreciation, even if it means transgressing conventions and offending others. In the presence of a close friend (dead or alive), the rest of the world is eclipsed. True friends have eyes only for each other and therefore act as if they are alone together. But for those considerably less than intimates (colleagues, neighbors, allies, and so on), exemplary civility, "decency and gracefulness," expressed in company sufficed to grease the wheels of everyday existence, "making those with whom one conversed easy and well pleased."[268] Moreover, like so many ethical resources schooled through and incentivized by pleasurable activities, the reserves of knowing sociability were likely to increase through use and to atrophy otherwise.[269]

Intimate Friendships
The goal of sociable relations was simply to maximize constructive relations within society while minimizing the occasions for attractions to destructive or antisocial behavior. Intimate friendships,

however, denoted far more powerful attractions and commitments than ordinary sociability. Both the *Shuowen* and the *Liji* describe intimate friendship, for example, in terms of shared commitments (that is, believing the trusted friend cares for what the other cares about), invoking a far higher standard than shared tastes, shared experiences, or shared interests,[270] these being seldom sustained over a lifetime.[271] Early on in their acquaintance, friends might observe certain proprieties (most of them negatives, for example, not being arrogant, not being careless, not asking too much of others, and so on). But an intimate friendship called for active engagement in a succession of subtle practices that enhanced mutual commitment, reaching a final point when the intimates felt free to overturn any of the conventional courtesies associated with mere sociability. Time spent with an intimate friend was like coming home and being yourself.

Notably, the available writings do not particularly focus on the voluntary nature of intimate friendship, unlike nearly all modern philosophical treatments of the same topic. Intimate friends instead feel "lucky to encounter" (*xing yu* 幸遇) one another, knowing that fate is not always so kind to people, howsoever worthy. Besides, attractions and commitments to intimate friends, even more than the casual, inevitably implicated the person in tight webs of relationships not entirely under his or her control.[272] Friendships, no less than fallings-out, seemed to "come to pass" as compulsions more than as calculations.[273] Post-facto analyses of mutual attractions mention selectivity and discernment as elements in the mysterious attractions that create and maintain intimacy, but such ideals resisted clear definition. Certainly, the conditions for intimacy were not satisfied simply by observing a series of formal courtesies, let alone the quasi-legal obligations that modern moralizing tends to associate with friendship; by definition, "market" (mercenary or self-interested) friendships contracted for profit were excluded from consideration.[274] Of the much-admired Huan Tan, it was said, for example, "He did not sell his friendship."[275] Fair-weather friends eager for commodified relationships were known to "grasp each other's arms, clutch each other's wrists, and, bowing to heaven, swear oaths of lasting friendship, pleading for favors and special treatment, heedless of the degree of commitment implicit in true intimacy."[276] These phrases become part of a long list of set expressions indicating duplicity and indifference, however.[277] By contrast, the relationship between intimate friends was supposed to be "understated" and "restrained" (*su* 素), based as it was on a deliberate disregard (*méconnaissance,* literally, "misrecognition") of their mutual indebtedness.[278] In consequence, intimate friendships were considered rare and not to everyone's taste.

Significantly, intimate friendship, like music, is in some senses a "preverbal" relationship.[279] "Preverbal" by no means denotes my belief in the "prelogical" survival of aspects of "primitive man" in a more advanced species, *Homo sapiens,* traveling triumphantly along a single evolutionary trajectory. Instead, I use "preverbal" to point simultaneously to several key facets of human experience noticed by writers of classical Chinese. "Preverbal" acknowledges that large swaths of human existence cannot be verbalized or explicated easily and well, regardless of intellect. It meanwhile refers to the related phenomenon that people often sense an affinity before a single word has been exchanged. Moreover, once an intimate friendship is in place, it is often signified and sustained less through words than through small gestures, a laugh, perhaps, or the slightest shrug, a shared mat or slice of fruit—the merest sounding of a "tone," as classical Chinese puts it, which marks the tenor of the back-and-forth as antiphonal improvisations.[280]

Needless to say, friends' sharing of a piece of music or a poem set to music employs the perfect medium to express perfect intimacy;[281] indeed, one of the most beautiful metaphors for thoroughly unscripted enjoyment of one another's company is "singing freely, with abandon."[282] As one text says, music creates "a whole that produces pleasurable exhilaration and the glow of mutual affection in everyone who experiences it."[283] Close friends are so exquisitely attuned to each other's moods and feelings that they can often anticipate them; hence their propensity to speak in a kind of shorthand to each other.[284] (See Wang Xizhi's letters to his friends, which Antje Richter has so movingly translated.) Or they may remain silent, partly, perhaps, overwhelmed by the sensory immediacy that characterizes the relation.[285] (As one famous letter puts it, "Given the true feeling between us, is it even necessary for me to say this, for you to understand it?")[286] Equally importantly, the word "preverbal" nods toward antiquity's unquestioned belief in cosmic regularities structuring the harmony of the spheres, a harmony in which people participate according to their sympathies and antipathies, their highest forms of social order and the polite arts working to augment the regularities and harmonies.[287]

Friendships can be postverbal, as well as preverbal, since the main purpose of exchanging words is to establish and keep trust in the friendship, whereas assurances of trust prove unnecessary if tacit trust already exists. For however crucial words and logic may be to a few legendary friendships in Chinese history (think Zhuangzi and Hui Shi),[288] the classical and classicizing texts present most close friends as chary of speech.[289] What is experienced cannot always be spoken, and what is said cannot always be experienced.[290] How, then, is a friend ever to employ ordinary language to communicate that sense of being in one's element when in the company of close friends and liable to depression when not? Indeed, the difficulty of portraying friendship, in literature or in the visual arts, is that the often repetitive, incremental, insignificant, and ephemeral acts that create or promote intimacy hardly make for exciting narratives, yet collectively, they turn a friendship into something that seems in hindsight "inevitable."[291] Hence the "aesthetics" of friendship and the consequent impulse to read for what is not said, as well as for what is. Like the silences in music, the voids in paintings and in architecture, the "unshowy," "undemonstrative," and reticent character of friendships—the very intensity of the companionable quiet can project incredible energy and lead to deeper engagements. Hence the sage's decisions to avoid declaiming.[292] As Kongzi comments in the *Analects*, "Words are merely for communication."[293]

The Benefits of Intimate Friendship
While candidacy for office and making a living (*zhi sheng* 治生) demanded sociability, constructing a life worth living (also *zhi sheng* 治生) required intimate friendship.[294] Those rare instances of intimate friendship, such as those glimpsed in the peculiar funeral scenes described above, while built upon the broad base of elite sociable relations, challenged those very relations in some respects. By consensus, friends became intimates not after a mercenary cost-benefit analysis, but through mysterious laws of attraction—as registered in the myriad passages in which *xiang* 相 ("mutual") is shorthand for *xiang gan* 相感 ("mutually attracting and affecting"). Supposedly this set of sympathies exerted a huge pull upon people's lives, though it seems to have been experienced mostly as the "way things

are," rather than as something demanding explication. Music, as noted above, was thought to be a most apt example of this especially subtle, yet moving "attunement between things or people of the same type" (物類相感) that "enters the spirit," transforming it profoundly, but by processes unknown.²⁹⁵ As readers will recall, the most deeply felt affinities between people were likened to "soundless music."²⁹⁶

Just how important intimate friendship was to the complex processes of self-transformation and social transformation known as "cultivation in the Way" can be judged by reference to *Analects* 13/28, which defines the cultivated gentleman as a person "exacting in his attention towards his intimate friends, and cordial toward his brothers." Huan Tan said, "Unless they are willing to tear out their hearts for one another," it "will always be difficult [for people] to carry out their proposals and bring their ideas to fruition."²⁹⁷ Similarly, Zheng Xuan 鄭玄 (127–200), the Eastern Han moralist, spoke for many when he remarked, "From the emperor on down to the common people, there is no one who does not need friends to accomplish things," most crucially, the development of his singular human potential.²⁹⁸ Some early Chinese texts claim that friendship is the very stuff out of which mature human beings form their second natures. Others assert that it is only friends who can lend the proper polish to the rough material of the human being.²⁹⁹ "Wise men made friends in order to expand their wisdom," as the proverb went.³⁰⁰ A family or community may foster a child's initial formation, but it is the knowing care of "unqualified friends" that ultimately lends that child, once mature, the confidence, the heart, if you will, to become the best possible version of her singular self.³⁰¹ "What is esteemed in human relationships is the just estimate of another's inborn nature, and helping him to realize it.... Lest you pervert or damage its innate quality, [the true friend] would rather see it find its proper place."³⁰² Indeed, in close relationships, the supreme gift conferred by the loving and perceptive friend is precisely this: seeing dignity and value in the other, the friend acts, like the beneficent Dao itself, to help the other attain her distinctive set of potentials.³⁰³ Accordingly, "individuality" (in the sense of distinctive or unique personal traits) figures prominently in the accounts of early friendship in China, even though "individualism" in the modern sense (that is, a sense of oneself as autonomous) is distinctly absent, and the lack of a sharp dichotomy between the self and others or self and society means that considerate actions on behalf of others were seldom construed as "selfless" or altruistic.[304] Friends benefitted and pleased one another and by attributing to one another a compelling persona enhanced each other's justifiable self-confidence. Hence the proverb, "It is human beings who broaden the Way, and not the reverse."[305]

Crucially, only intimate friends could have the needed insights into each other's defining features and characters, so only they could ascertain each other's desires and unspoken commitments. Because of their privileged access, friends were apt to be the best people to admonish one another for faults and be well and truly heard.[306] Few relationships aside from close friendships allowed frank admissions about a person's shortcomings, since mere sociability enjoined tolerance and forbearance instead. A good friend was expected to criticize and contest, by way of assisting and comforting. Thus, intimate friends could improve their lives largely because they recognized deficiencies and urged remedial courses.[307] Close friends did not necessarily know everything about each other, nor did they need to share all things, but, in the end, close friendships were sustained by a warm appreciation of the very qualities that represented sources of pride or vital concerns;[308] ergo the desire to see one's friends realize their own distinctive powers, energies, and capacities. And so the early Chinese texts stress friendship as the path toward "self-understanding" (*zhi ji* 知己) and "self-possession" (*zide* 自得), that is, knowing oneself so thoroughly that constructive actions bring great personal satisfaction. Thus, the early and middle-period texts hold that *not* to engage in the transformative relations associated with intimate friendships is to miss the most wondrous form of human experience and likely remain a less than complete human being.

Clearly, to find an intimate friend is to be blessed with great good fortune. True friends consider themselves stupendously lucky to have encountered one another, and given that this ineluctable attraction binds two or more parties tightly, ideally for the remainder of their lives, intimate friendship hardly qualifies as completely "voluntary" in the modern sense.[309] To cite a good example: many of us today would resist the notion of a father "choosing friends" for his son, but this did not strike early readers of classical Chinese as atypical or oxymoronic. As David Konstan notes with respect to friendship in the classical Mediterranean, "An achieved relationship does not necessarily mean one that depends essentially upon free or personal choice.... The role of election in friendship, though commonly insisted upon in modern accounts, appears to be historically variable."[310] As many of the finest modern philosophers have noted,[311] the common wisdom casting many, if not all "good" actions as "freely chosen" contradicts good evidence culled from every historical era. Wisely, the writers of classical Chinese did not posit intimate friendship as a "private relationship" that can be taken up or dropped without doing harm to one's sense of self and of belonging. Their preferred descriptions of intimate friendship imagined both heady commitments and added pleasures from sociable acts. Exemplary figures, male or female, inextricably bound and ceaselessly improved by mutual attraction, could ideally achieve that exalted state in which they felt their interactions with the beloved to be unhindered ("unfettered" and "unencumbered" and "unenslaved" were some of their terms), until the final transformation that is death.[312]

Friendship of this higher order represents a sublime fusion or merging of the friends, which creates a new entity that is more than the sum of its parts. As trust grows in any intimate friendship, the close friends tacitly vow, "I give you power over my future, and I trust you not to abuse it or diminish me in any way." This unqualified trust represents the binding force of the friendship that persists, sometimes, even in defiance of ritual or status considerations, oblivious to family or community interests.[313] Accordingly, friendship in early and middle-period China shares some of the same intensity ascribed to "romantic love" in the modern world. Indeed, in the classical texts under our consideration, romantic love is but one form that intimate friendship can take.[314] The difficulty is that the bonds of friendship may differ so in their qualities and in the reasons for which they are established that the friendships cannot themselves be categorized or constrained.

Friendship as a Social Model
These are large claims, of course, but I would argue that the overwhelming thrust of the early and middle-period texts at our disposal presupposes them. Let me now add two other assertions to the list: that whenever other social relations grow especially close in early and middle-period China, they tend to be constructed less on the model of family than on the model of friends. But with friendship

implicated so strongly in conceptions of social and political order, the severing of friendships posed problems, especially among elites and during times of social and political change, that extended far beyond the merely personal and that could place erstwhile friends in jeopardy, along with their families.

Friendship and Close Social Relations
Yang Xiong said, "A friend to whom one does not give one's heart and soul is but a superficial acquaintance."³¹⁵ In classical Chinese, the earliest notions of *you* 友 generally refer to a category of sustained and sustaining relationships best captured by the adjective "intimate" or "close friends."³¹⁶ In the bronze inscriptions, for example, the term *you* 友 refers to those who regularly take part in common repasts where the inscribed ritual vessel holds food³¹⁷—with the same vessel used for offerings at the ancestors' temple.³¹⁸ A whiff of the sacred lingers over notions of close friendship, then, since the character *you* first appears within the context of sacrifice, conceived as timely offerings presented in anticipation of future support.³¹⁹ Already by Western Zhou times, the idea of *you* meaning "friendliness" and "fraternal love" had become one of the most important ethical principles.³²⁰ Certainly, sharing common aspirations with others imparted a measure of strength to too-fragile human lives,³²¹ and such sharing was probably the more ardently desired by ambitious members of the elite when they left home to seek their fortunes at court.³²²

Over time, the term *you* (often joined in compounds with other graphs, for example, *pengyou* 朋友 or *liaoyou* 僚友 or *yousheng* 友生) tended to be diffused and watered down until it embraced a huge range of merely affable relations: colleagues, classmates, allies, or even mere pals, initially dubbed *peng* in classical Chinese or *jiao* ("contacts"). This may explain why, in late Eastern Han, Zhang Shao 張劭, famed for his extraordinary friendship with Fan Shi 范式, was careful to draw a sharp distinction between unqualified "friends unto death" (*si you* 死友) and mere "friends for life" (*sheng you* 生友), with the latter distinctly inferior.³²³ Importantly, outside committed friendships of complete trust, social relations were not construed as the product or function of the laws of attraction and sympathy; rather, they were deliberate and potentially risky constructions by one or both interested parties.³²⁴ To cite but one vocabulary change, the phrase *shi you* 士友 (colleague in office) does not appear in the extant pre-Han and Han texts; it surfaces in the fifth and sixth centuries, so far as we know, and most memorably in the seventh century *Liangshu*, in connection with Ren Fang's 任昉 active cultivation of a large clientele, gathered with a view to enriching Ren's own coffers and enlarging his political capital by pulling other men of cultivation into his orbit.[325] It is doubtful that literary conventions map neatly onto social realities, however, because the extant texts are more prescriptive than descriptive, and even the writings purporting to describe the world outside the text faithfully are highly selective and hence distorted representations of the elite realities of early and middle-period China.[326] (See above.)

Many early and middle-period texts, perhaps most notably the *Changes*, have two main subjects, trust and time, both related to the premier question of how to contract intimate friendships. Such texts contend that mutual aid and comfort can be expected only from "old" friendships that have stood the test of time.[327] With rare exceptions, only "old" friendships — if close enough — assumed heroic proportions, given that the "test" of true friendship was to see it thrive, so that might nurture the friends properly through life's reversals;[328] mere colleagues and acquaintances, by contrast, could be dangerous.[329] Many odes celebrate the joys of friends and mates who intend to grow old together. For example, Ode 31, "Jigu," contains the lines "With you I made the solemn pledge / That with you I would grow old" (與子成說, 與子偕老) and Ode 82, "Nu yue ji ming," the lines "With you it is right / With you I shall grow old" (與子宜之, 與子偕老).[330] Before and during the early empires, imputing great "age" to anything generally packed more rhetorical punch than a host of references alluding to the allure of the new and fresh. (Compare the attitudes toward music old and new.) As Cao Pi's (187–226 CE) "Discourse on Writing" tell us plainly, "Ordinary people value what is far away and feel contempt for what is close at hand."[331]

Fairly early in Chinese history (certainly before unification in 221 BCE), other serious and committed social relations come to be analogized to friendship. Most scholars of early China (Aat Vervoorn being a praiseworthy exception) have assumed that the "horizontal" relation of friendship, especially when uncalculated, was usually considered less important than the "vertical" hierarchical and calculated social relations of parent-child, ruler-subject, sibling-sibling, and husband-wife. Supposedly friendships were thought to pose a greater threat to the other, more predictable human relations.³³² Few early sources known to me support such a view, however. As Aat Vervoorn observes, the reason friendship often appears last in the lists of social relations is not because it is the least important or the most problematic, but because it is the culminating human experience built on those other relations, whose contours and obligations are more clearly defined. After all, the "Notes on Learning" in the *Liji* makes discernment in choosing friends the penultimate stage on the route to "great personal perfection."³³³ Admittedly, a few polemics declaim the priority of family over friends (raising the distinct possibility that many others did not),³³⁴ yet multiple texts stress that close friendship, epitomizing the virtues of fidelity, integrity, imagination, and flexibility required to sustain it, represents an admirable and authentic way of operating in the world.

It should not surprise us, then, that friendship as an ideal plays a large place in the rhetoric devoted to the other relationships, including the political.³³⁵ Eventually, friendship would serve as the new paradigm for the ruler-minister relationship in the early empires, where powerful sociopolitical bonds were no longer based solely on kinship.³³⁶ Meanwhile, friendship would be extended to describe good relations between siblings,³³⁷ as well as that between husband and wife and master and pupil, leaving only the parent-child relation less liable to description through tropes of friendship,³³⁸ despite the routine coupling of friendliness with filial duty in the compound *xiao you* 孝友. (The aid and comfort that true friends provided one another occasionally helped filial children better serve their parents.)³³⁹

For Xunzi, friendships are one of the three relations said to be constitutive of good order.³⁴⁰ Yuri Pines has catalogued the Zhanguo rhetoric that at once exalted the ideal ruler-minister relation and equated it with "true friendship."³⁴¹ By this rhetoric, the effective ruler was exquisitely responsive to his best advisors' remonstrations and correspondingly inclined to treat the latter generously (that is, by rewarding them with grants and nobilities).³⁴² To cite one early text:

> So well did the kings of old select men for office that their people all returned to the basics and pursued [good fortune] on their own initiatives. They were careful of their characters and they piled up good deeds, one by one. As they knew that good fortune [that is, an official career] did not depend upon another's favors, they did not customarily travel around to make contacts or ask for preferential treatment [unlike today, a more debased era]. Clear to their

very cores and settled in their bones, men enjoyed a quiet but blissful self-possession. Hand in hand, they guided one another to the straight path. They criticized one another honestly. Thus treacherous sayings and wicked behavior died down on their own.[343]

By certain measures, principally by sharing his wealth as well as his labors or dangers, the wise ruler could introduce a kind of parity between himself and the lowliest of his subjects, his foot soldiers (not just the ministers who are Pines's subject).[344] In return, his grateful subjects, touched by such gracious condescension, would give him their allegiance, as the early military texts explain (see also Chapter 3 in this book).[345] And there is much pleasure to be had in governing a realm with the help of friends, the *Zuozhuan* tells us.[346] (The *Zhuangzi* scoffed at such a rosy fiction, of course.)[347] As we know, the very word *you*/"friend" is part of the compound 兄友, signifying close sibling relations. And in a striking departure from Western tradition in which friendship is a pale semblance of erotic passion, in early and middle-period China, conjugal love was especially celebrated when the bedmates felt they were intimate friends, the latter providing a stronger and more complex commitment.[348] Even the ideal teacher-student relation was sometimes analogized to that between close friends, as in *Shuoyuan* 3/9, presumably because of the legendary Kongzi's fondness for Yan Hui, among other disciples.[349] As Aat Vervoorn astutely notes, the word for "intimate friend" (*you*) refers to a *type* or *quality* of relationship, rather than to a specific category or role of persons, with the result that "friends" and "kin," not to mention a host of other close relations, need not function as mutually exclusive terms.[350]

Let us conclude with one brief look at a charming early piece: it is an "intimate friend" (analogized to a mate or soul mate) whom the speaker seeks out in Ode 165 entitled "Hewing Wood." One verse is often cited when the topic is friendship crops up:

> Seeing, then, that even a bird
> Searches for its mate's voice (*you sheng* 友聲),
> How much the more must men
> Search out their friends and kin (*you sheng* 友生).
> For the spirits are listening....

The rest of the poem describes setting out wine, delicious meats of every sort, and grain dishes to entertain friends and relations, making the collective pursuit of pleasure perform double duty as the solemn fulfillment of a commitment to sociability[351] and to the merrymaking at the feast, which reinvigorated and renewed communal ties.

There have never been fixed rules to teach people whom to trust when they enter into the potentially mutually beneficial, yet highly vulnerable relationship we call "intimate friendship." At the same time, the personal costs borne when keeping our guard up in front of colleagues, neighbors, and associates are high, and higher still the costs of keeping close track of a host of virtual strangers. Habits of trust, reinforced on suitable social occasions, are vital to any sustainable vision of a flourishing life, even if trust is easily frayed and not easily repaired.

On Severing Intimate Friendships

Noting that it was sometimes impossible to end a friendship, Aristotle nonetheless advised his followers to try, should one friend become convinced that the other was not a good person.[352] By contrast, in early China, as we have seen, ritual propriety advised the avoidance, if at all possible, of any public break with a person once deemed an intimate, and it did so on two eminently practical grounds: how badly a long-standing or intense tie to an unworthy character would reflect upon the friend who belatedly wanted to break it off and how bereft the unworthy person would be of opportunities for humanizing contact if left entirely to his own devices. In an era before the "blind" civil service examinations initiated in the eleventh century in China, when a man's qualification for office-holding depended to a very great extent on dossiers compiled by local dignitaries on the lookout for lapses, the first consideration was especially important.[353] That helps to explain why civilized sociability presupposed the arts of tactful withdrawal, no less than those of tactful invitation; equivocation was an admirable form of speech with a recognized role to play. Chinese thinkers hoped that faithful adherence to the ritual prescriptions would allow the erstwhile friend to "retreat without leaving a bad taste in [either] mouth,"[354] letting the friendship diminish in intensity while avoiding a too-obvious rupture. As one text admonished, "No ugly utterance should come from the true gentleman severing a social relation" (君子之交絕不出惡聲).[355] "Fair-weather" friendships often ended abruptly, but "friends" who forsook those in trouble merely advertised their own lack of humanity. One passage deplores the reprehensible ways of "friends" motivated by self-interest:

> When men are in power, their clients and retainers gather in droves, but when they no longer enjoy access to power, their guests disappear.... They say of the Honorable Zhai of Xiagui that his clients and retainers filled his gates during his first term as Commandant of Justice, but as soon as he was demoted, he was able to set up nets for catching sparrows outside his doors [there being so little human traffic in his neighborhood]. When Zhai was later reappointed to the same post, his former clients and retainers tried to come back, but Zhai blockaded the entry to his house and remarked, "Only in matters of life and death can one fully appreciate the reasons for one's social contacts. Only after experiencing both riches and poverty can one know one's true attitude toward such contacts. And only with a career's ups and downs do the conditions for forging [trustworthy] social contacts become clear as day."[356]

Thereafter, "setting up sparrow nets" became a common metaphor for "experiencing abandonment by the very people whom one mistakenly took for dear friends."[357] A second story from the same source conveys much the same message, this time in relation to the semilegendary Lord of Mengchang 孟嘗, a political patron to thousands. When Mengchang lamented the disappearance of the old "friends" who once formed his retinue, Feng Xuan 馮驩, a would-be client, chided him for his refusal to see that patron-client relations are more like market transactions than like trusted friendships; that Mengchang's "friends" went away "was just the way things are."[358] The true fault lay in Mengchang's own failure to assess his men realistically.[359]

For careerists to "cultivate contacts to secure their post" (養交安祿) was already a well-worn trope prior to imperial times, as was the consensus that such contacts represented a travesty of real friendship.[360] The market-driven model whereby self-interested contacts pose as true friends continued to come under attack for centuries, as is clear from Wang Fu's 王符 (ca. 76–ca. 157) *Qianfu lun* and Liu Zongyuan's 柳宗元 (773–819) "Account of Song Qing."[361] Meanwhile, readers of classical Chinese were riveted by the poignant essays and letters composed by people of integrity who felt the need to sever their relations with former friends publicly, at considerable risk to themselves. A small selection of famous examples of those pieces, ostensibly "private," but fashioned for a wider circle, illustrates the huge range of responses to the problem of when and how best to sever relations, when necessary.³⁶²

Some men of worth, of course, grew so disillusioned that they preferred not to enter into any new potential friendships at all. But as Ying Shao's *Fengsu tongyi* and other texts demonstrate, much blame generally attached to such uncompromising behavior.³⁶³ Other men of integrity, by contrast, chose to anticipate the need for a break, reasoning that a later break would in all likelihood prove far more disastrous to both parties. For the party initiating the break, the key to softening the blow was to emphasize heartfelt regret. A sufficiently well-crafted "dissolving connections" letter laden with expressions of authentic emotion could avert blame for an author's abruptly putting an end to the intimacy, as we see from the story of Yuan Qiao 袁喬, recorded in the *Jinshu*. Yuan Qiao had long been friends with Chu Pou 褚裒, another worthy man. Once Chu's daughter became empress,³⁶⁴ Yuan foresaw that he, a mere minister, one day might out of sheer carelessness "enter into an irreverent relation with the empress' father," when in his cups or joking. As Yuan explained,

> This old friendship ... must come to an end as the demands of ritual shift. The joys of sitting together with legs splayed far apart [in relaxed fashion, ignoring ritual propriety] will be replaced with the passage of time. Although I might want to sing freely with you, letting my words flow out unrestrained, and throw off the burdens of decorum, would I be allowed to act in such a way [now that your daughter holds this position]? Things constantly shift and change. As they say, "Move one inch on the sundial, and it's a wholly different day." ... I hope that you, general, are happy and unburdened by things; that you serve with principle always in mind; that you rely on the virtuous and employ the good. As I hold this brush [to write this letter dissolving our relation], I can hardly give full expression to the mingled melancholy and nostalgia I feel at this time.³⁶⁵

All who read this letter, the history tells us, deemed it entirely "in accord with the rituals," perhaps because the letter's fulsome expressions of regret poured out in response not to waning regard, but to an unusual turn of fate that had artificially constrained the old heart-to-heart easy manners.

Although few friendships were formally severed because of fears of lèse-majesté, those dangers were real, as the sources attest. By the early empires, men in high position would have had ample opportunity to ponder the moral of the tale told of Mi Zixia 彌子瑕, the beloved male favorite of the Lord of Wei in antiquity. As legend has it, Mi was once strolling in the garden with his ruler when he bit into a peach (then a rare luxury) and, finding it particularly juicy, offered it to his lord and ruler. The Lord of Wei, in the first flush of erotic passion for Mi, took the offer as a sure sign of Mi's devotion. But recalling the incident after Mi's charms had faded, the Lord of Wei deemed the offer of a partly eaten peach insulting, and he duly punished Mi severely for it.[366]

Formal letters severing friendships betoken the real dangers of maintaining an intimate friendship rooted in an implicit quasi-equality at a time when the real powers and status of the parties were unequal or shifting. Not coincidentally, in many instances, the letters were sent by the less powerful party to the more powerful in an attempt to reassert a sort of parity in honor between the parties. While this motivation is hardly universal, the letter writer's adamant refusal to accept unequal status could sully the reputation of the letter's recipient, if the letter writer managed to avoid self-repudiation by appealing to higher values, especially more important forms of trust.[367]

By Eastern Han times, essayists and letter writers in the governing elite were ready to discuss the awkwardness shown by many of their peers in handling friendships and making contacts at court, whether intimate or not. Accordingly, in the later rhetoric on severing relations, standard literary tropes lament the highly public arenas in which heightened societal expectations about the nature of "true" and intimate friendships played themselves out.[368] Many texts frankly allude to the discomfort their writers felt when they or others of the requisite social standing paraded their virtues as trustworthy and sociable people; on the one hand, they reckoned the costs of a failure to do so to be too high for themselves or their families, while they were equally mindful that any parading of virtues left their motives open to suspicion.[369]

Another perspective from which to see Eastern Han instances of "severing relations" is the analysis offered by the upright official Zhu Mu 朱穆 (100–163) in an essay entitled "On Upholding Tolerance and Magnanimity," during the reign of Shundi (r. 125–144).[370] Zhu Mu's essay famously argued that the Confucian injunctions "to offend no one" and "to give no cause for complaint," lest one endanger oneself and one's family, could never be reconciled with the Confucian injunction to offer "praise and blame" as needed; obviously enough, criticisms riled the criticized. Zhu Mu's second essay "On Severing Unofficial Relations" took up a related problem, for there, an unnamed interlocutor asks Zhu whether, in refusing to maintain the usual contacts with others, he does not risk provoking their resentment and impeding his career. Zhu Mu stoutly declares that he, at least, can "bear the scorn" of others, since the current crop of officials is no good; they merely "steal their reputations," because they neither attend faithfully to their duties nor regard their ruler with proper awe. So, Zhu Mu concludes, to continue to observe the usual court practices of entertaining and being entertained risks damage to his own character and, ultimately, his future prospects for service to the court. In other words, to accept conventions and enter into close relations with flawed men would actually "abrogate ritual" and "undermine the common good." Until he finds promising friends of worth, he can do little except try to mend his own flaws. Zhu Mu notes that "the Ancients" were content to display their "awesome demeanors" in the formal settings afforded by court or community banquets. (Here Zhu may have been thinking of the example set by Kongzi, who had left courts, despite his ardent wish to serve).[371] The essay ends with a resentful poem expressing Zhu's intention to break off relations with an unnamed former friend, whom Zhu likens to a rapacious owl, "gluttonous and greedy . . . stinking and rotten." In the poem, a rapacious owl accuses the phoenix, the most exemplary of all mythical creatures, of lacking virtue, saying that "once the phoenix aspires to a different realm," the owl swoops down for the kill. One wonders whether Zhu Mu wrote this after defending himself against slanderers at court.

Cai Yong 蔡邕 (132–192), a later court official, was a fervent admirer of Zhu Mu and inclined to much the same grievances. Legend has him traveling to Zhu's family home after Zhu's demise, there to copy out Zhu's manuscripts; we know that Cai composed an essay expanding upon Zhu's points.[372] Cai's essay "Correcting Contacts" emphasizes the extreme caution with which the wise and good person initiates friendship, lest he later come to regret the association. Since the cautious person of true virtue, like Confucius himself, neither "waits upon the rich and influential" nor "holds the poor, the young, and the low-ranking in contempt," he need never fear going without friends. In the end, he will simply have dodged the disingenuous, who are apt "to desert old friends" in crises anyway.³⁷³ Holding back has an additional benefit, according to Cai: his assessments of other men's characters are likely more accurate, given his disinterestedness when observing. And only estimable men like him can hope to avoid false friendships, or at least quietly end an inappropriate acquaintanceship before intimacies complicate matters.³⁷⁴ Importantly, only close friendships forged between exemplary men prove beneficial to broadcast, says Cai,³⁷⁵ in a reproof of his contemporaries' penchant for facile characterological analyses.³⁷⁶ But how those in power may best differentiate genuinely worthy men from sycophants, cronies, and partisans remains one of the chief problems in governing — a problem recognized long before Cai's era (See Chapter 1).³⁷⁷

During times of political upheaval, in particular, even the most upright of men could feel stymied when breaking off relations with an intimate friend, knowing that he must not appear to be acting out of pique, but rather exemplify for both his peers at court and for later generations the ideal manner and timing. Otherwise, he jeopardized his own reputation as a cultivated man of superior insight and possibly the family fortunes, as well. (Might the upsurge in use of the term "dear friend" [*qin you* 親友] in the Six Dynasties literature point to the ubiquity of faithless friends in corrupt and unstable courts?) No formal break between men at court could ever be truly private, an act of individual agency or autonomous choice, but letters and stories as a forum or public space afforded onlookers a glimpse of recent events.³⁷⁸ For that reason, Ji Kang's 嵇康 (223–262) famous "Letter to Shan Tao" is marked by lofty intent and a grand style. Supposedly, this letter was sent to Shan Tao 山濤 shortly before Ji's execution, at the age of forty. Ji had sought to avoid court service several times, sensing (rightly) that his marriage ties with the imperial Cao family would endanger him at a court dominated by the Simas, would-be usurpers.³⁷⁹ When Shan Tao, presumably between a rock and a hard place, invited Ji Kang to become Shan's subordinate at court, Ji roundly berated Shan as one "who, in fact, did not get it!"³⁸⁰ According to Ji, exemplary figures in antiquity had but one thing in common: they had found a way "to follow their own commitments" (*neng sui qi zhi zhe ye* 能遂其志者也). Ji Kang admits to, even boasts of, a wide range of character flaws: he is arrogant, slovenly, and blunt to the point of rudeness. In addition, he cannot hold his tongue, and he is accustomed to finding fault with both the conventional models inherited from the distant past and the "new men" of ambition who are so widely admired in his own age. By Ji's own account, these propensities make him quite unsuited to office-holding, and so he prefers to live out his days in peace, "nourishing life" and avoiding potentially polarizing activities and occasions,[381] especially because he now, belatedly, has "come to realize that it is possible for a few men with lofty principles to exist," so long as they secure protection from the powerful.[382]

Ji's last assertion holds the key to any reading of his letter, which is a marvel of ambiguity. It may drip with sarcasm, in the belief that there are no decent men at court, it may represent a last-ditch, naked appeal for Sima protection, or it may be simply a justification for the break with Shan Tao, whom Ji had taken for a friend in days less fraught. As a friend, Shan Tao should know better to think that he, Ji, would be tempted by the prospect of mere fame and court distinctions. "If you insist on my joining you in the ruler's service, expecting our rise together to be a source of pleasure and help to one another, one day you may find that the pressure has instead driven me quite mad. Only my bitterest enemy would go this far [to harm me]!"

Did Ji Kang really think that Shan Tao had sunk so low as to "drift along with the vulgar habits" 流俗 of the other courtiers, craven in his desire to curry favor with the Simas, his patrons? That question has intrigued readers for millennia. Several scholars, Lü Lihan 呂立漢, Thomas Jansen, and Wang Yi, to name but a few,[383] have doubted the traditional understanding of the break given in the *Wenxin diaolong* less than a century after Ji Kang's letter and subsequent execution. How is the reader to square Ji Kang's ostensible self-denigration with his marked disdain for Shan Tao? Was Ji Kang determined to insult Shan Tao?[384] If so, why would Ji Kang have assured his son, shortly before his execution, that the son "would never be alone," so long as Shan Tao lived?[385] Was Ji not rather anticipating that he would soon be charged with a capital crime and so be trying to save his good friend Shan Tao from implication in his crimes? Or was Ji Kang anxious to avoid serving as Shan's subordinate because he foresaw and deplored Shan's clumsy attempts to silence him (well-meaning or otherwise)? Or was Ji Kang mainly seeking to elevate his reputation as a lofty recluse, a "lone pine tree,"[386] heedless of any possible future repercussions for himself and Shan Tao? Ji's letter is full of half-developed arguments that, pushed, could support any of these readings.

Perhaps this explains why later generations have read this letter so obsessively, whereas a second letter by Ji Kang has not garnered a fraction of the attention.[387] Are clues to be gleaned from another anecdote, which has Ji Kang, on the eve of his execution, calmly plucking the zither strings to the tune of the "Melody of Guangling" and going to his early grave refusing to teach the composition to anybody else, so that the arresting melody died with him?[388] Certainly this anecdote dramatizes Ji's lack of sentimental feeling for those who would live on, marking him as a lonely man habituated to a "selfish" way of life that was anathema to strict Confucians, who prided themselves on valuing the right sort of friendships with honorable men, living or dead.[389]

Reading any of the essays and letters about severing intimate friendships, one cannot but be struck by the high-stakes games involved in close friendships. Whereas it was silly and often dangerous to imagine that one could confront the world and accomplish anything without the help of close friends, it could prove no less perilous to promote close friends, since that left the one friend vulnerable to being implicated in the other's putative misdeeds, as many biographies attest.[390] Nor can one fail to notice that changes in political life inevitably transformed relations between friends, and seldom for the better. Perhaps "poems of parting" (*gaobie* 告別) represented the only entirely safe genre for expressing disengagement from friends, given the blurred lines between intimate, acquaintance, and colleague. Imminent departures became convenient covers for real emotional ruptures.[391] Does this help explain why tens and thousands of such poems, many little better than doggerel, were composed and recited or sung down through the ages?

We moderns are hardly alone in observing the high drama attached to the severing of intimate relations, with oh-so-careful phrases detailing the circumstances in which fond hopes are dashed; severing friendships had become the stuff of jokes by the third and fourth centuries CE, as is evident from one playful tale of "cutting the mat" told about Guan Ning 管寧 and his friend Hua Xin 華歆, where a potential breach nonetheless entails the usual anxieties about status and wealth:

> Guan Ning and Hua Xin were together in the garden hoeing vegetables when they spied a piece of gold lodged in the soil. Guan went on plying his hoe, as though the gold were but a potsherd or a stone. Hua seized the piece of gold,

but he then threw it away. But on a second occasion, Guan and Hua were sharing a mat while reciting a text together when someone passed by the gate in a splendid carriage and wearing a ceremonial cap. Guan did not interrupt his recitation, but Hua put aside his manuscript, so that he could go out and take a good, long look at the carriage. Guan cut the mat in two, and then sat quite apart from Hua. He said, "You are no friend of mine!"

Commentary on the passage insists that Guan was merely joking, because Guan was known to mock Hua's shortcomings playfully.[392] The word *wan* 玩 ("play"), as we will note in Chapter 6, not only denotes the tactile pleasure of rolling something smooth and round in the hand and the mental pleasure of rolling something over in the mind. It describes the pleasurable release felt after communicating thoughts to like-minded people of taste and refinement.[393] Here, Guan seems to be toying with Hua, though the story can just as easily be construed as a straitlaced and hurtful reproof. It is hard to tell, precisely because habits between friends are unreadable, unless we witness their smallest gestures and "hear the tone" of their exchanges for ourselves. No matter. People are wont to feel used or betrayed only when a friend does something that seems uncalled-for, something whose outcome may not prove pleasurable.[394] At that exact point, shared interests, shared views, or shared experiences no longer suffice, for the friendship no longer inspires trust. Distrust clouds what was done and felt, foreclosing any friendly constructions that might be imputed to a joint past.[395] Thus, many a close friendship in retrospect appears to have been the ultimate treachery.

One final question seems worth asking about the collapse of intimacy: in severing a relation, was it worse to express hatred or assume lofty indifference toward former friends? A brief look at two pieces of evidence from the Western Han, a famous letter by Yang Yun 楊惲 (d. 54 BCE) and the purported history of relations between Zhang Er 張耳 and Chen Yu 陳餘, suggest that both avowals could be equally disastrous for the body and soul. The letter by Yang Yun (dated ca. 60 BCE), which broke off his relations with a former friend Sun Huizong 孫會宗, is often thought to show the role that unsuccessful remonstrances can play in ruining friendships in real life. Yang, who once commanded noble rank, a high position at court, and great riches, had been demoted to the rank of commoner, ostensibly because he had incited the enmity of a fellow officer at court, but more likely because he was wont to offer his ruler impolitic reminders of his failings. Yang retired to the countryside, where he continued to live in ostentatious luxury. After some time had gone by, Sun, as a good friend, wrote Yang, urging him to live on a more modest scale. For only if Yang would show profound contrition for his former misdeeds would he likely regain the emperor's trust and hence his former titles and powers.

Yang's reply to Sun's entreaty opens with the humilifics customarily deployed by elite strangers — but never close friends — when addressing each other. But Yang's rhetoric strikes one as over the top, even by the standards of the day:

> Moved to pity by my gross stupidity and immorality, you have been good enough to write me a letter teaching me my shortcomings and how to correct them. Your heartfelt concern is truly magnanimous. I am disturbed, however, that you may not have examined deeply enough the circumstances of my own case [which ended in demotion].... [After my disgrace,] I considered, in view of the magnitude of my errors and my less-than-ideal conduct, that I had better end my days as a farmer. So I took my wife and family to the country, where ... we plow, tend mulberries, water the orchard, and so manage to produce enough to pay the taxes we owe to the state.
>
> One cannot put a stop to human emotions. Even sages do not try. Therefore, although rulers or fathers command the greatest honor and affection, after they die, the period of mourning for them always comes to an end, sooner or later. And it has already been three years since I incurred the disgrace [and three years corresponds to the longest period of mourning allowed].
>
> My family and I work hard in our fields, and when the holidays, summer and winter, come, we boil a sheep, roast a young lamb, bring out a measure of wine, and rest from our labors. My own family came originally from Qin, and so I can make music in the Qin style. My wife is from Zhao, and she plays the zither very well. We have, moreover, several maidservants who sing well. So after I have had enough to drink and my ears are beginning to burn, I gaze up at the sky, thump a crock to keep time, give a shout, and sing this song: ... "Man's life should be spent in pleasure. / Why wait in vain for wealth and honors?" At such times, I flap my robes round in delight, wave my sleeves up and down, stamp my feet, and dance about. Admittedly, this is a wild and unconventional way to behave, and yet I cannot say that I see anything wrong with it....
>
> As Confucius said, "Those who do not follow the same road cannot lay plans for one another." Why, then, do you come round with a career statesman's ideals and use them to censure a person like *me*?[396]

After going on in this vein about his idyllic life away from court, Yang peremptorily concludes the letter to Sun by severing his friendly relation: "I fully understand from your letter where your ambitions lie. Now, when the Han ruling house is at the height of its glory, I hope you will diligently pursue them [at court], without wasting any more of your time conversing with me!"

Yang's indifference to societal mores did more than ruin his friendship with Sun. Soon after hearing about the contents of Yang's return letter to Sun, Emperor Xuan (r. 74–48 BCE) of Han decided it was high time to execute Yang for his "satirical remarks." So whatever pain Sun experienced upon receipt of Yang's curt reply, Yang's formal letter severing relations in effect signed his own death warrant.

Turning to the tale of Chen Yu and Zhang Er, we see bosom buddies who once "vowed to die for one another" eventually become archenemies who let their hatred eat away at them until each indulged in malicious attacks and counterattacks in hopes of killing the other. Zhang managed to eke out no more than a single year's existence under the protection of the Han founder after destroying Chen Yu. What are we to conclude, when we read this story in light of Yang Yun's tale? Since loyalty between friends was believed to be so crucial to ethical development, allowing each party to glimpse in himself and his friend some admirable potential that would "enlarge and magnify" both, which was the more destructive emotion, feigned indifference or outright hatred? Indifference was arguably worse, insofar as it dampened spirits and diminished the zest for life. On a more mundane level, however, medical practitioners warned that hatred would eat away at the vitals, leading to an early death, either by "natural" or by violent means.

Pleasure, Music, and Friendship

The modern philosophical category of "friendship" is configured (like a great many modern concepts considered important) through such normative dichotomies as public/private, cause/effect, agent/patient, and coerced/voluntary, while a love of music tends to be relegated to the realm of the "aesthetic" and "impractical." Modern academia's demands for clear summaries, totalizing syntheses, and unambiguous hypotheses cannot capture the delicate constructions in classical Chinese rhetoric positing the nature of intimate friendships or resonant music, painstaking rhetoric phrased in distinctive ways through analogies to the cosmic sympathies epitomizing order and beauty.[397] Governing elites conceived the social world "between heaven and earth," no less than the sphere of the gods, as an immanent domain whose deepest structures were intelligible to thinking and feeling people, despite the inherent messiness of contingency and change (aka, the "mystery" or *xuan* 玄). One of the strengths of the antique thinking is that it generally eschewed universal rules or totalizing systems when judging the actions of particular people. Instead, it preferred to examine the courses of action adopted in specific social relations within historical contexts and the possible motivations that underlay them. Noteworthy is the plurality of approaches and criteria routinely invoked when judging the range of attractions and allegiances ascribed to exemplary figures.[398] In such accounts, the pitfalls of bonding in sociopolitical alliances become clear, clear as the weakening of the moral fiber that follows listening to the wrong sorts of music in bad company. By contrast, the countermodels of intimate friendship and musical influences seem at once wondrous, inexplicable, and yet vital to human flourishing.

I have coupled music with friendship early in this book not only because I believe the two cannot be uncoupled, but also because so many pre-Han and Han (not to mention later) writings cast theories about music, due to the graphic pun with "pleasure," as the source of all ancient rhetoric about pleasure. And yet I have shied away from giving a systematic account of the canonical "Record on Music" ("Yue ji" 樂記) chapter in the *Record on Rites*, the obvious inspiration for most secondary studies, in part because the *Liji* is a relatively late compilation, in part because foregrounding it might lend the impression that talk about pleasure in music was confined to treatises whose main subject was music.[399] I have also intentionally put texts together that are not usually, these days, placed in conversation with one another—texts drawn from histories, from Classics, from masterworks, and from less official sources—because such cross talk was far more common in the period under examination than today. This constitutes a refusal to draw a hard-and-fast line between the rhetoric assessing political considerations and the loftier ideals ascribed to music and friendship, in the belief that the ancient texts move seamlessly from one subject to the other.[400]

One of my favorite American writers, Wendell Berry, describes a tight-knit local community with much the same lyrical quality of close friendship that appears in anecdotes from early China. In Berry's novels, the characters come to learn that they are "part of a place already decided," "part of a story begun long ago and going on." From that place and story they travel through "happiness and hardship and longing and satisfaction and death and grief, and somehow become innocent again." For they come to find that "to be loving something is to take comfort from it." These neighbors, friends, and lovers "wish to help what could only be endured," and they see, too, that "in a place like that you don't need to say much." This, it seems to me, captures the quality of close friendship, which obviates the need for much talk, rooted as it is in a "particular love for particular things, places, creatures, and people, requiring stands and acts, showing ... practical or tangible effects."[401]

Figure 3.1. Handscroll entitled "Illustrations of the Classic of Filial Piety" (detail), 26.4 x 71~ 139.4 cm, ink and color on silk, National Palace Museum, Taipei, Taiwan, R.O.C. Traditionally (mis?)attributed to the painter Ma Hezhi 馬和之 (fl. 1131–1189) and the colophon-writer Emperor Gaozong 高宗 (1107–1187).

A classic scene of instruction, in which the disciples surround the master (in this case Confucius), so that they can receive his oral teachings. Here Zengzi (right), an exemplary disciple, kneels before Confucius sitting on the dais at center. We should imagine the same type of oral transmissions going on with Mencius, Xunzi, and Yang Xiong and their disciples, during the age of manuscript culture. Han-dynasty pictorial stones from Zhucheng (Shandong) also depict nearly the same scene of instruction, with the master seated at the center of his disciples, though they place the master and disciples within a building, rather than outside.

CHAPTER THREE

Mencius 孟子

on Our Common Share

of Pleasure

> What is there to fear, if you share your blessings with others?
> —*Zhuangzi*, "Tiandao"
>
> So I say it, again and again: pleasure is shared. —Lucretius

At least since the beginning of the third century CE, all readers of the Chinese masterwork the *Mencius* have approached its sophisticated philosophical arguments through its first chapter, which is devoted to the topic of pleasure. In Book 1, Mencius (act. 320 BCE), typically cast as the most faithful follower of Confucius, claims to be not "fond of disputation" (that is, rhetorical argument), even as he crafts a single, subtle, and complex argument about the crucial role of pleasure in sociopolitical constructions of the common good.¹ There when Mencius converses with rulers great and small in a series of dialogues, three main motivations seem to propel his highly nuanced rhetoric: he would nudge the rulers he meets (each with their own distinctive concerns) to adopt the belief that it's always in their self-interest to commiserate with their subjects, who will return the favor by granting their legitimate authority and working enthusiastically on their behalf; he would maintain his position as self-styled "friend" and valued advisor to those rulers, because that wins him his own economic and psychic security; and third, he would establish criteria by which to assess future rulers, depending on whether they serve or undermine the common good. (Certainly, later readers tried to extract his criteria from his writings.) Mencius throughout Book 1 labors mightily against the common perception, as true as Mencius's day as in our own, that anyone in power will surely want to consume as many luxuries and services as possible. To counter that widespread perception, Mencius offers a vision wherein true sovereignty (defined as relative freedom from self-induced disasters) depends upon the ruler's forging gracious compacts with his subjects, who then support him through thick and thin.

Because the air of condescension evoked by Mencius can be so off-putting to moderns (I speak as one occasionally put off), it may be necessary to recall that Mencius works in nearly every episode through the personal, and if he is to nudge his far more powerful superiors in a constructive direction, it would likely prove counterproductive were he to scold, coerce, or threaten them outright with ruin. Instead, Mencius must, through dialogue, exhortation, and parables, awake in those rulers what we might call a "social conscience" — never an easy task for a persuader and one presumably made harder during the Zhanguo ("Warring States") period, when, in the war of all against all, displays of brute power tended to translate into awesome majesty and sincere admiration. Note that the dialogue form is perfectly suited to the task that Mencius sets himself, insofar as it allows him to take up seemingly minor comments and use them as pretexts for extended discussions relating to his project. Meanwhile, in dialogue, he may portray himself as the perceptive friend who sees into the soul of the ruler with whom he converses, divining aspects of the ruler's character that the ruler himself fails to credit in himself ("The heart is another man's / But it is I who have surmised it").[2]

Because modern studies of the *Mencius* rarely, if ever allude to the topics of pleasure and desire,[3] this chapter restores to this important classic the integrity of its original approach, recasting Mencius's discussions of value neither as disembodied, abstract formulations of bloodless virtues nor as denatured accounts of human nature, but rather as earned, self-aware, and ultimately humane calculations encouraging bodies and hearts that are full of desire and responding at an unusual level of passion. This reading of the *Mencius*, which returns to the basics, develops a familiar line of thought found in the *Analects*, whereby "taking pleasure in Heaven" (our given allotments) and in fellow-feeling constitutes the whole of doing good, which itself represents the best kind of human knowing.[4] (As readers will recall, twice in the *Analects*, Kongzi, that is, Confucius, names fellow-feeling, *shu* 恕, as the one thread running throughout his teachings.)[5]

Every important part of Mencius's arguments on pleasure and desire is laid out in Book 1 of his collected writings. For that reason, what follows will survey, one by one, the arguments pursued there before proceeding to analyze the relation between Book 1 and the rest of the writings attributed to Mencius, in particular Mencius's ideas of sharing, empathy, and commiseration as the ground for all humane action. If pleasure is the starting point for Mencius, the adoption in Book 1 of the dialogue genre is hardly coincidental, since the dialogue is the genre best suited for eliciting and directing the focus of particular individuals to specific views and desires.[6] In contrast to the Socratic dialogues, which continually ask, "What is X?" the dialogues in the *Mencius* nearly always circle around the questions "What do you like?" and "What do you want?" Mencius's conversations attempt to help the listener/reader discover the true relation of ordinary desires to more civilized desires for the good. Drawing out his partners in dialogue so that they identify their own needs and predilections is the first step toward getting them to accept the needs and desires of others. The ordinary desires experienced by every human being, desires for food, sex, music, companionship, security, and respect, are the key resources that can be transmuted or redirected—*if* the person's whole being has been sufficiently aroused—into the single overpowering desire for community and cooperation premised on mutual care and the cultivation of integrity.[7] Not coincidentally, the ruler provides the perfect example of how a person's thoughts about pleasure-seeking factor into his ordering of priorities, since the ruler's actions are less likely to be constrained by others, and the ruler has within his grasp a seemingly infinite array of pleasures, as well as the power to dictate how others may enjoy themselves. In educating those in power, Mencius potentially educates all.

Mencius's Book 1: Entering the Realm of Pleasure

In Book 1, Mencius's aim, as we will see, is to encourage the rulers with whom he converses to recognize the essential humanity underlying all their desires, then "push" (*tui* 推), that is, "enlarge" (*da* 大) them to the fullest extent possible.[8] With the "great man" (a useful rhetorical term that is suitably ambiguous, because it indicates either the "ruler" or the "noble in spirit"), enlargement comes simply through crediting other human beings with having the same human desires and on the basis of those shared sympathies seeking ways that allow others to satisfy their desires so that all end up at once civilized and gratified.[9] The process outlined in Book 1 has as its goal the increase of the great man's pleasures to the degree that he knows he has prudently provided pleasures for others in his community, no less than for himself.[10] This acknowledgment of the importance of desires in human lives helps the person come to accept responsibility for human failings.[11] In turn, the acknowledgment of the shared seeking and shared failings of humanity will represent the crucial first step on the proper path toward achieving the best the person is capable of, since it enables the moral imagination.[12] By Mencian logic, ardent longing to realize that goal — longing sufficient to sustain a studied commitment to humane conduct over time — opens a broad and level path to those striving for the same sense of the wholeness that the cosmic order ("Heaven") models so effortlessly. "Integral wholeness is the way of Heaven, and longing for that is the way of human beings."[13] That sense of wholeness carries several connotations simultaneously: first, it refers to keeping oneself from harm, in an unmutilated state; it moreover implies a steady commitment to acting with integrity. Supposedly, wholeness of body and spirit imparts to the person in power a well-nigh irresistible charisma that attracts others to follow him in his causes.[14] And there is an even larger claim to which Mencius works his way eventually: the process of making oneself more whole, if Mencius is to be trusted, strengthens the sympathetic connections between the person's enlarged sympathies and the entire universe, for Mencius reports of himself, "The whole of phenomenal existence is complete within me. There is no greater pleasure than to find, on reflecting on my person, that it has integrity.... This is the shortest path to real humaneness."[15]

In Book 1, the opening passage of the *Mencius* appears to say very little about pleasure, aside from the conventional pleasantry employed by King Hui of Liang to express his delight that Mencius has troubled to pay a visit to his kingdom. The king inquires, most politely, what Mencius would have him do to profit his realm. Quite irrationally, it seems at first reading, Mencius launches into a lengthy tirade, protesting that the king should set an example for his people. If he asks, "How I can profit my realm?" each of his subjects will soon start asking how to profit himself, and before long, regicides may be perpetrated. For, Mencius reasons, "If profit is put before rightness, there can be no logical satisfaction" to his subjects' desire for ultimate profit short of total usurpation, since the ruler's expression of his desires influences his subjects' formation of their desires. "All that matters is that there should be benevolence and rightness. What is the point of mentioning profit?"[16]

Mencius's critics have had a field day with this passage. Since "profit" (*li* 利) in classical Chinese, as in modern English, means "benefit," Mencius's argument is criticized as weak, if not downright puerile. The critics have missed Mencius's point entirely. Mencius knows precisely what he is doing. As he remarks elsewhere, "Today all those who are in the service of princes boast that they are able to extend the territory of their princes and fill their coffers for them, but the so-called 'good subject' of today would have been dubbed a destructive force attacking the people in antiquity,"[17] given the aggressive character of such claims.[18] What is peculiar about evaluations of human worth is that they so seldom reflect thinking about what is needed to nurture life, despite the importance of this subject to every single person.[19] Mencius's first dialogue intends, therefore, to plunge the listener/reader immediately into the heart of the problem, and the reader, naturally enough, reacts much as King Hui must have done — taken aback at the pedantic quality of this advisor's hectoring.[20]

The critics have failed to discern the thrust of Mencius's diatribe. Mencius, like other classical masters of the pre-Qin era, exhibits a deep sensitivity to the fact that descriptive vocabularies are hardly "neutral" mental constructions; concepts shape human behavior and hence the sociopolitical realm. In reality, Mencius in one short dialogue has accomplished quite a lot. He has established a sharp distinction between socially constructive and destructive types of pleasure-taking by way of showing that single-minded pursuit of *li* 利 can easily devolve into antisocial behavior and thus is a sustainable goal neither for governance nor for personal behavior. At the same time, Mencius indicates that the pursuit of appropriate duty (*yi* 義), a term inextricably tied to notions of social acceptability, handily secures the same benefits for the ruler while avoiding the negative consequences of short-sighted greed for profit. He has deftly compelled a change of topic from talk of profit and whatever nurtures outsize ambitions to the serious consideration of several moral topics needing further explication: topics such as the feasibility of limits on our actions, the urgency of our desires and those of others, and the real dangers — moral and physical — that attend entertaining desires in ourselves that we forbid to others while tending to overlook the actual human costs that certain policies will likely exact. To hide in comfort behind seemingly "objective" cost-benefit analyses is to fall short of humanity. Suddenly there looms the specter of injustice, retribution, and societal chaos unleashed by "treating people like horses and hounds" — or worse.[21] The unequivocal implication is, "If you persist in acting in those terms, you will soon die a violent death," and yet, already in this opening passage, Mencius has raised a more palatable prospect for contemplation: that if the ruler learns from previous mistakes, he "will thrive due to his [salutary] anxiety, and only perish if he [continues in his old ways of thinking himself] perfectly secure in his pleasures," which becomes a recurrent theme.[22] (Remember: to threaten the ruler would likely prove counterproductive to Mencius's project.) In addition, Mencius has unobtrusively, if crucially, forged a common identity among the listener/reader, the omniscient recorder/spectator of Mencius's dialogues, and Mencius's interlocutors themselves, blurring time, space, and character.[23]

Because such topics can usefully be explored only through a more overt pleasure discourse, the next reported interview (1A/2) finds the thinker again in the company of King Hui of Liang, who is idly taking in the view in his pleasure park. (The park setting is important, not only because hunting parks provide the conventional setting for companionable outings and feasts, but also because parks epitomize high status and exceptional privilege, excessive luxury and immense wealth.) The king wonders if a man of true goodness and wisdom would ever deign to feel the ordinary sort of enjoyment that he feels — an aesthetic appreciation heightened by what we might call "pride of possession" — when he views his well-stocked parks and ponds. By implication, the king does not consider himself to be a good and wise man, since his thoughts are not particularly elevated. Mencius hastens to correct the king. "Only if a man *is* good and wise will he be able to enjoy" sights such as these, he insists. "Otherwise, the king would not enjoy them, even if he had them." Were the king an immoral tyrant, this would preclude true enjoyment, Mencius remarks, for true enjoyment comes only when the person shares his pleasures with others, as Mencius will demonstrate, with a "proof" from history.

Long ago, King Wen of Zhou had a vast complex with three sites, the Magical Terrace, Pond, and Park, built near his capital. The people completed the three sites "in no time," according to Ode 242.[24] They did not find the work of building the park and terrace burdensome, because they were given access to the fabulous structures and the creatures inhabiting the area. The ode is cited by Mencius because it confirms the view that so long as public works such as these are shared, the ruler becomes the ultimate source of the people's abiding enjoyment. In consequence, "King Wen received Heaven's mandate [to rule the empire], and the people took pleasure in his possession of magical virtue." King Wen's triumphs, as every novice student of history knew, undid the last king of Shang, whose notorious cruelty was epitomized by a closed park. Legends denouncing his lakes filled with wine and forests hung with meats pictured an island constructed in a large pool filled with alcohol and stocked with skewers of roasted meats (some, they say, taken from the flesh of honest officials). With past paragons such as King Wen summoning the counterexamples so vividly to mind, Mencius can convince King Hui of Liang that "it was by sharing their enjoyments with the people that the men of antiquity were able to enjoy themselves." Knowing full well that they could never really have enjoyed their perquisites "all by themselves," since selfish acts would incite their subjects' envy, the most excellent kings of yore through various boons, grants, and concessions secured the people's loyalty, unified the realm, beginning from their own small tracts of land, and thereby prevented the grave harms attending social isolation and further violence.[25] Mencius leaves the king to wonder what pleasures cannot usefully be shared, if such large and costly projects as hunting parks, usually deemed major factors in the downfall of rulers, could be made to fortify instead of to weaken his realm.

In the next exchange (1A/3–1A/4), King Hui of Liang proceeds to consider a resource that is strictly limited: stocks of grain. Being strictly limited, it presents a harder case, for the grain that feeds one person is taken from another. The king complains that he has been willing enough to move his farmer subjects at great expense from one district to another, as natural disasters necessitated, so that they might get enough to eat, and still he is not widely regarded as a model ruler. Why not? Beginning once again with the ruler's likes and dislikes, Mencius answers with a seeming non sequitur, "Your majesty is fond of war." Borrowing an analogy from war, a subject that he knows to be of absorbing interest to the king, Mencius asks whether those who retreat a mere fifty paces on the battlefield can justly mock those who flee a full hundred paces.²⁶ Obviously not. The unspoken implications are two: that the king himself has been proceeding by half measures and that the king does not know how to employ the resources at his command properly. "There is fat meat in your kitchen and there are well-fed horses in your stables, yet the people look hungry and in the outskirts of cities people drop dead from starvation. This is to show animals the way to devour humans."²⁷ What is required, Mencius says twice, is wholehearted attention (*jin xin* 盡心) to redress the problems at hand, mainly by attending to the cares of the poor and disadvantaged.²⁸

Mencius then imagines a better way for the ruler to manage his resources:

> If you do not interfere with the busy seasons in the fields, then there will be more grain than the people can eat. If you do not allow nets with too fine a mesh to be used in large ponds, then there will be more fish and turtles than they can eat. If hatchets and axes are permitted in the hill forests only in the proper seasons, then there will be more timber than they can use. When the people have more grain, more fish, and more turtles than they can eat, and more timber than they can use ... then those who are fifty can wear silk ... those who are seventy can eat meat ... and families with several mouths to feed will not go hungry.... When those who are seventy wear silk and eat meat and the masses are neither cold nor hungry, it is impossible for their prince not to be a true king.²⁹

For "who among us does not want wealth and rank" or the benefits they bestow?³⁰ By careful planning and provisioning, the lucky subjects of the true king may come to know, most particularly in their declining years, luxuries fit for a king. There will be silk to wear and meat and fish to eat before a warming fire. The people will finally know what it is to be satisfied, and in their gratitude, they will surely support the king, so that he may have his way as well. However, should the ruler fail to provide the opportunities that allow his people to savor the creature comforts that he values so much, how can such a king dare to call himself "father and mother of the people"?³¹

The king remains silent; he does not take the bait. Instead, he begins to relate his great sorrows (1A/5). His realm has met defeat in the east, the west, and the south. His eldest son perished while on campaign. All he wishes in his old age is to wash away the shame and anguish through a victory, but how can he possibly accomplish such an enormous task within the short span of his remaining years? Mencius replies that the king's territory, while small, is "sufficient to enable its ruler to become a true king." The king need only practice benevolent government, reducing punishments and taxes and teaching the people to get the most out of their lands and their families, and Liang can yet be victorious. For in a world where the king's rivals "push their people into pits," brutally ignoring their interests, "the benevolent man has no real peer."[32] The king's rivals take their subjects away from their work during the busy seasons, making it impossible for them to minister to their own household needs. Comparing the king's people to rice sprouts, Mencius advises the king to consider this analogy (1A/6). If his people are to grow sturdy, they require sufficient moisture, with moisture a standard metaphor for the king's civilizing influence, his boons, and his favors. If the people get what they need—and they are satisfied with little enough—the people will turn to the ruler "like water flowing downward." "Who can stop them" from doing so?[33] Another silence ensues.

Abruptly, the scene shifts to the munificent court of King Xuan of Qi (1A/7), who asks Mencius to describe the heyday of his country several centuries ago, under the hegemon Duke Huan (r. 685–43 BCE).[34] Mencius refuses to discuss the proposed topic. "None of the followers of Confucius spoke about the history of the hegemons.... I know nothing about them. If you insist on my speaking, perhaps I may be permitted to tell you about becoming a true king."[35] King Xuan asks, somewhat anxiously, "How virtuous must a man *be* before he can become a true king?" Mencius replies that by definition, anyone who brings peace to his people is a true king. King Xuan has all the necessary qualifications, particularly because he leads a powerful kingdom. King Xuan wonders how Mencius can be so sure. Mencius reports what he has heard at court: one day, the king, catching sight of an ox being led to slaughter, ordered his men to spare the animal, since he could not "bear to see it trembling like an innocent man going to the execution grounds." He commanded his underlings to use a sheep for the sacrifice instead. On the basis of this story, Mencius concludes that the king has the very heart of compassion that is the precondition for becoming a true king: he is not complacent in the face of suffering.[36]

The king, mulling this over, remarks: "I looked into my heart, and I failed to understand it [my motivation in this incident]. Your description of it has struck a chord in me, however. What makes you think that my heart really accords with the Way of the True King?"[37] Mencius insists that the king need only extend to his own people the same feelings of empathy and compassion he has *already* felt for the ox; he would then quit confusing a *refusal* to act on the people's behalf with an *inability* to act.

> Treat the aged of your own family in a manner befitting their venerable age and extend this treatment to the aged of other families. Treat your own young in a manner befitting their tender age and extend this to the young of other families. Then you can have such complete mastery over the realm that governing well will be like rolling it on the palm of your hand. In other words, all you have to do is take this heart that is here, in the one case, and apply it to the other cases.... There is just one thing in which the Ancients greatly surpassed others, and that is the way they extended what they did.... It is by weighing a thing that its weight can be known and by measuring it that its length can be ascertained. It is so with all things, but particularly so with the heart. Your majesty should take the measure of his own heart.[38]

Mencius has the king imagine that the duties he owes to his family members are like those that his subjects owe to theirs. It is only right, then, that the powerful king help his lowly subjects experience a sense of duties well performed. "Why should it be different in your own case alone?" If the king will but "take the measure of his own heart," he will soon understand how obvious the lesson is.

This passage provides a fine illustration of a point registered in Chapter 1: that early philosophy is in better shape than most modern ethical theories, principally because it does not advise the person, in Kantian fashion, to cordon off moral from practical considerations when deliberating. Moreover, since modern ethical theories generally deal with universals and hence disregard a particular action's precise effects on the moral psychology of any single agent, they would see little difference between killing the ox or a sheep.[39] Mencius, on the other hand, makes much of the distinction he finds implicit in the king's decision, however poorly formulated or dimly understood by him, because this affords him a convenient pretext to score a significant rhetorical point: that the king's budding exercise of his inborn capacity for empathy vitally depends on the specific context or environment in which it occurs (a point to which Mencius will return repeatedly in later dialogues), not upon mere happenstance or the conscious calculation that the common folk readily ascribe to him.[40] Extrapolating from that, the morality of any action cannot be determined by an abstract principle, but by a complex determination that factors in, among other considerations, how performance of the action will affect the agent's future sense of himself and others.[41] For whenever a person acts, that act creates new realities for all concerned and hence a new opening for the common good.[42]

That Mencius has been remarkably successful in prompting the king to reflect upon his own motivations becomes evident when the king, for the first time, responds by assuming some blame for the present sorry state of his realm. "I have a failing," he admits.[43] Though the king refuses to name his principal failing, his overweening ambition, he believes that in order to achieve that ambition, he must start wars, imperil his subjects, and incur the lasting enmity of the other local lords. To counter this, Mencius replies, "Do you find satisfaction in such things?" "Of course not."

> "Are fats and sweets not enough to satisfy your mouth? Are light, yet warm fabrics not enough to satisfy the body? Are beautiful colors and enticing women (*cai se* 采色) not enough to satisfy the eye, and the sounds of a musical performance not enough to satisfy your ears? Are your attendants and officers too few to carry out your orders at court, and your slave boys and serving maids not enough to supply your needs?"
>
> "No, it is not that."
>
> "Well, then, it is obvious what your great ambition is. You wish to expand your territories ... and view the Central States as your own.... But to seek this desire by the means you employ is just like barking up a tree to catch a fish."
>
> "Is it as bad as all that?"
>
> "It may be even worse! ... You will not harm yourself by climbing a tree to look for fish. But if you seek to fulfill your ambition by acting as you do, after you put all your heart into the pursuit, you are sure to reap disaster in the end.... Perhaps you should go back to the basics.... If you do that, who can stop you [from attaining your heart's desire]?[44]

But what, pray tell, are those basics (*ben* 本) to which Mencius refers? Mencius has named two, the extrapolation from one's own needs and desires to those of others and a concurrent willingness to recognize the basic facts about the larger social order in terms of both the "constants" (predictable outcomes of certain actions) and the current less-than-ideal situation:

> If, in governing your kingdom, you applied the insights you have gained from examining yourself to the cases of other people, then all office seekers in the realm would want to find a place at your court, all tillers of land to till the lands in the outlying areas of your realm, all merchants to enjoy the refuge of your marketplaces, and all travelers to go by way of your roads. And it would cause all those who hate their own rulers to come lay their complaints before you.[45]

Mencius's language here in classical Chinese is highly patterned and rhymed — just the sort of phrasing that will most likely prove persuasive (as any good rhetorician knows), for people, grateful for the relative cognitive ease with which such passages can be mastered, are apt to find their logic familiar and hence commonsensical.[46] Mencius takes understandable pride in his ability to persuade others, but the stakes, as he sees them, are higher. For the ruler's recognition of these basic facts about the human condition alone can move him to adopt an agenda capable of correcting at least the very worst abuses and indignities, to give the correct terms for them, which agenda, thankfully, puts the king in a more favorable situation vis-à-vis his rivals.

Mencius's analysis continues: "Only a truly noble person can have a constant heart, despite a lack of constant means of support."[47] The common people cannot be expected to be constant in their virtue if they are not secure in their livelihoods. The unkind ruler who punishes his poor subjects when they disobey his laws is merely setting a trap for them, for they are driven, in their desperation, like animals into the "net of the law." Accordingly, the clear-sighted ruler "determines what means of support the people should have" and then ensures that "their means are sufficient" for their parents, wives, and children. Only then does the best ruler manage to "nudge or even drive them toward goodness," at which point the people will find it easy to follow him.

Fleshing out his "basics," Mencius again offers the suggestion that the king rule in such a way that those who reach the age of fifty can wear silk and those who are seventy can eat meat. The king can do this by not requiring labor service during the months of planting, tending, and reaping. Nor should "the grey-haired have to carry heavy loads on the road." "When the aged wear silk and eat meat and the masses are neither cold nor hungry, it is impossible for their ruler not to become a true king." Mencius has cleaved rather closely to two *Analects* teachings: "The noble person cultivates in his or her self the capacity to ease the lot of other people" and "To demand much from oneself and little from others is the way for the ruler to banish discontent."[48]

Let us draw back a bit to appreciate what Mencius's approach has done. Aside from one odd outburst at the outset (surely calculated), Mencius has engaged in strikingly little carping in his opening dialogues. Still less has he trafficked in the standard moralizing talk about curbing desires and abstaining from pleasures. Having concluded diplomatically that "you can never succeed in winning people's allegiance by trying to conquer them with goodness,"[49] Mencius has constructed a single community of interests between the ruler and his subjects, extrapolating from the rulers' dislike of being snubbed, slapped, lectured, and exhorted to ascribe the same disinclinations operating among his people.[50] After all, Mencius's main job is to "get the cruel man to listen to reason," to realize that "he dwells contentedly and unwittingly amongst dangers and disasters, taking pleasure in the very things that will destroy him."[51] Again, Mencius nicely conveys the king's own misconstrual of pleasure's real utility in governing, in the process distinguishing the consuming from the sustaining pleasures and defining the latter as shared pleasures. Again, Mencius ends, in friendly fashion, encouraging his prince to feed, educate, and moralize his people.

A would-be advisor to powerful men accustomed to having their way, Mencius knows that he can never accomplish this task through persuasion if he continually confronts men with their failings, because this only makes it harder for those he would advise to believe themselves capable of undertaking moral action, not once, but again and again.[52] The opacity of the human heart is such that all people, including the king, his advisors, and the lowliest of his subjects alike, share the human experience of having curious, even bizarre reactions well up from mysterious places.[53] Apparently Mencius feels no need even to establish definitions agreeable to all parties straightaway.[54] Instead, through indirection, Mencius's conversations attempt to help the listener/reader ascertain the true relation of ordinary and sometimes inexplicable desires to more civilized desires for the good.

For much the same ethical and rhetorical reasons, Mencius does not seek to contain the desires of the powerful, even when convention labels those desires as flaws, faults, weaknesses, and shortcomings (*ji* 疾). All human beings at every rank share the desire to be "honored and exalted" (*gui* 貴) — today, we might talk of being afforded basic human dignity — and that specific desire becomes the chief motivation for moral action, so long as the person deems himself capable of acting morally and then works at it.⁵⁵ The development of the requisite human emotional and imaginative capacities rests, most importantly, on two grounds: first, the ability to imagine others as having intentional states comparable to our own, and then the allied ability to imagine ourselves as having intentional states different from those we have now.⁵⁶ In other words, human beings require community, as well as the much-vaunted autonomy, in addition to courage and curiosity, with community preceding and fostering a degree of autonomy and of individuation.⁵⁷ As Mencius puts it elsewhere: "All people share the same desire to be exalted. People simply fail to appreciate the fact that every person has the wherewithal within his or her person to be exalted."⁵⁸ To be exalted, to be a dignitary, is to be held in high regard in one's own estimation and in that of others. The inherent nobility of the person aiming to do good — the quality that is "exalted within" — lends dignity to even the simplest and most casual of social gestures. "To be resolute in acting with empathy and commiseration (*shu*), and to seek a sense of shared humanity in social relations, nothing comes closer to perfection than this."⁵⁹ So where a ruler seems unaware of his potential to tap this sort of authority, he must be given better teachers — teachers such as Mencius himself.⁶⁰ As Mencius bluntly puts it, it is not worth talking to a man "who has no integrity... and who is not trustworthy with his friends."⁶¹

By the end of the first half of Book 1, Mencius has established the definition of the true king (a sage by any other name): it is he who finds a way to satisfy his own ambitions while satisfying the ambitions of those below. He achieves his heart's desire by extending to all his subjects the natural compassion he feels for helpless animals and for his own person, so that "those below" live well and can attain their natural desires, as well.⁶² In successive dialogues of Mencius, then, the supremely human desires that others tend to identify as the source of all human misery and shame figure as the best means by which a person in power can attain greatness. Believing the sage to be the epitome of compassion capable of deploying his charismatic attractions to establish himself and the social order (that is, "one who turns disasters into blessings and repays offenses with acts of kindness"), Mencius begins to build the case that will, in a grander vision found in a much later passage, link desire, goodness, and political greatness:

> The desirable is what we call the good. To have it in one's own person is what we call being "true and authentic." To possess it fully in oneself is what we call "excellence, fineness, and beauty," and to have it shine forth in full possession is what we call "greatness." To be great and to be transformed by this greatness — that defines the sage,[63] and when those sagely qualities are deftly applied, in unconscious virtuosity, this defines being and acting in divinely efficacious ways (shen 神) [in governing].[64]

The second part of Book I reiterates these themes while making the rhetoric of pleasure even more prominent, as in the following exchange:

> Zhuang Bao went to see Mencius. "The king received me and told me that he was fond of music. I was at a loss what to say."...
>
> "If the king has a great liking for music, then there is perhaps hope for the state of Qi."[65]

Another day, when Mencius was received by the king, Mencius inquires if he is really fond of music. The king blushes and says, "Yes, but I am incapable of appreciating the music of the Former Kings [exemplars of antiquity]. I like only the popular music of today." To the astonishment not only of the king, but also of later readers, Mencius replies: "The music of today or of antiquity — that makes no difference."[66] The king begs to hear more.

> "Which is greater, pleasure-taking by oneself or in the company of others?"
> "Pleasure-taking in the company of many...the more the merrier."

Based on this seemingly minor, but crucial concession, Mencius leads the king through a protracted thought experiment:

> Let me explain a thing or two about pleasure-taking. Let us suppose for the moment that you were having a musical performance here, and when the people heard the sound of your bells and drums and the notes of your pipes and flutes, they all with aching heads and knitted brows said to one another, 'In being fond of music, why does our king bring us to such straits that fathers and children do not see one another, and siblings, wives and children are parted from one other?'...

> Now suppose that you were hunting here, and when the people heard the sound of your chariots and horses and saw the magnificence of your banners, they all with aching heads and knitted brows said to one another, 'In being fond of hunting, why does our king bring us to such straits that fathers and sons do not see one another, and siblings, wives, and children are parted from one another?' The reason would be simply that you had failed to share your pleasures with the people.
>
> Now suppose you were having a musical performance here, and when the people heard the sound of our bells and drums and the notes of your pipes and flutes, they all looked pleased and said to one another, 'Our king must be in good health. Otherwise, how would he have music performed?' Or suppose that you were hunting, and when the people heard the sound of your chariots and horses and saw the magnificence of your banners, they all looked pleased and said to one another, 'Our king must be in good health. Otherwise, how could he undertake to go hunting?' The reason again would simply be that you shared your pleasures with the people.[67]

On the basis of the foregoing suppositions, Mencius concludes, "Now, if you share your pleasures, enjoying them with the people, you will be a true king."

In the time of Mencius, "music" meant "ensemble performances," bells and drums, pipes and flutes. As readers will recall, the two words for "music" and "pleasure" (*yue* and *le* 樂) are graphically identical (see Chapter 2) and, in classical Chinese, either homophonous or phonetically close. The metaphor of the musical performance allows Mencius to teach the listener/reader a series of important lessons about the group dynamics of pleasure-seeking and pleasure-taking. Following the standard tropes from texts such as the *Zuozhuan*, Mencius links the harmonies of musical performance to the exemplary balance of dispositions displayed in social situations by men of cultivation.[68] For Mencius to reassure the king that a love of "the music of today" works just as well to promote a more capacious attitude as a love of "the music of antiquity" when it comes to the grand musical entertainments that Mencius's rivals, the Mohists, condemn, is not so much a claim about melodies or lyrics as a statement about the shared responses to music, and, by implication, to every occasion for pleasure. (Compare the statements by Xunzi in Chapter 4.) That all people, unless they are tone deaf, seem to like some form of music or another attests our shared humanity. That musical performances are best appreciated in the company of others reminds us that humans are social beings. That laments change quickly to odes of joy shows us that life is volatile, at best, and precarious, at worst; hence the need for serious reflection about priorities. To affirm that in music, as in pleasure, "the present pleasure/music comes from the past pleasure/music" is to invest current fashions *and* long-standing traditions with added glamour.[69] And finally, Mencius subtly conjures the commonplace that solitary pleasures can be more dangerous, since no natural social checks exist on them. So while all Chinese thinkers presume the influences that the powerful exert on those below (as noted in Chapter 1), Mencius draws attention to the fact that the influence is entirely mutual, since the ruler is no less dependent (psychologically, politically, and economically) upon his people than they are on him. Both sides must happily cooperate if the realm is to flourish, in acts "wrapped in the experience of mutual pleasure," and that cooperation can be built only from the ground up, through personal trust.[70] Such an insight is not always registered in the early rhetoric of pleasure, except by those who would emphasize the precariousness of the ruler's position—but in his dialogues, Mencius quietly alerts the rulers to the potential for disaster without making this his central message, stridently delivered.

The trick is for Mencius, the master rhetorician, to figure out how to drum up as much enthusiasm in the king for listening to his own conscience as he already feels for musical performances at court.[71] In making tradition relevant to the discussion and closing the distance between the king, with all his flaws, and the hallowed kings of old, Mencius makes it much easier for the king to imagine himself taking his place in the select group of the paragons of humanity.[72] Mencius urges the king to see his apparent weaknesses (his fondness for wine, women, and song) as the chief source of his moral strength. The king, sharing these weaknesses with his subjects, can hear the people's voices as if they were his own. Having attended to those voices, the king may then find that the people's expressions of pleasure in the king's well-being echo and amplify his own awareness of his delight. To draw the king in still further, Mencius methodically pursues this line of argument, taking up three sources of pleasure available to occupants of the throne: the building of pleasure parks, the defense of one's honor in response to minor affronts, and the enjoyment of the fine prospect from a winter palace, moving from the physical to the psychological and the aesthetic. Mencius concludes,

The man who is deprived of a due share in such pleasures is bound to speak ill of those in authority. It is wrong, of course, to speak ill of those in authority because one has been deprived of a share in such pleasures. But for one in authority over the people not to share his pleasures with the people is equally wrong. The people, for their part, will take pleasure in the pleasures of the ruler who takes pleasure in their pleasures and feel concern for the cares of the ruler who feels concern for their cares. He who takes pleasure and feels concern on behalf of the realm is certain to become a true king.[73]

Again Mencius has invested the king's subjects, as well as the king, with dignity and humanity. For that reason, even the king's progresses through the land — typically identified as a waste of scarce resources that fosters "drifting, lingering, rioting, and intemperance" — can become occasions promoting the serene functioning of the administration of the realm. "In spring the purpose is to inspect the ploughing; in autumn, to inspect the harvesting."

> If our king does not travel,
> How can we have rest?
> If our king does not go on tour,
> How can we have help?
> Every time he travels,
> He sets an example for the local lords.[74]

Apparently, the same activities known to sap the people's energies can educate and energize subjects as well, so long as the ruler at every opportunity considers the welfare of his people to be no less important than his own and arranges his affairs accordingly. The ruler need only open the granaries and have his musicians play music as he passes through. Mencius ends this flight of fancy with a suitably sober classical gloss on a citation from the *Odes*, "What harm would there be in curbing our lord?"

Ironically, however, what Mencius has suggested in everything to date is that the lord need *not* curb himself in any substantial way; his account delineates few, if any obstacles that would impede the ruler from attaining his ultimate desires. By such means, Mencius persuades his prince not only to order his musicians to compose the sort of music that better expresses "the harmony between ruler and subject," but also to accept Mencius's proposition that "to curb the ruler is to express love for him."[75] The king is eager to seek Mencius's advice on other knotty problems (1B/5). So although Mencius has carefully broached the topic of limits, he has sweetened the pill: since no pleasures are detrimental per se, except those that deprive the people of a reasonable level of comfort and security, any ruler who takes the trouble to plan that he has sufficient resources to feed the people thereby satisfies his own tastes for pleasure and still attains his supreme ambition to rule his corner of the world. Even the king's wasteful outlays promote the proper functioning of his administration. This is indisputably a win-win situation.[76]

Mencius conveys this profoundly (overly?) optimistic message in yet another dialogue with King Xuan of Qi. The king reports to Mencius that all of his advisors would have him tear down a magnificent Hall of Light, the gorgeous setting for ritual gatherings (1B/5). The king "loves money," and money — lots of it — is required for the upkeep of such follies. Mencius evinces no problem with the king's love of money. Long ago, the exemplar Gong Liu loved money, too. "He stocked and stored / He placed provisions in bags and sacks." "You may love money, but so long as you grant the people this same love, how can it interfere with your becoming a true king?" The king then raises a further objection: "I love women." Mencius responds with yet another gratifying example from antiquity: Tai Wang loved women. "He brought with him the Lady Jiang / Looking for a suitable abode." "You may love women, but so long as you grant the people this same love, how can it interfere with your becoming a true king?"[77]

Mencius makes it all sound very easy indeed.[78] The king may enjoy his palaces, his parks, his well-stocked ponds, his money, his dogs, and even his seductive beauties so long as he sees that the basic economic and ritual needs of his own people have been met. But when Mencius, by indirection, suggests that the king has neglected his duties, the king — clearly uncomfortable with such lines of argument in successive dialogues — "turned to his attendants and changed the subject" (1B/6). In the next encounter (1B/7-8), Mencius returns to the origin of the king's discomfort, lest the king continue to be indifferent to his duties; rather than let the king off the hook, he proposes two stark alternatives: either the king is a true king — one who wisely ensures that his people will taste enjoyments, too — or he is a "mutilator, a cripple, and an outcast."[79] Having established his point with that remarkable contrast, Mencius then drops the topic of the king's responsibility for a time, confining his remarks to advice on the way the king should punish the evildoers in his realm: see that the wicked are put to death "by the whole state," Mencius advises, so that king and subjects share this activity, as well. The tone of this injunction reminds the king that major undertakings require expert advice in building consensus — and that Mencius is an expert.

So while Mencius refuses to give straight answers to the urgent questions whether Qi should invade its neighbor Yan, or Teng ally with Qi to its north or Chu to its south, or take action if a neighboring state fortifies a border town, Mencius knows the true king to be one whose invasion will always be perceived by the conquered as a rescue. The king grumbles about his disloyal subjects, who are too numerous to punish. Mencius, having prepared the king to accept more responsibility for the worst abuses and indignities his people suffer, ultimately renders a harsh verdict: "In years of bad harvest and famine ... the old and young were abandoned in the gutters ... though your granaries were full. ... This shows how callous those in authority have been and how cruelly they have treated their people." Turnabout is fair play, although Mencius allows the lion's share of the blame to be placed on the king's officials, rather than the king. "It is only now that the people have seized the opportunity to pay back the treatment they themselves received" at the hands of their ruler and those who discharge his commissions (1B/12).[80]

Book 1 of the *Mencius* ends with a curious story in which the Duke of Lu, who originally thought to seek Mencius's advice, is dissuaded from such a course by his current male favorite. Mencius comments, "When a man goes forward, there is something that urges him on, and when he halts, there is something that holds him back. When it is not in his power either to go forward or to halt, it is due to Heaven" (1B/16).

When all is said and done, is it fate or human agency that determines Mencius's success at the courts of the Central States? Is the king refusing to act, or incapable of action? The answer for Mencius is clear, since Mencius has earlier distinguished a refusal to act from an inability to act, but Mencius is too polite or too wary to say it.

While Mencius returns to such questions in other writings attributed to him, it is important that every major proposition that he will seek to prove has been laid out by the concluding dialogue of Book 1 of the *Mencius*. Mencius has equated "true kingship" with "sharing one's pleasures with the people," and he has gotten the kings with whom he has conversed to see themselves, however briefly, as they see others. Human nature may be a bundle of conflicting desires, yet good rule is entirely feasible, since it rests on a series of conscious decisions that do not go against human nature.[81] "Tyranny" has been traced to the refusal to allow other people's desires to be satisfied. And the very qualities that convention treats as weaknesses—a love of hunting and arms, of palatial surroundings, of money, music, and sex—have been identified as the stuff from which true kings can be made, insofar as these pleasures require social interaction and illustrate the companionable ease to be had by sharing pleasures. Good rule is principally a matter of seeing the connections between one's actions and their likely effect on one's dependents. Some items can be shared directly (silk, fish, or spectacles). Others require the king to create the conditions in which those below can procure these goods for themselves (as with women, music, and education). Above all else, the people desire good rule, for by definition, it maximizes their opportunities for receiving and giving pleasure. And through good rule, the king can make himself the ultimate object of his subjects' desires. "No one will be able to stop him," if only because "the appearance of a true king has never been more overdue than now."[82] It is easy, in other words, for a ruler to create the desired effect and so attain his own supreme ambitions when so many long for him to institute good rule and when all human desires compel men and women to travel along essentially the same paths. What counts is this: "that those who are morally well-adjusted look after those who are not,"[83] because those in power realize that the people—not the rulers—"are of supreme importance."[84]

Beyond Book 1: The *Mencius*'s Later Chapters

The contents of Book 1 have been summarized in some detail because Book 1 sets the stage for all subsequent talk about morality in the *Mencius*. Alertness to the pleasure discourse permits a keener and subtler appreciation of the parts of the *Mencius* that tradition has favored.

Let us look first at Mencius's impressive description of the "floodlike *qi*" 氣 that the good man summons up in times of crisis, found early in Book 2. An interlocutor asks of Mencius whether he would be willing to tell him something about the "heart that cannot be stirred." Mencius replies:

The will is commander over the *qi*, while *qi* is that which fills the body. The *qi* halts where the will arrives [meaning, the *qi* holds to the position determined by the will]. Hence, the saying, "Grasp your will firmly and do not abuse your *qi*. . . .

The will, when blocked, will move the *qi*. On the other hand, the *qi*, when blocked, will move the will. Now stumbling and hurrying affect the *qi*, producing palpitations of the heart. . . .

The flood-like *qi* [circulating in the body] is, in the highest degree, vast and unyielding. Nourish it with integrity and place no obstacle in its path and it will [expand to] fill the entire space between Heaven and earth. . . . Deprive it [the *qi*] of rightness and the Way, however, and it will collapse. Because it is born of accumulated rightness, it cannot be appropriated by anyone through a sporadic show of rightness. Whenever one acts in a way that falls below the standard, it will collapse.[85]

This response will make no sense unless readers understand that *qi*, configured energy, has physical and ethical dimensions, and desires induce alterations in the body that shape, in turn, our interactions with the phenomenal world. Thus, the moral will — the desire to do good — operates in a person like the surge of adrenalin a stalwart warrior feels on the battlefield: once the will is sufficiently determined to carry out a particular action, the person can perform seemingly impossible feats. The steady desire to act morally, like other desires, propels the good person to pursue the Way and live well.[86] But just as constant exercise is required to build up the warrior's muscles prior to battle, so, too, the constant employment of our moral faculties for judging right from wrong is the prerequisite for responding spontaneously, resolutely, and effectively to the moral crises we encounter while making our way in life.[87]

Why is *ren* 仁 — the supreme virtue touted throughout the *Analects* of Confucius — so little in play in the 2A/2 discussion in the *Mencius*? *Ren*, the ability to extrapolate from one's own condition to that of others, is reduced in 2A/2 to one virtue among the four that define what it is to be human, the others being having a capacity for shame, a willingness to yield, and a sense of right and wrong.[88] *Ren* is premised on ethical understanding, "knowing what lies far away by knowing what lies near to hand; knowing what is in another's heart by knowing what is in one's own heart."[89] Doubtless, *ren* is important, but for Mencius, to measure the relative value of other people and things and to feel another's pain is comparatively easy, insofar as such judgments involve primarily the head and the heart, but not necessarily the concentrated will.[90] The challenge is to forge ahead and do what is right each and every time, like the best soldier in battle, heedless of the social status of those whom one engages.[91] The vast majority of situations do not require the person of good intentions to rethink all ethical priorities in order to ascertain the proper course of action.[92] They require the courage to do what one already knows one should do—a "measuring of the heart" and a "commanding of the *qi*,"[93] unfazed and "unstirred."[94] This sort of moral courage derives, first, from being filled to the brim with the desire to do good, and second, from the serene confidence that good acts performed in a just world will be celebrated and in an unjust world conduce to beneficial order.[95]

In 2A/6, Mencius demonstrates the counterintuitive proposition that "no man is devoid of a heart sensitive to the suffering of others." (It is this heart that King Xuan of Qi exhibited when he decided to save the ox that trembled as it was led off to be sacrificed.) The superb example that Mencius cites is that of a man who sees, "all of a sudden, a young child on the verge of falling into a well." Mencius believes that the man would certainly be moved to compassion initially, not because he wanted to benefit himself in any way by "getting in the good graces of the parents, by "winning the praise of his fellow villagers or friends," or by "stopping the cries of the child." Nonetheless, Mencius never promises that the man will actually go on to save the child; he is not so rash as that. Self-interest may intervene and prevent him from acting. The man may dislike his neighbors or prefer to curry favor with enemies of the child's parents. Or it may be that he is too lazy or timid. Mencius does not claim overmuch, in other words. What he says is consistent with his analysis in Book I, which defines "moral potential" in terms of the impulse to alleviate the suffering of others. Whether that first impulse will ever be translated into action depends on the person's economic situation, his ruler's suasive example and intervention, other external conditions, and finally, the accumulated force of the person's previous decisions that have strengthened or crippled his early capacity to become fully human.[96] But Mencius's example implies that the initial impulse to act to save the baby involves emotion and rationality, belief and desire, and these cannot be usefully separated when conceptualizing this kind of situation.[97] Mencius then adds that the maker of arrows does not begin by being more unfeeling than the maker of armor. It is the daily conditions in which he works that make him keener to witness death than to protect life.[98] Habit will play a large role in how one decides to act in response to a scene.

Understanding the rudiments of Mencius's pleasure writing can also help us unpack puzzling passages in the text, for example, the linked maxims in its fourth chapter:

> If a man in a subordinate position fails to gain the confidence of his superiors, he can never hope to govern people well. There *is* a way to gain the confidence of superiors, but a man will surely never gain their confidence unless he proves trustworthy to his friends.
>
> There *is* a way to be trusted by friends, but a man will surely never be trusted by them unless he pleases his parents.
>
> There *is* a way to please parents, but a man will surely never please his parents if he himself turns toward a course that lacks integrity and wholeness (*cheng* 誠).
>
> There *is* a way to become whole and true, but a man will surely never have integrity and wholeness in his person if he fails to be clear about what is good in a given situation (*shan* 善).
>
> Thus we see that wholeness is the way of Heaven, and longing for that integrity is the proper way of human beings.... The man who is completely true to himself invariably moves others, and the man who lacks integrity is never capable of moving others.[99]

On first reading, the obscurity of this passage borders on the Delphic: how can the treatment of one's parents be a decisive factor in one's relationships with friends, for example? A closer look reveals that the strength of each relation where pleasure is given and received is predicated on the relative steadfastness displayed in fulfilling the prior commitment. As Mencius explains elsewhere, "For a man to give full realization to his heart is for him to understand his own nature. A man who knows his nature will know Heaven [his ordained lot in the form of his own basic nature and proclivities]. By retaining his heart and fostering the potential in his nature, he will serve Heaven."[100] Put another way, a steady desire, a steely determination to pursue the good, on his own for his own improvement, must have possessed the person before there can occur a suitable extension in the form of the desire to do good to others, introducing improvements to their situations.[101] The only alternative is a corrosive mistrust that ultimately weakens all parties.[102] For that reason, the wise leader persists in the desire for wholeness and integrity, which is achieved through successive acts signifying, rehearsing, and recording the relative worth of different things, people, and impulses.

> As for a man's relation to his physical person, he cherishes all parts of himself, he nurtures all parts of himself.... Nevertheless, the parts of the physical person differ in value and importance. The body possesses what is greater and what is lesser.... Muddled, certainly, is the man who takes care of one finger to the detriment of his shoulder and back without realizing his mistake.... Those who nourish the lesser parts are petty people, and those who nourish the greater parts are great by definition.[103]

Whereas the courage to act well must be absolute, the ranking of those desires that allow a person to attain the greatest quantity and quality of experiences fostering the good life admits of degrees, it being a measured response to contingency derived through a process that is at once the most sophisticated of human tasks and the most gratifying of human endeavors.[104] To strive for and to attain wholeness is not only to "return to one's [best] self,"[105] but also to make oneself more graceful and attractive in the eyes of all. Fortunately, this form of adornment is available to the very least of us: "Seek and you will obtain it; shunt it aside and you will lose it. The seeking is invariably of benefit to the getting, for what is sought [a sense of wholeness and its true value] resides within the self."[106]

In the context of this discourse on pleasure, modern readers can begin to fathom why Mencius has such extreme reactions to the rather commonplace assertions about the endowed human nature made by Gaozi 高子 and Master Gongdu 公都. Mencius accepts Gaozi's belief that "the appetite for food and sex is human nature," but he rejects out of hand most of the other ordinary propositions that Gaozi advances: that accepting social obligations does violence to one's own nature; that human nature has no natural inclinations toward either good or bad behavior; that natures vary significantly by person; that a sense of duty must be learned in social settings and thus is not universally endowed at birth, unlike a sympathetic heart; and so on. Mencius's arguments seem at points to hang tenuously on a very slender thread: that communal rituals of all sorts make palpable a range of full-blown commitments that a person's birth in a particular culture at a particular time and place inevitably entail, commitments that spring from (literally, "germinate" or "sprout" from) early senses of compassion, shame, respect, and a determination of right versus wrong.[107] Mencius hopes to convince his partners in dialogue, including the listener/reader, that upright, trustworthy, and empathetic conduct toward others is hardly something external to human nature and so is not potentially alienable from it.

Common sense and everyday experience tell us that Mencius is flat wrong. Is it not the case that we must be taught to do good and to prefer it to evil? Are there no people who by their very natures are ornery and destructive? In the following extended dialogue, Mencius refuses to concede the point:

> A man of Ren asked Master Wulu 屋廬子, "Which is more important, the rites or food?"
>
> "The rites."
>
> "Which is more important, the rites or sex?"
>
> "The rites."
>
> "Suppose you would starve to death if you insisted on the observance of the rites, but would manage to get something to eat if you did not. Would you still insist on their observance? Again, suppose you would not get a wife if you insisted on the observance of *qinying*, the ritual where the groom goes to the home of the bride to fetch her, but would get one if you did not. Would you still insist on the observance of the ritual?
>
> Wulu didn't know how to answer. The following day he went . . . to give an account to Mencius. Mencius said,
>
> "What difficulty is there in answering this argument? . . . If you compare a case where food is important with a case where the rite is inconsequential, then the greater importance of food is hardly the only absurd conclusion you can draw. Similarly with sex. Go and reply this way to your interlocutor: 'Suppose you would manage to get something to eat if you took the food from your elder brother by wrenching his arm, but would not get the food if you acted otherwise. Would you do it [wrench his arm]? Again, suppose you could get a wife if you climbed over the wall of your neighbor to the east and dragged away the daughter by force, but would not if you acted otherwise. Would you do it [drag her away by force]?'"[108]

Mencius asks at what point such actions cross a moral line, a line marked by discomfort in those with "well-disposed hearts."[109] Then, to shore up his case that our potential to do good is innate and unlearned, Mencius relates the story of Ox Mountain, a hill near the capital that has been clear-cut. Understanding that most people tend vastly to underestimate their own ability to do good and so refuse to undertake to do what is right, Mencius asks his potential adherents to imagine their own lives not as they are, but as they might be. In the case of Ox Mountain,

> People see the barrenness and presume that it never had any resources or potential (cai 材). How could this be the true nature of a mountain? As for humanity, how could it be that humans lack the heart to discern humanity in others and to commit to doing good? But how can humans realize their potential to become a thing of beauty if they are struck down day after day?[110]

Ox Mountain, in Mencius's description, is a system under stress, one that is so depleted or denuded of resources that it soon cannot summon the necessary reserves to continue functioning well.[111] His conclusion: however ugly the face of humanity may appear to be at times, if Ox Mountain is any guide, every person is born with the potential to do good and to grow into a thing of beauty. As Mencius has already informed us, "things are unequal is part of their nature. Some are worth twice or five times, ten or a hundred times, even a thousand and ten thousand times, more than others."[112] Now, Mencius admits, "There *are* cases where a person is twice, five times, or countless times better [at doing something] than others, but this is only because the others have failed to make the best of their innate endowments."[113] "Should a man act badly, it is not the fault of his basic stuff," since he is endowed by Heaven with everything it takes to fashion a dignified life commanding the respect of others, including the capacities for weighing values and applying self-restraint.[114] For Mencius, the important lesson circles back to this: the starting point for doing good is little more than the capacity for enjoyment — delighting in the verdure of a grove, savoring roast meat, to take two examples — and a candid acknowledgment that all humans seek to gratify their tastes for pleasure.[115] Deliberation and the conscious pursuit of sustainable pleasures follow from that lesson, because "assuming one's proper form"[116] makes the person comfortable in his or her own skin.

To agree with these assertions is to go a long way toward accepting the astonishing proposition that Mencius advances time and time again in the writings attributed to him: each human has the endowed potential to become a sage. Mencius says of Shun, for example, "He is a man and I am a man. Why should I be in awe of him?"¹¹⁷ Both the ruler and his subjects, in assessing their accomplishments, may presume an essential parity between the past and present and between high and low. The king, for his part, in thinking of each subject, had better remind himself that "he is a man and I am a man." For those sorts of observations invariably lead the person to a still more important question, "What sort of man am I?"¹¹⁸

In Mencius's terms, "There is no person who is not good, just as there is no water that does not flow down."¹¹⁹ Any person is capable of becoming good, though people can be made, through mismanagement and insecurities, to mischaracterize their basic proclivities. That is what Mencius means by saying that all persons *are* good. Mencius insists, "A sense of compassion, of duty, of the rites, and of right and wrong — these are not welded onto me from the outside; they are in fact something that I have in me. Surely it is simply that I have failed to think about them and long for them" to be developed in a sustained fashion.¹²⁰ Because ruler and subject alike share the appetites for pleasure-seeking and pleasure-taking and the consequent abhorrence for suffering — whether their own or that of others — Mencius finds us all equally human.

If all human beings are equal in having sufficient moral potentials, how is it possible by the Mencian system to justify rulers getting far more in the way of goods and services than their equally human subjects? The answer is simple: rulers and their advisors *deserve* more, because they alone "work with their hearts and minds," determining what is in the best interests of the common good, instead of working with their hands alone (*lao li* 勞力). Those at court, to be legitimate, must anticipate the needs of many people of different stations (something that ordinary people without much training cannot always manage), plan policies to help others achieve their aims, and set good examples of justice and probity for those below. Insofar as they must apply, ideally, a wider domain of skills in their profession, they merit greater compensation.¹²¹ (The same argument, of course, suffices for CEOs today, often with less justification.)

Contrary to the charges of his fiercest critics, Mencius was not under the delusion that the mediocre rulers he advised would suddenly "become Tangs and Wus," dynastic founders of exemplary goodness and wisdom.¹²² That the objects of one's desires will naturally vary with education, heredity, past experience, and present opportunity is obvious, says Mencius. Still, the rewards to be had from inciting heads of state to desire to do good are potentially limitless, insofar as just administrations promise to usher in a world full of stable pleasures wherein each enjoys doing what is suitable to his station and temperament:

> Now if your administration applied insights derived from your best self [ren] to governance, it would give rise to the following situation in the realm: the officers in the realm would all want to be standing guard in your court; the farmers would all want to be ploughing in your outlying districts; the merchants would all want to store their goods in your majesty's markets; the travelers would all want to go by way of your roads. Moreover, all in the known world who are sore aggrieved by their rulers would want to flock to you to make their appeals. Were that the case, who could stop them [from contributing to the enlargement of your resources]?[123]

In this best of all possible worlds, the people will naturally be enveloped in joyful music, "opening with the bells and ending with the jade flutes."[124]

Wanting, wanting, and more wanting — Mencius calls up a litany of desires. The moral person is one who "yearns all his life" because he never ceases to feel the yearnings that others experience.[125] "What need is there for a false front in a situation?" Mencius asks.[126] Mencius is playing with differing notions of sovereignty and legitimacy, but he assures us of one thing: that yearning for moral legitimacy in all human relations confers infinitely more blessings upon the striver than status and wealth ever can, especially when worldly blessings are unmerited and liable to be jeopardized by misconduct. Confucius himself, by Mencius's accounts, serves as an exemplar precisely because he is a sage with an unparalleled capacity for longing. The Master is avid for the middle Way; he moreover longs to return home to his beginnings in both the physical and moral senses.[127] And "when he could not have the principal object of his desire [to be an advisor to kings]," he did not stop longing, we are told. "He simply longed for the next best" person to carry out his vision of a just and noble state.[128] Going still further back in history, Shun likewise "was the best and most noble of sons," for "at the age of fifty, long after their deaths, he still yearned for his parents."[129] Doing good means doing good tirelessly, and an admirable persistence comes from seeing a clear connection between the object of longing and how one may achieve it.[130] In general, the sages are presented as men who may have made a number of mistakes,[131] but whose sense of humanity and longing for deeper human connections was never wanting. And it follows as night follows day that the sage, as a result of his steady quest, becomes completely himself and a figure widely emulated. "That is the definition of a great man."[132] Mencius, who never doubted his own sageliness,[133] commented, as we have seen, "The whole of phenomenal existence is complete within me. There is no greater pleasure than to find, on reflecting on my person, that it has integrity [*cheng* 誠]. Trying your best to treat others as you wish to be treated yourself, you, too, will find that this is the shortest path to real humaneness [*ren* 仁]," with insights about shared humanity likely to end in suitable action.[134]

To those who are receptive, Mencius brilliantly makes the case that the greatest moral error that a person can make is to see human society as a zero-sum game — to think that "if one's aim is wealth, one cannot be benevolent, and if one's aim is benevolence, then one cannot be wealthy."[135] For "when it is clear to all that those in authority understand human relationships, the people will be affectionate.... One who can put his heart into his actions should therefore be capable of renewing his state."[136] Furthermore, "he who loves others is always loved by them, and he who respects others is invariably respected by them."[137] Morality, being rooted in reciprocity, is supremely easy to understand: "Do not do to others what others would not choose to have done to them. Do not desire for oneself what others do not desire for themselves. There is nothing more to it."[138]

Bad things happen to good people, a truth of which Mencius is hardly unaware. He therefore describes a person of great virtue who received outrageous treatment from others. When that person finds, upon examining his conduct, that he has responded invariably with benevolence and courtesy to instances of ill treatment, he sadly concludes that he somehow, unbeknownst to himself, has failed to do his very best. "The benevolent person never harbors anger, never nurses a grudge overnight.... All he does is love others."[139] He even loves those who harm him, not because some Chinese counterpart to Christian charity requires this of him, but because he, with his heightened sense of shared humanity, feels the pain and perplexity beneath the evildoing. Mencius's noble man has cares enough, since he sympathizes or empathizes with everyone, but he escapes all the "unexpected vexations"[140] that originate in egotism. Such generosity of spirit, as it happens, has its reward: it simplifies the process of deciding what to do. "Shun was a human being. I also am a human being. I should be like Shun. That is all; there is nothing more to be done!"[141] "The trouble with a man does not stem from a lack of strength, but from his refusal to make the effort.... The way of Yao and Shun is simply to be a good son and a good younger brother. If you wear the clothes of Yao, speak the words of Yao, and behave the way Yao behaved, then you *are* a Yao."[142]

In Mencius's opinion, "No one has ever erred through following the example of the former kings" in their loving responses to the world around them.[143] Mencius's prescriptions for the common good readily recall kindred teachings in the *Analects*, most especially these:

> 14/45: Kongzi's disciple Zilu asked about the qualities of the noble person. "The noble person cultivates his or her person by comforting others." "Is that all?" "To cultivate one's person by comforting everyone else is something that even the legendary sage-kings Yao and Shun found difficult."

> 12/2: "Behave when away from home as if in the presence of an important guest. Deal with the common people as if you were officiating at a solemn sacrifice. Do not do to others what you would not like yourself."

Given these broad commonalities between Mencius and his model, Confucius, there may be more to be learned if we compare and contrast Mencius's approach with that of another Confucian master, Xunzi, who lived a century later (see Chapter 4). After all, all students of Chinese history since the Southern Song (1127–1279) have been trained to think of Mencius as the best-known adherent of Confucius, dueling with Xunzi. For devotees of True Way Learning cast Mencius as the one "true Confucian," downplaying Xunzi's enormous influence, placing too much attention on their different definitions of the human nature (*xing* 性) endowed at birth—creating a fierce debate on an issue that drew little interest in earlier times, judging from the extant texts.[144] The two opposing slogans ascribed to Mencius and Xunzi, respectively—"Human nature is good" versus "Human nature is not a pretty sight to see"—supposedly encapsulated their views adequately.[145] And as beginning students quickly discover, Mencius is famous (and famously criticized by Xunzi) for making it seem too easy and "natural" for an authority figure to learn to do good. (Of course, Mencius seldom discusses the moral condition of anyone but powerful rulers and their chief advisors.) Nonetheless, the two thinkers share far more than is generally conceded, since both imagine it to be perfectly possible for human beings to attain human goodness. Ultimately, both Mencius and Xunzi agree that were human beings to have no desires, there would be no basis for community or morality, let alone morality in the pursuit of the common good. Some of the modern secondary literature concedes that this is Xunzi's position, but under the sway of the late imperial version of True Way Learning, which highlights the problematic nature of human desires (see Chapter 1), few discern that Mencius's position on this matter is virtually identical to that of Xunzi. Larger areas of overlap also should be noted, as when Xunzi defines the hexagram "Togetherness" 咸 as "Resonant Feeling" (*gan* 感),[146] then proceeds to argue that all elevated social feelings find their basis in the strong mutual feelings pertaining between loving husband and loving wife; out of desires as natural as those for cuddles and for sex there can develop a refined sense of sociality in several discrete registers, including affection for and dedication to one's ruler.[147]

So whence the differences in their approaches?

Mencius is content to let the ruler decide policies on behalf of his people. The prudent ruler, by Mencius's account, consults good policy advisors like him, who will guide him to the path where he may gain most, if not all of his moral awareness. Not surprisingly, given that Mencius so often is in conversation with rulers, his rhetoric highlights the heroic aspects of a ruler's decision to undertake moral action, likening the decision to battle-worthy knights girding themselves for battle. He suits his rhetoric to his ruler, certainly. ("I understand words," says Mencius, immodestly, and his rulers often greet his carefully phrased analogies and cases with an admiring "Well spoken!"). The beginning and end point — in particular, the initial personal decision to pursue the common good and the benefits from having done so — constitute the main Mencian directive. "Wish for compassionate behavior, and lo, it is at hand!" Mencius might say, whereas Xunzi emphasizes a lifelong learning process by which one gradually comes to realize one's full potential as a human being who is "whole and intact" and so capable of independent judgment.

Perhaps for this reason, Mencius pays surprisingly little attention to parsing events in history as storehouses of powerful empirical evidence about human propensities. Mencius's appeals to history, more often than not, portray impossibly distant culture heroes (distant in kind and remote in their perfections) and how they acted with exemplary (again highly implausible) heroism. Xunzi, being far more interested in the long process of moral cultivation, painstakingly shows how good social institutions and good teachers alike school the person to see that practical goodness or efficacy (that is, the objective interests of a particular agent) maps closely onto moral justice, so long as the realm is well managed. Thus Xunzi looks to actual historical situations in the recent past to discover ordinary behavioral patterns. Drawing upon these reasonable conclusions, he hopes to persuade reasonable people inclined to careful deliberation.

Mencius may posit strong connections between personal morality and social mores, but his ruler-centered narrative hardly goes beyond the statement that the ruler's model of compassion or indifferent cruelty exerts an extremely powerful influence on the thinking and motivation of his subjects,[148] while their collective behavior tells on his psyche and situation in turn. Mencius urges his rulers to make policies with a view to benefitting the common people, certainly, but the writings ascribed to him evince little interest in devising a coherent system of reliable incentives, whereas Xunzi regards the implementation of such a system as the primary key to leading people in the direction of moral (that is, socially constructive) action. Impatient with Mencius's reliance on moral intuitionism and idealism, Xunzi carefully details the roles that good institutions and leaders play in creating the proper environment to aid and encourage each person first to identify and then to embrace the pursuit of steady, if incremental, moral and social improvements.

Mencius is content to have Heaven function as the ultimate justification for morality; hence his attention, in discussions with Gaozi on human nature, to inner/outer distinctions,[149] designed to mark off what is innate and thus not liable to change as an endowment from Heaven (for example, one's lot in life) from what is "external," that is, learned in society or imposed by it and so subject to change.[150] In stark contrast to Mencius, Xunzi makes human welfare the sole criterion by which to judge human morality. Relatedly, Mencius's definitions of morality and the right are for him absolute truths, matters of Heaven's sanctions and cosmic principles, whereas Xunzi's efforts are directed to building sociopolitical institutions that will allow people to ascertain both their own best interest and that of the larger society. How some of those differences play out with respect to pleasure theory forms the subject of the next chapter, devoted to Xunzi.

Pleasure, Politics, and the Common Good

Ideas, expressed or stifled, have consequences in the real world. Take sharing in the wake of Mencius's pronouncements and the ideas of sharing now. In large part because of Mencius's teachings, rulers in early China tended to be judged by three main criteria: whether their actions conduced to social order and thereby instilled trust among their subjects; whether the farmers (nine-tenths of the population in Mencius's day) could make a decent living, given taxes and the prevailing terms of land tenure; and whether the rulers provided a modicum of security to disadvantaged populations: the aged and those bereft of family support such as widows, widowers, and orphans, plus groups afflicted by natural disasters.[151]

Mencius all but promised Lord Wen of Teng that his small state could not only survive a series of brutal interstate conflicts, but even triumph over larger kingdoms, so long as his subjects gave him their undivided allegiance ("looking to him as home" or *gui* 歸) and trusted him to provide for their best interests.[152] Mencius's reasoning was sound, for the ruler of Teng needed his subjects to serve loyally in campaigns and to produce an abundance of grain and silk when not engaged in war. Farmers too hard pressed would simply flee, in Mencius's day, possibly to a rival state. Accordingly, Mencius urged the Lord of Teng to institute a well-field system, designed to give visible form to the mutual dependence and trust between lord and subject. The lord's arable land was first to be divided into tracts of nine hundred *mu*, which were then subdivided into nine equal plots of 100 *mu* in something like a modern tic-tac-toe board. Each farming family was to till and reap the harvest from one of the nine plots so that they could grow enough to eat. All eight families in the unit were to work together to till and reap the one central plot, for the maintenance of their lord. "Those sharing the same well-field unit "in comings and goings were close associates; they looked to each other for mutual aid; they supported each other in times of crisis, and thus the commoners treated each other as members of one family and they were at peace."

At another point, Mencius proposed a similar plan, based on four farming households tilling one hundred *mu* of presumably better land, which he said would allow all four families to reap enough to live above the margin of subsistence. Let us take a look at the key passage:

> I have heard that the reasons why King Wen of Zhou was so good at seeing that the aged were taken care of were these: he regulated the land allotments; he taught his subjects to plant trees and to raise domestic animals; he guided their wives and children to take care of their elders. At the age of fifty, an elder with no silk robes may feel the cold, and at the age of seventy, an elder with no meat dishes may feel hungry.... King Wen did not allow his aged to endure either cold or famine. That is what I am talking about. Whoever in the empire is good at taking care of the aged now will find that his people regard him as benevolent, and that man will be given his subjects' allegiance.[153]

Mencius was no less quick to put explicit caps on fair rates of taxation on humble farmers.

Examples abound from the Han empire of honoring the Mencian belief that only mutual trust and care prevailing between rulers and ruled promotes social stability across the realm and security for the ruling house, operating to mutual advantage. Edicts stipulated that people over the age of seventy could not be prosecuted for serious crimes unless they were directly responsible for violent assaults ("murder and inflicting wounds"); that males and females over sixty who had no sons need pay no land tax or sales tax at the market; that orphans and other disadvantaged populations (the blind, cripples, and those who lived alone) were not to be called up arbitrarily for state service; that the bearers of a dove staff (given to those aged seventy or above, depending on the reign)[154] were to benefit from imperial privileges, such as the right to travel on the imperial highways; and that amnesties were to be issued on important ritual occasions, reducing legal penalties or absolving whole populations of their crimes.

In addition, regular (in a few cases monthly) gifts of grain, meat, wine, or silk were to be given to the aged, sometimes in substantial quantities, from at least 179 BCE. With some regularity, bolts of valuable silk were given "to the poor," no doubt in pleasing ceremonies, those communal rituals applauded by Mencius. One edict even granted gifts to "those who have lost their jobs."[155] Still more radically, at least one edict decreed that such groups were to be asked to give their opinions of the local administrators in their areas.[156] State loans could be forgiven, and sometimes imperial land was parceled out to help such people. (On the northern frontiers, "ever-normal granaries" and "lend-lease" bureaus were set up, the first to stabilize grain prices and the second to release farmland to would-be homesteaders.)[157] Public birthing houses were erected, and something very like poorhouses were built in the capital.[158] During times of natural disasters, relief efforts were to be superintended by the state, whenever possible. For example, under Emperor Cheng (r. 33–7 BCE), a flotilla of 500 huge boats was dispatched from Chang'an to aid in rescuing flood victims.

Nearly always, in arranging for the social welfare and common good, leaders employed the language of Mencius.[159] Here is one sample of that language in an edict issued by Emperor Xuan: "Widowers, widows, orphans, those who live alone, people of advanced years living in poverty or distress, these We pity. Earlier, an edict was issued to loan them public fields as well as seed grain and food. In addition, let the high officials of 2,000 bushel rank with all due severity instruct their subordinates to treat such people with solicitude, lest these groups be disappointed."[160]

Perhaps the most stupendous instance of a social-welfare measure, an instance of "lending state land," occurred under Emperor Cheng: The expanse of the fabled Shanglin Park, stocked with rare animals as well as imported plants, had brought the throne an estimated annual profit of seventy million in cash, a sum said to pay the entire annual defense budget for the far west. Yet Chengdi ceded to his capital's urban poor large tracts of parkland in three directions, the east, south, and west (the so-called *san chui* 三垂),[161] presumably "to extend favors to the people" in a manner analogous to that ascribed to the antique sages.[162]

The difficulties of carrying out such regular disbursements in the early empires, with their poor transportation and communication facilities, can hardly be overestimated, yet repeated edicts (for example, those of 179 BCE) aspired to maintain them as customary, until the mandated disbursements gradually yielded to irregular boons to mark auspicious events or to mitigate natural disasters. By the most cynical view, the Han emperors (even those said to disdain Confucian teachings, such as Emperor Xuan) thought it useful, at least, to identify disadvantaged groups as "objects of imperial pity and solicitude" (朕所憐也) and repeatedly to profess their fellow-feeling by donating not a small part of their resources to sustain their populations. As the legalist Han Feizi noted, no court could count on its people acting well unless the people thought doing so served their interests. But the perennial cynicism misses something, surely. If we apply the historian's habitual question, *cui bono* (who benefits?) to the throne's policy making, we soon discover that it was conceivably to the mutual advantage of both parties for those in power to treat the least powerful with dignity, thereby winning allegiance and trust from "those below" and a sense of self-worth, to boot.

True, the early emperors' paternalistic rhetoric rankles the modern ear, for the ruler was to protect his children like "father and mother," as if they were but "babes in swaddling clothes." In light of repeated cutbacks to social services by "enlightened" democracies, however, it may make sense to rethink our automatic recoil from paternalism. Many ethicists would argue that the main task of our moral development is "to move the circles progressively closer to the center, so that we regard our parents and children like ourselves, our other relatives like our parents, and strangers like our relatives."[163] If so, how would we reconfigure Mencius's injunctions to King Xuan of Qi to begin to reverse the current disinclination to recognize the basic facts about our current sociopolitical situation? The answer is just as obvious as in Mencius's day: by seeing the worst abuses and indignities inflicted on others as savage blows to *our own* lives and by pushing forward an agenda to correct the most damaging elements of the situation. Paternalism, altruism, enlightened self-interest, or ethical action—call it what you will—increasingly looks like the more pragmatic, not to mention the more moral option than studied neglect. How different, really, is Mencius's paternalism from the "nudge" advocated by today's social engineers and widely accepted?[164] Meanwhile, chances are still good that the person who credits another with dignity is more likely to receive decent treatment herself. It is no accident that modern political theorists calling for improvements to the current Chinese political system typically cite Mencius as their chief inspiration.[165]

Mencius's persuasions were hardly exceptional. The *Zhuangzi* epigram at this chapter's opening echoes his sentiments, as does the excavated Guodian "Chengzhi wenzhi" 成之聞之 manuscript, which speaks similarly of the signal advantages to be gained from sharing and empathizing with others who are less blessed:

Thus the noble man places value not in lavish goods, but in sharing things in common with the people. If he possesses understanding yet stands in line with others, the people will want his wisdom to achieve success. If he has wealth, yet shares it with the low-ranking, the people will desire that his wealth grow. If he is honored and yet able to yield to others, the people will want his honor to be ever higher.[166]

Some would argue that our very capacity to think at all, and especially to think accurately, depends on our thinking in community with others to whom we communicate those thoughts. (This is the language enshrined in the United Nations' "Universal Declaration of Human Rights" proclaimed by the United Nations General Assembly, in Paris on December 10, 1948, and the ethical thinking that philosopher Allen W. Wood locates in Kant's writings.)[167]

Like Confucius and Xunzi, his fellow Confucians, Mencius was a man of passion and of scruples, and no bloodless parser of the human experience. Mencius aspired to exemplary conduct without any blemish whatsoever: "washed by the Yellow River and the Han, bleached by the autumn sun, so immaculate that the whiteness is unsurpassed."[168] Admittedly, Mencius at his worst seems barely to move beyond the tribalism that divides the world into humanity ("people like us") and the rest ("people not fully human by virtue of their not belonging to our tribe"). And Mencius's denunciation of a rival teacher as a "shrike-tongued barbarian from the south"[169] assuredly detracts from his moral standing and dims our belief in his optimistic portrayal of the moral potentials embedded in the very fiber of the human project.

Still, were it necessary to sum up Mencius's distinctive contributions to theories of pleasure and, more generally, to Chinese civilization, surely the most important would be his emphatic reiteration of what seems today (to some, at least) a crazed belief: that even the least moral person shares a significant degree of humanity with the sage-kings of yore. Perhaps seeing his lapses in this light can be salutary, if we conclude from it that even most extraordinary human beings may sink to appalling pettiness, and thus we share a starting point with them after all. Perhaps, too, the charitable reader of the *Mencius* will come, with its help, to the basic insight that all of us need to trust that there is a common world, and that agreement *might* happen, for this is a necessary concession, if the politics of the common good is to take place.[170]

Finally, given Mencius's admiration for Confucius, this chapter inevitably considers the effects, conscious and unconscious, that great teachers exert upon their pupils, with their helpful conversations ideally like "falls of timely rain."[171] While Mencius speaks of pleasure, he himself performs in such a way as to heighten his pupils' pleasures, whether they be listeners or latter-day readers. He recites odes, he relates stories of the famous and the forgotten, he devises parables, and above all, he empathizes so strongly with his interlocutor's flaws that they succumb to Mencius's flattering vision of themselves. Having heartened others by his exalted vision of what they can become, Mencius has improved their chances of reinterpreting and fashioning their own inchoate yearnings into a highly self-conscious pursuit of personal integrity and compassion toward others.

Thus, on his best days, Mencius must have aroused in his boon companions and maybe even a ruler or two the strong desire to enlarge their hearts so as to extend to others the fellowship they enjoyed with him. The "family resemblance" between parts of the Mencian dialogues and the *fu* 賦 rhapsodies, the dominant form of versifying in the Han (206 BCE – 220 CE), hardly seems serendipitous, in consequence.[172] For both the Mencian dialogues and the Han *fu* celebrate the intensity of experience associated with pleasure-taking, while they exhort the ruler to regulate his desires and imagine likely outcomes. The alternation of moral and immoral, what is wished for and what is found wanting, serves as point-counterpoint in these two sets of writings.[173] And always, *fu* writers, in company with Mencius, presume the power of heady language to move men in both dangerous and constructive ways. "And when an awareness of pleasure arises, how is one to stop it? And when it is irresistible, you begin to dance and wave the arms before you know it!"[174] Xunzi, the subject of the next chapter, penned intriguing *fu*, in addition to his much better-known treatises on good governance, and the foremost *fu* writer of the Han era was the classical master Yang Xiong, the subject of Chapter 6, whose *Fayan* championed the boldness of Mencius's vision of humanity.

CHAPTER FOUR

Xunzi 荀子

on Patterns of Brilliance

Xunzi (d. ca. 230?), unlike Mencius, is a highly systematic thinker, intent upon defining his key terms and overturning settled habits of mind. (The usual comparisons cast Xunzi as a Chinese Aristotle.) Through beautifully wrought essays delineating the role of rites and music in learning and a model for good governance in society, Xunzi builds the case that his readers can fashion themselves as virtual works of art, marvelous to behold and pleasing to emulate, thanks to their conformity with due proportion — works imbued with nearly miraculous power to improve the sociopolitical realm and their own lives.[1] Xunzi, who won fame for discounting the intervention of unseen powers in human events,[2] instead deserves greater renown, to my mind, for his signal lack of deference to hereditary status; he is far less interested in defining humans by their rank at birth than by the drives and commitments they acquire through acculturation. Xunzi, so far as we know, is also the first thinker to imagine a person deprived of worldly powers who nonetheless commands greater authority and autonomy than the occupants of thrones; he is the first, not coincidentally, to portray the nobility of failure.[3]

This chapter will show how well talk of pleasure lent itself, in the hands of an acknowledged master, to consideration of a wide range of compelling questions. It has three sections: the first addresses the role of desire in the *Xunzi* (a role similar to that found in Mencius's rhetoric, even if Mencius is more suggestive than precise on many points); the second lays out four sociopolitical institutions that in an ideal polity would help to guide human desires toward aspirations for the higher pleasures; and the third unfolds Xunzi's rationale for saying that the best man (the "noble man" in the parlance of the day, that is, he who is dedicated to the loftiest expression of humanity) cannot be stripped of honor, no matter how abysmally he is treated by the powers that be, for he has made himself a charismatic figure whose ultimate worth is inestimable and inviolate.[4]

The impulse to seek pleasure is far less a deplorable aspect of human nature than the most suitable basis from which to construct sound theories about good (that is, healthy and constructive) personal and social orders — Xunzi argues this elegantly in nearly all the extant chapters ascribed to him.[5] Xunzi's accommodation of desires may in part explain why he, of all the Zhanguo persuaders, undoubtedly left the largest mark upon Han political, social, and ritual theory.[6] At the same time, Xunzi is probably the figure *least* associated in either popular or scholarly accounts with an interest in pleasure-seeking and pleasure-taking, aside from basing his chapter devoted to music on the graphical coincidence of the Chinese words for "music" (*yue* 樂) and "pleasure" (*le* 樂). Yet fully two-thirds of the *Xunzi* chapters take up the subject of pleasure substantially.[7] Let the following account of Xunzi's preoccupation with nourishing (*yang* 養), satisfying (*man* 滿), refining (*xiu* 修), and gradually expanding (*duo* 多) the human capacities for pleasure attest to the pervasiveness and the sophistication in the classical era of the pleasure theme, which bound together seemingly disparate conversations about human nature, court institutions, and superior cultivation.

Consistent pronouncements on pleasure, shored up by precise definitions, compose a multipart argument in the extant *Xunzi*.[8] Logically, the argument begins with a complex theory of desire as motivation, which argues that the innate desires (*yu* 欲) and dispositions (*qing* 情), as constituent parts of human nature, cannot possibly be uprooted from human beings. Besides, there is no need to excise them, because these desires may be employed for good as well as for ill. Xunzi therefore not only denounces the many calls to "reduce desires" made by the rival persuaders of his day, but also casts one metadesire — the fundamental desire to achieve and sustain pleasurable sensations for as long as possible — as the principal impetus to personal cultivation and social improvement. Since humans by nature seek pleasure, it follows that the only substantial distinction between the noble man (*junzi* 君子) and the petty is the level of discernment (aka "wisdom") that the former has at his disposal when confronting competing objects of desire. By Xunzi's account, only the noble man as a matter of course succeeds in identifying his own long-term benefit and his greatest means of satisfaction, and only he wholeheartedly commits himself to it. Whereas the petty man pursues every momentary gratification he can, diminishing or even ruining his chances of long-term security, the noble man carefully calculates the set of prudential actions most likely to produce and sustain the highest degree of pleasure over the long run, then pursues that course with absolute single-mindedness. Consequently, the Way of ritual and music, in Xunzi's reiterations, constitutes the only conceivable path by which to provide reliable satisfaction for the jumble of desires and longings (*si* 思) that inhere in even the better (that is, more satisfying) version of the self, one that is more whole and integrated, and full of integrity (*cheng* 誠).[8] The noble man, is confident that "those who secure their benefits [by the Way]" as a rule take pleasure [in life] and feel at ease."[9] Satisfaction derived from the increase of pleasure and security, in turn, allows the noble man to pursue the Way in the full knowledge that it maximizes his opportunities for pleasure-taking.

Partly because the noble man finds himself less frustrated and frantic than lesser folk (because nothing need thwart him in walking the Way), and partly because he is more focused and steady (since he is fixed on ultimate, rather than fleeting, pleasures), the noble man's fidelity to the Way strengthens over time. The contradictory impulses and drives with which he was born gradually give way to the single, self-aware endeavor to achieve an integrated body and spirit capable of taking maximum pleasure in life and then giving it to others. With that integration achieved, the noble man's virtuous acts (*de* 德) invest him with a compelling grace and charisma (also *de* 德). His social and cultural accomplishments make him a fully ritualized person who is "beautiful and fine, whole, and integral."[10] As an outstanding exemplar, he naturally attracts others' interest and emulation. It is the singular grace with which he handles his social relations (his final achievement) — more than the keenness of his mind that some of his accomplishments presuppose — that ultimately provokes widespread admiration. Finally, and no less importantly for Xunzi, the autonomy with which the *junzi* pursues his ends so as to realize his highest potential marks him out for consummate status. As a man who radically transforms himself and others, he is said to partake of divinity (*shen* 神).[11] Above all, he has escaped the slavish condition (in the twin senses of subordination and mediocrity) to which the vast majority remain subject.

Needless to say, the desire to acquire a perceived good expected to bring pleasure is not the only conceivable motivation for acting honorably and well to be found in the whole of the *Xunzi*. Xunzi, for example, speaks of the negative emotion of fear spurring some decisions, as when fear of dishonor or execution for treason prevents a warrior on the battlefield from fleeing an enemy. But desire plays a crucial role because of its ties to approval. *Ke* 可 ("approve") has three senses in classical Chinese: "I approve it"; "it is socially acceptable, customary, or mandated, and thus approved"; and "it should be approved as a norm." In the *Xunzi*, all three logically separable meanings are deeply implicated, each one with the others, which means Xunzi is no Kantian lauding certain drives (exclusively moral motivations) and condemning others (practical motivations). For example, in most cases, societal approval or disapproval of some object or course of action affects personal approval or disapproval of it. As Xunzi observes, people are often motivated to act well for a mixture of reasons, and regardless of their initial motivations, they may grow accustomed to acting well, especially if society and its ruler encourage them in that path. What began as hypocritical conduct or a thin veneer of politeness can become ingrained, and meanwhile, all good actions, by definition, improve society by some degree, even if all three notions of approval are perfectly aligned only in the ideal realm led by a sage-ruler.

None of the standard labels applied to Xunzi, including that of Legalist, captures the dominant role that pleasure plays in Xunzi's system. Xunzi's pleasure theory, as presented here, has major implications for the study of classical thinking, among them that Xunzi takes very seriously the sorts of arguments about pleasure posed in the *Zhuangzi*, the subject of Chapter 5 in this book; that Xunzi, in highlighting the noble man's habit of attending to what makes him singular (*shen qi du* 慎其獨),[12] is nonetheless far less focused on interior states than on effective social constructions; and that Xunzi articulates an aesthetic theory from which later thinkers in the Han, such as Yang Xiong 揚雄 (53 BCE –18 CE), the subject of Chapter 6, drew inspiration.[13]

Desires and Deliberation
Xunzi grounds his entire intellectual project on the commonsensical proposition: "All men possess one and the same nature. When

hungry, they desire food; when cold, they desire warmth. When exhausted from toil, they desire rest. And they all desire benefit and hate harm."[14]

Because this list represents the most ordinary of desires experienced by everyone—the urgent, indiscriminate desire for food or sex, as well as yearnings for a specific food or person—Xunzi can assume that all will readily acknowledge the force of such desires and dispositions.[15] People want what they need to survive and to flourish, but what a person wants at any given moment depends upon what the person believes he or she lacks currently. I do not crave more food after a huge meal, and so I will not go to look for it. In this way, drives for food, sex, or shelter can serve as shorthand examples of the entire range of possible desires and dispositions, in that far more complex desires depend equally upon perceived needs and lacks. Indeed, many a passage in the extant *Xunzi* casts the inherent human disposition (*qing* 情) to evaluate desirability and the resultant desire (*yu* 欲) for an object, person, event, or circumstance as the chief catalyst for a host of human activities. The compulsive quality attached to these dispositions and desires seems intuitively obvious, as well; as Xunzi puts it, "Now, as we all know, people intend to have what they desire, regardless of the great distance and obstacles [between them and the objects of their desire]."[16]

The particulars of Xunxi's pleasure theory rest upon the initial premise that it is the function of the sensory receptors endowed at birth (the ear, eye, nose, and body or skin being all conceived as loci for touch), as well as the faculties of the heart and mind residing in the single organ of the *xin* 心, to crave pleasurable contact with external phenomena. The sensory, the emotional, and the cognitive—all these are inextricably bound up in the craving. Only with death does this craving stop. So long as the person lives,[17] "The eye likes colors and sex appeal; the ear likes sounds, the mouth likes flavors, the heart [*xin* 心], as seat of thinking and feeling, likes anything of benefit and profit [*li* 利], and the body in its constituent parts likes ease. These are all born of nature's dispositions [*xing qing* 性情].... They do not await a conscious effort before being born."[18]

Furthermore, when several choices are available, it is always toward the most pleasurable experiences that the senses propel the person: "It is the essential condition of man that his eyes desire the richest colors, his ears the richest of sounds, his mouth the richest of flavors, his nose the richest of aromas, and his mind the highest degree of relaxation and repose. The propensity for these Five Extremes is unavoidable."[19]

Thus, the sensory organs do much more than merely register external phenomena. They also assess the phenomena they encounter and create a hierarchy of values, at least in a preliminary way. The *xin*, as the locus of thinking and feeling, further sorts through the initial impressions, assigning them unreflectively or even unconsciously to the categories established by prior experience, unless confronted with a truly novel situation that does not fit. As Xunzi puts it, "The eye distinguishes black and white, holding some colors to be beautiful and some ugly; the ear distinguishes tones and sounds, holding some notes to be pure and some muddied; the mouth distinguishes sour from salty, sweet from bitter; the nose distinguishes the fragrant from the rank."[20]

Given that the evaluative impulses and sorting mechanisms are inborn, but also liable to shaping by memory and experience, the person inevitably seeks certain satisfactions outside, in hopes of remedying any perceived lacks. A human being has no "value-free," "neutral" response to phenomena, in other words. Instead, a preliminary assessment of a given phenomenon's value to a person disposes that person to desire or to shun the phenomenon. In Xunzi's schema, these dispositions may be classified by level of intensity: if a person finds a particular phenomenon pleasing, she may feel a liking or preference (*hao* 好) for it, a heady frisson of delight (*xi* 喜), or a more lasting and profound sense of pleasure (*le* 樂). Confronted with a less than enticing phenomenon, she may feel dislike or distaste (*wu* 惡), a spurt of anger or frustration (*nu* 怒), or a more profound and lasting sense of pain and loss (*ai* 哀). These six basic dispositions (*qing*), all inborn, color the initial, unmediated inclinations to act.[21] All action, however deeply considered, starts from such basic "spontaneous reactions which are summoned rather than considered."[22]

In addition to the sensory organs, with their range of dispositions, humans — unlike the beasts — come equipped with another faculty, the feeling and thinking heart (*xin* 心), whose function it is to process contacts made through the sensory receptors and correlate those impressions with stored memories of preceding encounters, resulting in more deliberate, that is, mindful and committed, actions to some ends, but not others. Such deliberations, made time and time again, inform and affect the dispositions prior to the self's taking action, we learn; habits of reflection can become second nature.²³ Thus, the initial impulses to act need hardly be synonymous with acts taken after mature deliberation (*lü* 慮), just as a child's instinctive reach for a dazzling light will be tempered in later years by the knowledge that flames hurt.²⁴ Certainly, the more intense the pleasure derived regularly from certain contacts, the clearer and more conscious the later commitment (*zhi* 志) to kindred experiences. Contrarily, the worse the experience, the stronger the reverse drive to avoid any repetition of less than pleasurable experiences. (In addition, Xunzi seems to believe that the negative emotions of fear, loss, and pain may, at the beginning at least, impress themselves more deeply on the conscious deliberations than the prospect of pleasure.)²⁵ As a result, over time, seemingly mundane deliberations can lead to quite sophisticated calculations gauging the likelihood that a particular phenomenon will or will not fulfill the person's needs and desires in the short or longer term.

What the sensory organs and the heart regard as gratifying or distasteful is to a large extent, of course, a matter of experience and custom, as Xunzi readily acknowledges.²⁶ In one of many passages that recall Zhuangzi, Xunzi says, "The frog in the well cannot speak... of the pleasures of the Eastern Sea."²⁷ In another, a hypothetical country bumpkin raised on the coarsest of grains, pulses, and vegetables is for the first time served the very finest wines and meats. His first reaction is to exclaim how strange these things are. But, says Xunzi, "since the wines and meats would be pleasing to the nose when smelled and appealing to the mouth when tasted, and their consumption would be reliably good for the body, then every single person, after giving them a try, would reject the inferior foods and choose the new instead."²⁸ Presumably, the more refined the experiences to which a person is introduced, the more likely that person will be aware that his choices may be consequential. When all the farmer knows is potatoes, to choose them to relieve pangs of hunger seems not only natural, but inevitable. But when the lucky fellow can choose between plain boiled potatoes or delicious lobster, he begins to notice some choices are better than others. The most fortunate man is he who has grown up with a sufficiency in the company of good teachers, friends, and parents in the cultured environment of the Central States, as we will see. Nevertheless, Xunzi insists, the picture he paints of the human condition and its potential for deliberate, ritualized social interaction holds true of all men and women in every age. Because Xunzi's "proof" for his propositions is quite complicated, it is worth examining in some detail.

In every society in every era, as Xunzi tells it, two aspects of the human condition work against the easy attainment of pleasures: externally, the basic search for satisfaction of the senses in a world of limited resources leads to ruthless competition over whatever is perceived to be valuable, which only intensifies in proportion to the scarcity of the acclaimed valuables (for example, the highest ranks; the most attractive women; the most succulent of pigs). Strife within and between social units in a virtual war of all against all sooner or later breaks out, unless some person or institution intervenes to regulate competition. Concurrently, there exists within each person another series of contests, which sometimes pit the physical needs for food, clothing, shelter, and sex against the desires for wealth, high rank, honor, long life, beautiful spectacle, social approbation, lasting fame, and a sense of self-worth. Incommensurate goals, some easy and some harder to attain, typically demand different sets of activities. Invariably, therefore, attempts to obtain all these desires simultaneously lead to frantic (*ji* 急) and fragmented (*bu cheng* 不誠) efforts,[29] no matter how correct the person's estimation of the value of each desired object. Pursuits must be prioritized, clearly, lest the resources of time, energy, and materials, innately limited, be dissipated to no effect or — still worse — predictably disastrous effects ensue, following disorder (*luan* 亂) in society and anxiety (*you* 憂) in the body.[30]

The words "frantic" and "fragmented" point to the basic insecurities that permeate the vast majority of human lives. Humans at every level of society find it surprisingly hard to "nourish and satisfy" the senses as precondition to fashioning a "good life." For to foster the senses, humans need not only material security and opportunities for sexual, social, and symbolic interaction, but also profound peace of mind. Lacking peace of mind, humans become too distracted to register their own sensations, even when the most exquisite objects of gratification lie ready to hand.

> If the heart is full of anxiety and fear, then the mouth may be crammed with grain-fed and grass-fed animals, but it will neither recognize nor appreciate [*zhi* 知] the tastes. The ears may be listening to the sound of bells and music, but they will neither recognize nor appreciate the sounds. The eye may alight upon embroidered patterns, but it will neither recognize nor appreciate the forms.

And the body may be clothed in warm, yet light garments and rest on soft mats, but it will feel no ease. In such a case, [even] when a person confronts the most excellent and beautiful [*mei* 美] things, he finds himself incapable of feeling pleased. Even were he to feel a moment's gratification, he would still not be able to allay [his anxieties and doubts permanently].... When things stand like this, does a person's pursuit of things nourish his life or put his longevity up for sale to the highest bidder?... He may wish to attend to his pleasures, but [anxieties and doubts] will assail his heart.[31]

For the foregoing reasons, provision for the immediate gratification of the basic biological needs is not enough to nourish the entire domain that constitutes the whole person. Insecurity diminishes the experience of pleasure, and the failure to experience satisfaction can lead, in extreme cases, to such a serious state of deprivation that death ensues. By Xunzi's account, anxiety, born of the insecurities engendered by the frantic pursuit of an inferior or partial good, constitutes as large a problem as acting in an excessive manner ("giving free rein to the desires") or racing after a variety of conflicting pleasures. For all three compulsions brought about by wrong-headed attempts to secure pleasurable experience essentially "harm life."

Fundamental to Xunzi's larger argument is his claim that the person preoccupied with cares and concerns, worries and doubts, can never truly experience pleasure. For Xunzi, a true, profound, and durable pleasure invariably entails the engagement of the evaluating heart *at the time*; the person must be fully aware of the experience and must consciously acknowledge that the experience satisfies his or her present needs. Unfortunately, even (or especially) the most satisfying of present experiences may serve only to awaken anxieties about the future. In the midst of sufficiency, the specter of a future loss of sources of gratification can sap present enjoyment. That paradoxical sense of death in life and destitution in plenty Xunzi, in company with the best of the late Zhanguo thinkers, ably communicated. As a phrase ascribed to Zhuangzi would have it, "People go through life with worry as their companion. To be careworn for so long before dying—how bitter that is!"[33]

Xunzi fully agrees with many rivals regarding the pervasive anxieties of humankind. But Xunzi, drawing quite different conclusions about human potentials and final goals,[34] adamantly refuses to advocate the reduction of desires, let alone their suppression. According to Xunzi, neither the desires nor the dispositions can ever be reduced, since they are part and parcel of the human endowment. "Humans are born with desires. When the desires are unattained, then humans cannot but seek to satisfy them."[35] More importantly, the desires to attain the conventional "valuables" of long life, wealth, social standing, and official rank constitute the first level of motivation for human beings just starting to cultivate themselves. It is wonderful that the impulse "to get what nourishes" the senses of sight, smell, sound, taste, and touch can trigger positive action, as well as regrettable acts that create conflict. For quickened by a love of patterns (*wen* 文) of inestimable beauty that is partly innate and partly learned, the heart can be stimulated to exercise its "inclination to make proper distinctions" (*hao qi bie* 好其別), with the result that it can also make highly self-conscious and commendable selections in pursuit of higher goals.[36] One of the most effective spurs to "the impulse to get what one wishes" is the example of the cultivated man whose patterns of behavior prompt admiration and emulation. That said, desires for rites and music prove to be of little or no help to personal improvement until a taste for them has been duly acquired through habit and long practice: "The rites and music provide models, but they do not persuade us [to practice them]."[37] Still, the utility of human desire must never be underestimated, since it strongly motivates people to set aside self-interested activities and short-term benefits in order to cultivate better relations as a way of achieving longer-lasting pleasure and security in that pleasure.[38] Only short-sighted thinkers such as Song Bian 宋鈃, who are "blinded by the concept of desire, do not understand what can be gotten" from the careful employment of desire as catalyst.[39] By such artful phrases, Xunzi indicates that he may be better equipped to show how to adapt the unchangeable aspects of the human condition — its desires, drives, and dispositions — so as to provide more substantial and lasting pleasures in sufficient measure to sustain, rather than deplete, each person.

For Xunzi's persuasions would convert the severe constraints on human existence into major advantages promoting the success of the search for long-lasting, deep pleasure. And once again, Xunzi's logic seems irrefutable, because it is well grounded in the most ordinary of learned experiences, that of delayed gratification. Given that the human condition is plagued both by an insufficiency of physical resources and by the normal afflictions of the psyche, even the least reflective person will, upon occasion, find his desires to satisfy his present needs offset by a yet more pressing desire to set aside sufficient "wealth in the form of surplus money and provisions" so that he will never want for anything in the future.

> Now, the human condition is to want to eat grain-fed and grass-fed beasts and to wear patterned and embroidered clothes.... In their lives, men may be known for raising chickens, dogs, pigs, and boars, and they may even raise oxen and sheep, and yet they do not dare to have wine and meat at mealtimes. And even with a surplus of knife-shaped and spade-shaped coins and stores in cellars and storehouses, they do not presume to dress in silk.... It is not that humans do not desire to do such things [as would supply immediate gratification]. It is rather that, taking the long view of things and calculating the consequences of their actions, they are apprehensive lest they lack the means by which to perpetuate their wealth.[40]

In this extended passage (only part of which is quoted here), Xunzi suggests that all adults accept the reality that they cannot at any given moment have everything they desire ("do not dare to have wine and meat at mealtimes... do not presume to dress in silk"). They see that the continuance and improvement of their comforts depends at least as much upon self-restraint as upon any other factor ("taking the long view of things and calculating the consequences of their actions"). (Already others had raised the example of the demands of ritual, backed by the natural desire for social approbation, dampening the urge for immediate gratification. A young man longing for sex is still unwilling to have sex in the presence of his elder brother or parents, as the "Five Conducts" piece excavated from Mawangdui plausibly claims.)[41] In other words, however rudimentary their level of moral development, human beings choose to moderate or (temporarily) ignore their basic desires if that will improve their future prospects for enjoyment. Take the simple matter of the household budget. Ordinary humans are not only capable of this type of reasoning; they regularly employ it in daily life. Thus, all are capable of conscious behavior modification, even restricting their most immediate drives for food and sex. Obviously, to obtain the less physical desires predicated on a cluster of skilled calculations (for example, those for a good name, honor, or fame) is more complex than devising a household budget; often, too, it requires greater self-restraint. But the more deliberate and reasonable the ardent pleasure-seeking, the more likely it is that the person can be taught to assess the future consequences of present actions and to modify his behavior so as to achieve his ends.

Specific measures to attain and prolong satisfaction will vary according to situation, but since all human beings in all times and places have operated under roughly the same sorts of constraints, to the person of discernment and experience, it should be relatively easy to determine, at least approximately, the wisest course of action, if not from personal experience, then from knowledge of others' past initiatives and their outcomes. A person can judge the merits of any particular experience by putting it to two simple tests: how it compares with previously known phenomena in satisfying the senses, emotions, and mind and whether it is likely to provoke no ill reaction. Witness the story above of the poor and ignorant peasant who found the fine meats "reliably good" both while eating them and afterward.[42] Through his rough apprehension of the connections, that peasant has learned enough about decision-making to assess aspects of the phenomenal world by a reasonable standard.[43]

One key to the persuasiveness of Xunzi's analyses is his word "reliably." If one wishes the experience of pleasure to "last long," "broaden," and "be secure,"[44] the pleasure-taking or pleasure-giving must remain pleasing over time. Repetition or prolongation of the valid pleasure act must not sate or disgust. (One likely registers the sense of having eaten too much after the consumption of three peaches, but there may be no limit to the pleasure to be derived from a wider range of friends or a deeper experience of friendship. Stolen fruit may taste sweeter at first, but will it continue to yield as much pleasure over time, given the risks involved?)

It goes without saying that a sustainable pleasure must be attainable. True, desire is "not dependent on its object first being obtainable," since it spontaneously erupts as one makes contact with external phenomena, but with the thinking person, "what is sought follows what is possible.... That we seek to satisfy our desires by following after the possible is an impulse generated by the faculties of the *xin*."[45] It is expectation born of memory — the sense "that certain notes will follow in sequence after certain others"[46] — that leads a person to elect one path over the alternatives. With that in mind, a person can go a long way toward arriving at a healthy hierarchy of pleasures, such that the determined search for pleasure will not frustrate, but satisfy over a lifetime.

In Xunzi's view, it remains only for him to show that ordinary men assume (except, most inexplicably, as regards their own lives) the operation of intelligible laws of cause and effect in the social sphere, no less than in the practical realm of household budgets.[47] Most men, for example, believe that those who contravene the social norms do not operate for long in safety. Bullies and tricksters do not feel secure enough in their possessions and powers to savor them ("Rudeness and aggression are techniques that bring constant danger")—though Xunzi worries that the egregious sophistries fashionable at contemporary courts will obscure such commonsensical conclusions.[48] In due course, the utter isolation of the unthinking and selfish person, his inability to keep allies, patrons, and friends, will impede his quest for greater power, authority, and self-esteem and make his life precarious.

Having witnessed this "constant rule" in operation, the observant man—the very man who may become noble through the deliberative process—weighs his own prospects by the same standard that he applies to others.[49] And so he chooses, consciously and consistently, to adopt the socially constructive behavior that society regards as "virtue" (de 德), since he will thereby be able to savor his pleasures securely and in good fellowship. The lesser man, by contrast, is far too inclined to trust to luck; he counts on being the single exception to every norm, an exception who will manage to escape the predictable consequences of his own despicable behavior.[50] Only in one respect, then—a propensity to trust in the rational law of social consequences, rather than in dumb luck—does the noble man secure in the enjoyment of his pleasures differ from the petty. As a key passage in the *Xunzi* says,

> With respect to the basic substance and capacities for understanding, the noble man and the petty are one and the same. What the noble man and the petty share is a liking for glory and a dislike of shame, a liking for benefit and a dislike of harm. It is surely only with respect to the means by which they seek these things that they differ.
>
> The petty man is quick to self-aggrandize, and yet he wants others to trust him. He is quick to deceive, and yet he wants others to regard him as their nearest and dearest.... It is difficult for him to create security through his actions.... And if he goes on in this way to the end, he will never get what he would like and he will always meet with some awful outcome.

The noble man, by contrast, is entirely trustworthy, for he also [like the petty man] wants others to trust him. He is entirely loyal and sincere, for he also wants others to treat him as their intimate. He certainly cultivates the right and the rules in a discriminating way, for he also wants others' approval.... So he easily achieves security and peace of mind.... And if he perseveres to the end, then he will surely get what he would like and, moreover, he will *not* meet with what he finds most distasteful and distressing.[51]

As Xunzi puts it elsewhere, "In general, every man wants safety and glory and abhors danger and shame. But it is only the noble man who can get what he *prefers*. The petty man just 'happens to meet' every day what he abhors [because he fails to see the causal connection between his evildoing and its outcomes]."[52] The decision to consider the moral and practical consequences of one's actions becomes far easier when one learns two things: that the attainment of conventional rewards, not to mention an enviable nobility of character, in the vast majority of cases is due to choices made rather than to accidents of birth or fate and that in all but a few extraordinary circumstances beyond the person's control, the pursuit of conventional long-term self-interest (for example, honor, rank, wealth, and physical security) and the pursuit of goodness are identical in practice. Put another way, the noble man by definition is simply one who accepts the direct cause-and-effect relation between the focus of his pursuits and habits, which reveal what he works at (*zuo* 作), and what he seeks to become. He therefore seeks "what in thought is easily understood [this commonsensical reckoning of cause and effect], and what in conduct easily brings security ... and what *in the end* inevitably brings about what he prefers."[53] He does not begrudge the effort to ascertain what is most likely to "secure glory and ease" without "harming himself,"[54] knowing that any path or profession involves a certain amount of trouble.

Once Xunzi has proven to the satisfaction of his listeners that the sensible person will invariably go after whatever he perceives he lacks (as is evident as soon as the belly senses a lack of food),[55] he goes on to postulate an analogous mechanism propelling the listener/reader toward more complex and profound goals, for instance, the attainment of more deeply satisfying social relations:

> Every man who wants to act so as to gain social approbation does so precisely because his [inborn] nature is not a pretty sight to see. It is the sense of his

own meager accomplishments that leads to a longing for greatness and the consciousness of his own ugliness that leads to a longing for beauty. It is the feeling of being cramped that leads to a longing for breadth [in physical or psychic terms], of being impoverished that leads to a longing for abundance, and of being base that leads to a longing for honor and eminence.

Whatever a person lacks in himself he feels compelled to seek. The rich man does not long for wealth nor the man of high standing for greater powers. What a man already possesses in himself he will not bother to seek elsewhere. From this we can see that the motivation for men's desire to do good derives from the uglier aspects of human nature. Ritual principles are certainly not a part of man's original endowment ... and so he forces himself to study them and he seeks to acquire them.[56]

There is reason to believe that repeated deliberations premised on such insights will gradually induce major changes in the dispositions themselves. (Compare Mencius's benevolent man operating on the principle that he already has within himself the "seeds" of righteous conduct, weak impulses that he need only bolster.) After all, people quite regularly accustom themselves to attitudes and circumstances that strike them initially as odd or irksome — things that are, in fact, the artificial products of culture and habitual practice. This is the basic learning process, from infancy. Discounting both the objection that virtue is "unnatural" and so harmful (registered by Laozi, Zhuangzi, and Yang Zhu, among others) and the contrary notion that virtue is entirely "natural" (a position attributed to Mencius by Xunzi, when he is tetchy), Xunzi would open us to a third possibility: that we may inscribe in our very core manifestly better propensities by virtue of everyday experience and formal schooling until they come to seem second nature to us.[57] In so doing, Xunzi strikes a fine balance among rival positions, knowing that human and societal evolution is neither wholly conscious nor wholly automatic and that crises often result from the unintended consequences of ill-conceived directives.

The question is how is a person to get from here to there. How is he to move from the original endowment, fraught with competing desires and inclined to rash judgments, to a second nature that is more uniformly balanced and calmer in its thinking?[58] And how to progress from a first dim awareness of cause and effect to the exalted condition that consistently yields the greatest satisfaction and pleasure by abandoning habits that deplete resources (*qiong* 窮)

and prevent a higher state of flourishing?[59] Those transitions even Xunzi does not explain well, doubtless because the path for each person varies so much, depending on the original endowment, opportunities and obstacles met in life, and the level of commitment mustered. Xunzi's studied use of the generally pejorative word *wei* 偽 ("artificial," meaning "contrary to one's [original] impulses") for the learning process highlights the voluntary, yet somewhat "unnatural" effort required to realize one's full potential.[60] Xunzi does say that the process is long and cumulative, rather than easy and quick; that it requires a keen ability to recognize what is categorically true and better; that the human senses and the evaluating heart are adequate to the task, though the process will proceed more smoothly and more quickly for those fortunate enough to live in a cultivated society, where good teachers, rulers, and friends abound.

Much of Xunzi's discussion hinges upon the double association of the character *si* 思, which means "to long for" as well as "to ponder." Insofar as a person longs for a particular phenomenon, he ponders the best method to attain it. Presumably, the more he thinks, the better he becomes at making connections and apportioning his efforts. Gradually, the person of increasing cultivation comes to view pleasure-taking and relative advantage in ever more subtle ways that insure pleasure's prolongation, enhancing the ability both to obtain the pleasure and to enjoy it.[61] As the heightened capacities of the senses and heart/mind cause radical reassessments of advantage and disadvantage and related readjustments to the desires, such readjustments over time prompt self-conscious advances toward a greater security of mind and a minimum of external threats. That the person's sense of his relation to the world has been substantially changed by the shift in habits of thinking and acting and also by the succession of experiences confirming the pleasures of goodness—that much is sure. When next he thinks, feels, or acts, he then acts upon the new insights with a newfound steadiness of purpose and excellence in execution.[62] Or as Xunzi says, "Seeing what is good, the noble man cultivates it, regarding it always as his best means of self-preservation." "He honors what will bring him security."[63] Just as the archer's aim and the musician's ability to stay on pitch improve through long practice, after the successful coordination of separate microskills, the ordinary person traveling through life may hope to attain a new level of proficiency in his deliberations and actions if he works steadily at the separate skills in unison.⁶⁴ Gaining proficiency depends heavily on the degree of commitment to continue one's practice, regardless of setbacks.

> [By definition] the good farmer does not give up ploughing because of flood and drought. Neither does the good merchant stop doing business because of occasional losses. [By analogy,] the noble man of breeding does not neglect the Way just because of poverty and hardship....
>
> The sage is defined by what he accumulates and lays store by. If a person ploughs, time and time again, then he is a farmer; if he carves wood or stone, over and over again, then he is an artisan; if he accumulates assets, then he becomes a merchant. And if he performs the rites and does his duty, consistently over time, then he becomes a noble man.... So whether a man becomes a sage or villain, a laborer, an artisan, a farmer, a merchant — it all depends upon nothing more than an accumulation of efforts to perfect his practices.⁶⁵

Because it rests mainly upon the resolution to persist, to attain goodness evidently lies well within the reach of anyone.⁶⁶ The sage, however superior he seems to the ordinary person, is simply a man who "sought it [goodness] and later got it, who acted to achieve it and later perfected it, who accumulated it and later was elevated"⁶⁷ to that rarified height. Without a doubt, each vocation entails some obstacles, and all require some persistence in the face of adversity. The farmer, for example, must toil, day in and day out, to secure his livelihood, and the merchant travel far from home. But oddly enough, the world deems the choice to farm or to trade eminently "practical," whereas the decision to school oneself to prefer the good and perform it is widely mocked as "impractical" or "eccentric." (Compare today's ridicule of the "ivory tower.")

Not at all, Xunzi insists. Xunzi counters his critics in this way: the wisest and most gratifying choice is to become a sage, since it requires less physical exertion, avoids the risk of infamy, and secures for the person a state of ease and contentment.

> One can either become a legendary sage [a Yao or a Yu] or a notorious tyrant [a Jie or a Robber Zhi].... To be a legendary sage is always to find security and fame; to be a notorious tyrant is always to create danger and infamy. If the legendary sages constantly enjoy their ease, whereas doing the work of a laborer, artisan, farmer, or merchant means constant trouble and toil, why do men work so hard at the one sort of endeavor and so little at the other?⁶⁸

It matters little that Yu first gained fame for the hard physical labor that he put into stabilizing the realm, for once he had finished his physical labors, he, like Yao and Shun before him, sat on the throne, from which vantage he could survey his subjects and allies with calm satisfaction. A commitment to the moral Way, on this view, requires no unusual measure of self-abnegation. To the contrary. That commitment entails, in the end, far less toil and trouble than most or all other goals while maximizing a sense of pleasure and ease. "The noble man seeks his own benefit and profit" (*junzi qiu li* 君子求利), and the determined search for what the thinking person knows to be pleasurable leads to right conduct, to health, and to all manner of honors in death, if not in life.[69] Thus Xunzi as classical master arrives at moral necessity via his initial premises about pleasure.

Much of the "glossy appeal" (*runse* 潤色) of Xunzi's rhetoric derives, no doubt, from his insistence that wanting physical pleasure and security prompts the next course of action.[70] But Xunzi was hardly alone in positing such a correspondence. By the Zhanguo period, proverbs likened proper discrimination in evaluating objects, persons, and events to "liking sex" with alluring women.[71] Xunzi's arguments, though much more logically rigorous, expand, yet alter the claims recorded in Book 1 of the *Mencius*: there, the ruler could reach sagehood through an imaginative "extension" of his love of music or his concern for a particular ox,[72] while the *Xunzi* gives the transformative aspects of the gradual process their due. It is worth considering, then, why Xunzi, along with a number of other masters, used the pleasure rhetoric to such robust effect to talk about two subjects that are quintessentially moral: deliberative choice and awareness.

Preachiness is likely to irritate the listener, and anecdotes from history and fable are frequently misconstrued. But pleasure is a subject that lends itself to discussing a conscious decision to act as the rational function of its predictable consequences, rather than as a moral imperative. Everyone has a sense of what constitutes "dissipation," if not their own, then that of others, though it may be harder to trace the predictable consequence back to a pattern of behavior when the person is ruling a realm, forming an alliance, or evaluating prospective officials. Yet everyone can agree that people are moved to action by considerations of pleasure. Reaching for a pear at lunch or a peach after dinner — those sorts of acts embody a conscious preference built upon an anticipation of pleasure ("I prefer the peach, because it will give me more pleasure now"), though the choice in no way implies that a peach is always good or a pear always bad.[73] Being surfeited temporarily with respect to a particular taste or feel does not preclude the enjoyment of something similar (the pear with dinner instead of the peach), nor does it inoculate against a resurgence of desire (for example, sex with yesterday's partner).

Timing and situation, as well as individual predilections based on prior experiences, play an enormous part in setting preferences in daily life. But through talk about the entangled notions of desire, longing, and pleasure, a rhetorician as skilled as Xunzi can induce a keen awareness that the stakes in deliberate action may be complex and shifting, that changing realities may call for frequent revisions to one's earlier calculations in light of unfolding events, and that seeing the whole picture eventually yields more predictable and weighty consequences.[74] As a chapter influenced by Xunzi, if not authored by him, says, "The deeper the person goes into the human Way in our society, the greater its potency and his, and the more rarified the music and what he takes pleasure in."[75] Attention to situation, timing, and likely outcome lies at the root of all morality, though it is not equivalent to the developed moral sense. It was the latter that Xunzi hoped to instill in the ruler and see institutionalized at the court and realm.

Institutions of the Ideal Realm in Support of Pleasure

Up to this point in his set rhetorical pieces, Xunzi has devised a theory that demonstrates pleasure as motivation and the necessity for forethought and self-restraint to secure all manner of pleasures, but he has not once required his audience to believe in the sublime efficacy of ritual to secure the highest pleasures. Xunzi recognized a widespread disinclination among his contemporaries to adopt Confucian ideals and behaviors; ergo his exact framing of his rhetorical question: "Who could have known that the patterns and principles of rites and duties are the way to nourish one's own dispositions?"[76] So when Xunzi moves from his general description of the human condition to advise the ruler to institute the ritual Way within the realm, he continues to urge this on almost entirely practical grounds. The ruler's own pleasure-seeking will be best served if the court's institutions provide a way for all — ruler and subject alike — to sustain their pleasures (echoes of Mencius). Notably, as with his analysis of the human drives, Xunzi's discussions of social policy are predicated upon an insight akin to that articulated by Fredric Jameson: "Pleasure, like happiness or interest, can never be fixed directly by the naked eye, let alone pursued as an end; it is only experienced laterally, or after the fact, as something like the by-product of something else," where the "something else" is often social approbation.[77] Xunzi, in his wisdom, recognized an added complication: simply to indulge the people will not necessarily bring them pleasure, yet the ruler ignores the people's desires only at his peril.[78] Given such pleasure calculations, the ideal state, for Xunzi, is one in which the opportunities for serene and steady pleasure-taking are enhanced for clear thinkers, since the court, through various policies, routinely signals its approval of socially constructive behavior (see below).

Critical to Xunzi's formulations on policy measures is the conviction that people are liable to be equally endangered by repression and by overindulgence of the basic sensory equipment as above.[79] Desires as such cannot be extirpated, for if they were, no motive would remain for pursuing the "good" pleasures. At the same time, overindulgence overwhelms the human capacity for rational calculation. Because most people, being self-indulgent, habitually snatch their pleasures without salutary regard for the consequences, the intelligent ruler, in order to help his subjects avoid injury, ensures that his subjects have a material sufficiency (so that their desires and needs are met) and that he himself furnishes a model of prudent self-restraint (so that they will disdain excess).[80] Grateful subjects will then be happy to repay their ruler's efforts on their behalf. In this way, the ruler's patterns of exemplary leadership induce a positive ripple effect within the sociopolitical order that, in turn, allows his subjects to flourish in their vocations. Thus is the proper balance between hierarchy and reciprocity struck and "the hearts of the people made good."[81] As will become clear, Xunzi's remarks on policy matters in an ideal realm, like those on human nature, circle round and round the entwined themes of pleasure and desire, encouraging the performance of charismatic virtue through ritual and music.

The famous opening passage of Xunzi's chapter "On Ritual" attributes the potential frustration of human desires to the scarcity of resources, but moves on quickly to assert that people can learn how to satisfy their desires without exhausting available resources through the ritual institutions devised by the sages:

Humans at birth have desires. If their desires are not satisfied, then they cannot but seek [some means of satisfaction]. If they seek, but there are no limits and degrees and barriers [to the seeking], then they will inevitably fall to contending. From contending arises chaos; from chaos comes exhaustion [of resources, including material resources and human energy]. The former kings disliked chaos, and so they institutionalized [ritual norms of] humanity and duty in order to provide a barrier, wanting at the same time to nourish human desires and to supply what humans seek. They formulated laws and institutions that prevented desires from being exhausted by things and things from being used up by desires — so that the two would be in balance and develop in tandem.[82]

In the analysis of ritual that follows, spectacles of various types (for example, parades, progresses, and military reviews), mourning, and sumptuary regulations are presented as the three main "ritual" institutions of the realm devised by the former sage-kings to forge stronger links in the populace's mind between pleasure and virtue. Thanks to these links, the people can correctly identify ritual performance and social institutions as "what nourishes" them in their needs.[83] (Chapter 20 presents a fourth institution, musical performance, as a particularly powerful sort of spectacle.)[84] By their very natures, human beings are predisposed to form societies, so unless they are "eccentrics or rogues," they can be led, without harsh punishments, to an awareness of the dire consequences of destructive behavior and social isolation.[85] If people live alienated from one another, refusing to render service to each other, "there will always be poverty... and conflict," if only because "even the able cannot possibly be universally skilled, and it is impossible for a single person to hold every office."[86]

Complex social bonds, supported through sociopolitical institutions, are to secure men and women in their rightful places. For that reason, Xunzi celebrates notions of shared, stable communities. These permit the establishment of long-lasting fame (as opposed to the celebrity of the infamous); these facilitate the fair distribution of scarce commodities according to societal contribution. This sort of community does far more than gratify the individual's senses, emotions, and mind.[87] Inculcation in community values can strengthen the resolve to engage in the deliberations that locate long-term pleasure in the most constructive forms of behavior. How does it do this?

The ancient kings' institutions "caused the various classes of people in the world to realize that what they desired and longed for is to be achieved via such institutions [rather than through antisocial behaviors]. This is why their incentives work."[88] Ritual performances, among the most important social and cultural institutions, constitute the sage-kings' primary "recipes" (*fang* 方) providing pleasure and music (*le, yue* 樂). Ritual allows "the Hundred Pleasures to find their targets," while putting "the ruler of men in the most authoritative position from which to benefit the world."[89]

Reviewing the institutions named by Xunzi, spectacle by its very definition performs important human functions: it appeals strongly and at once to the senses of sight, hearing, and smell (and sometimes taste, as well) while addressing human cravings for symbol, for safety and security, and for belonging. Moreover, as is evident from the munificent rulers' progresses and his musical performances, the two most captivating spectacles mentioned by Xunzi, spectacle renders palpable the good society's balance of hierarchy with reciprocity. For at the same time that a spectacle invests an "awesome authority" (*wei* 威) in its sponsor and patron, most often the ruler or his court, all those with any connection to a given spectacle — whether they prepare it, participate, or merely watch — come away with a heightened sense of shared community. Thanks to spectacle, the satisfaction of inner human needs confirms the powerful reality of a larger imagined community.[90] These congruent observations are critical to Xunzi's final analysis.

For Xunzi, it is axiomatic that the ruler appear to the public as a model of generosity. Otherwise, the ruler's will cannot be identified with the common good. So arises the widespread institution of what I call "display culture," which offered to early rulers a distinct set of advantages. For Xunzi, in particular, "sharing pleasures" through magnificent forms of court-sponsored display kept the throne at the center of the public gaze while obviating the need for onerous self-restraint on the ruler's part.[91] At the same time, the considered use of pleasurable display accomplishes three tasks vital to the preservation of the realm: In "making a standard [pun intended] of virtue,"[92] it lets "those below" know what are the mandated "general" or "common" interests; it advertises the utility of actively promoting those interests (though it provides no adequate mechanism to distinguish apparent conformity with communal interests from real commitment); and it gratifies those already inclined to identify with the prevailing hierarchies. Notably, with spectacles observing sumptuary regulations, it became possible for the socially constructive behavior reinforced by public display to give pleasure to the patron ("The noble man takes pleasure in his public expressions")[93] no less than to crowds of onlookers. Meanwhile, it made the social and court hierarchies seem both inevitable and just. The intended result: to generate in as many people as possible, up and down the ranks, the distinctly pleasurable sense that they had shared in the great honor conferred by the *edifying* spectacle — be it largesse distributed by the ruler, community conferrals of honors upon the aged, village banquets, or mourning processions — even if excluded from direct participation and its tangible rewards. In an era millennia before movies, television, the Internet, or social media, such edifying spectacles, it was thought, could bestow suitably scaled pleasures upon people of wide-ranging standings and ranks while highlighting the ruler's awesome authority as giver of spectacles, imparting a strong sense of community.

Mourning, as a special form of antique spectacle cued to sumptuary regulations, gratified the senses equally. Its real significance for Xunzi, however, was that, in paying due heed to and substantiating the immaterial dead, it drew attention to human commonalities, since all humans experience death, underscoring the reliance of one generation upon the next. It tied the mourning circles composed of the living and dead to the individual's fundamental sense of well-being, assuring the living that they would live on after death in reputation and in ritual, if not as purposive agents, then through the regular commemorations offered to the honorable dead by those in their debt. From the solemn mourning rituals, members of the community would learn to figure post-facto assessments about others into their pleasure calculations, thereby broadening their bases for judgment, exercising their powers of imagination, and undergirding their propensity to consider the long-term consequences of actions in relation to whole communities, past and present.[94]

Sumptuary regulations were meant to operate in a manner analogous to the penal code, in that they could deter aspirants from pretending to higher ranks and prerogatives than the laws allowed, lest their pretensions to undue eminence imperil self and society. Simply defined, sumptuary regulations stipulated the number and quality of the items for consumption (everything from landholdings, to houses, to banquet utensils) those of a certain status were privileged to use. To judge by Xunzi's comments, sumptuary regulations are important to the health of a society for two main reasons. First, since it is "human nature" or the "human condition" to experience mimetic desires (that is, desires to possess something prompted by the realization that someone else already possesses the object of desire), the drive to attain more material goods will surely devolve into destructive competition among individuals and groups, given the limited supply of the potential sources of pleasure, unless legislation and custom place limits on the innumerable urges to consume. For all people, being endowed with a wide range of contradictory desires, will seek to gratify their impulses all at once, unless such folly is overridden by some more powerful drive or prohibition. Equally importantly, sumptuary regulations, if fairly devised and enforced,[95] help the ordinary person to identify for himself those aspirations most likely to bring his person and his family long-lasting pleasure. For when those most worthy of emulation (defined as those who contribute most to the social order) possess the best material goods in the greatest quantities, as should happen in an ideal realm, the link between prudential deliberations and just rewards becomes clear as day to all. Thus, when the ruler has proffered noble rank and riches to induce men to take "the path of goodness,"[96] those of even mediocre insight in all likelihood will hasten to imitate worthy models, initially propelled by self-interested desires to attain a comparable level of material enjoyment.[97] That pursuit will, in turn, improve their insights.

At this point, Xunzi hopes to have an ideal administration intervene, for a ruler and his officials, armed with a better understanding of the reasons for human conflicts, can hope to motivate entire populations to act in conformity with the ruler's dictates by means of proportionate material rewards and punishments.

> When the king sets about regulating the terms [of official service and of proper social roles], if the terms and the actualities to which they apply are fixed and clear, so that he can apply the Way and communicate his intentions to others, then he may guide the people with circumspection and unify them.... With a humane disposition [renxin 仁心], he may explain his ideas to others. With a receptive disposition eager to learn [xuexin 學心], he listens to their words. And with an impartial disposition [gongxin 公心], he then makes

his determinations.... He honors what is fair and upright and disdains petty wrangling over benefits.... No laws and institutions can stand by themselves [without the right men to judge how best to administer them].... In general, each human affair has its own rationale, yet the totality of human affairs abides by the same set of ways.[98]

When "those below" discern an obvious advantage to acting in accordance with the ideal ruler's will (increasingly equated with the "common good"), they will happily undertake to obtain their desires for wealth and status by practicing the court-sponsored social virtues, for therein lies the path of safety. Properly governed and instructed, the commoners willingly forgo the chance to act upon certain cravings in the short run, so long as they have a reasonable prospect of achieving their larger goals in the long run. Those fully accustomed to this complex negotiation of desire have the potential to develop a second nature more reliable and discriminating than the first, a second nature capable of the serene pleasure-taking that was so admired in classical times.[99]

Xunzi's "Enrich the Realm" chapter, to take one elegant spin-off from this, envisions the ruler's court as primary distributor of status items, including manuscripts on gorgeous silk and well-seasoned bamboo (Figure 4.1). Regular ritual dispersals of high-quality material goods (many of them inscribed) act as carrot to the penal code's stick. Of the two tools at the court's disposal, it is primarily the carrot that motivates subjects to follow their superiors' lead, generating good order throughout the realm. Naturally, the wise ruler is eager to garner (through taxation, tribute, and gift exchanges) sufficient wealth in the form of grain and finished goods — not so that he can hoard this wealth, but so that he can freely dispense it among his subjects, in effect making their homes his own storehouses. Such gifts and awards are more than worth the cost, however great, since the regular disbursements bind the ruler's subjects to him tightly. Absent such a close bond between ruler and subjects, rebellion and regicide may ensue. For this reason, according to Xunzi, the old Mohist arguments in favor of frugality must be condemned as "exaggerated reckoning," that is, a counterproductive focus on the fiscal bottom line in willful disregard of the fact that material goods represent one of the best incentives at the ruler's disposal. No wonder the wise ruler keeps a close watch on the manner in which such powerfully

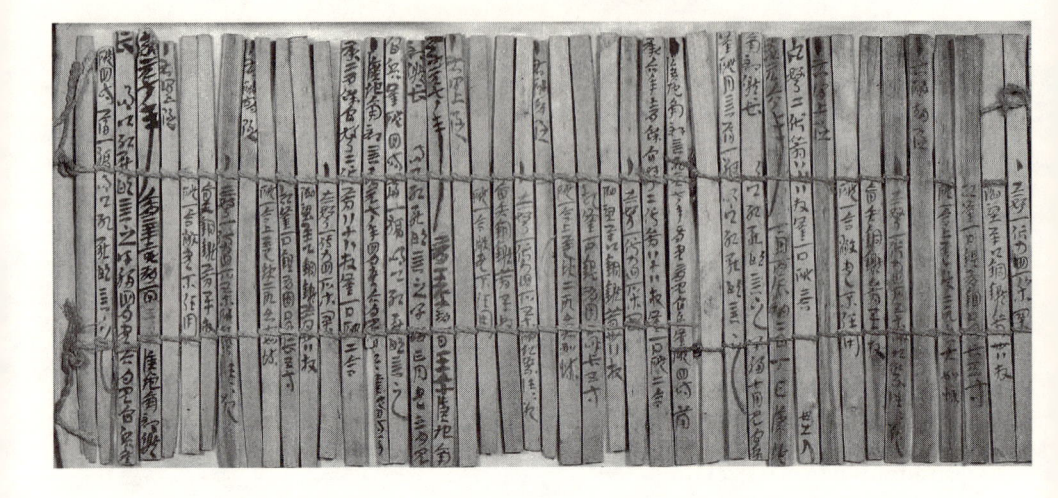

Figure 4.1. An official document of Eastern Han composed of strips (in this case, wooden rather than the more common bamboo) tied together with string cords. Excavated at Juyan 居延 (present-day inner Mongolia), site A27. Length 23.1 cm; width 91.6 cm. Weight 243.63 grams. Item 128.001. For the complete transcriptions, see http://ndweb.iis.sinica.edu.tw/woodslip_public/System/Main.htm

Here we see a late Eastern official document entitled "Yongyuan qi wu bu" 永元器物簿, which records the military supplies for the Yongyuan year corresponding to 93 CE. While this particular text is composed of wooden slats (more common in the northwest than elsewhere), the format where strips are held in place by strings or cords is identical to that used for bamboo slips in early manuscript culture. Needless to say, once the strings break, the slats or slips fall out of order, and that alone works against the possibility of verbatim transcription when copying the text, as does the propensity for people with the requisite learning to memorize texts and recite from memory or interpose their own comments or renditions of the text in their hand copies.

desirable items as ritual objects, including manuscripts, are meted out to those of lower rank.

With the judicious application and regular enforcement of rewards and punishments, in compliance with sumptuary regulations, a person decidedly less inclined to be moral can even be schooled to set aside selfish considerations in the interests of the larger social good, for sumptuary regulations advertise the paradox that he who intends to profit most from his own actions had better profit his ruler and his fellow countrymen first. As Xunzi says,

> Who could have known that the fearless rush to confront death on the battlefield is the way to nourish life [while cowardice spells death]? Who could have known that expending resources [on behalf of the community] is the way to build up wealth? Who could have known that to be respectful and yielding is the way to improve one's own security? ... [But sociopolitical institutions can insure that] a person whose only goal is survival will surely meet death and one whose only goal is profit will surely meet loss, and that a person who mistakes indolence and ease with real security will surely risk dangers, and a person who mistakes short-term gratifications for long-term pleasures will surely perish.[100]

With institutions reinforcing the message that practical and moral considerations will align in his realm, the ruler can teach his officials to choose the path of virtue consistently, if only so they may benefit themselves. Then he and his representatives can in turn convince most of his subjects — no matter how average, amoral, or grasping — to consider the interests and advantages of other members of their community. This creates a very useful operating principle, since nearly all forms of pleasure-taking entail a dependence, albeit temporary, of one person upon many others. (Indeed, rituals of all types enact such relations.) So when the people go about their daily lives trusting (correctly) to the justice of a system that allocates resources in proportion to societal contributions, each person feels more confident that he will receive his due share of the goods that have long been valued by ordinary men, as well as by the wise: long life, wealth, rank, a good name, social standing, beauty, honor, and freedom from debasement.

With the support of such institutions and customs, every person in an ideal realm "rests secure in what he or she finds pleasurable" (*an le* 安樂), utilizing talents in the ways he or she knows best[101] while trusting to the ruler's "art of calculating what [rewards and privileges] suit each respective station."¹⁰² Lyrical passages in the *Xunzi* describe the ideal realm as one in which serenity is achieved by all, with "every commoner finding security in his place and taking pleasure in his locality."¹⁰³ In that ideal world, "The ploughman takes pleasure in tilling his fields, the soldier feels secure in his labors, the many clerks enjoy their measures and laws (*fa* 法), the court exalts ritual, and the high-ranking ministers cooperate in their deliberations," as Xunzi puts it in his chapter "On Music and Pleasure."¹⁰⁴ Where "a benevolent person occupies the top position," upholding the right institutions,

> The farmer uses every ounce of his strength to till the fields as best he can. The merchant uses all his powers of close observation to create as much wealth as he can. The artisan uses all his know-how to craft the very best implements. And the court counselors [*shidaifu* 士大夫] on up to the aristocrats employ humaneness, liberality, knowledge, and skill, so that they execute their responsibilities to the fullest extent.¹⁰⁵

The well-governed polity, then, requires more than that each component of society doggedly function like a small cog in a large wheel. Good governance is predicated on each person experiencing a zest for his calling because it is so well suited to his capacities and predilections. All people, "from the Son of Heaven on down to the commoner," want to "maximize their capacities, attain their goals, and take secure pleasure in their activities."¹⁰⁶ Like a fish in its element (a metaphor employed by Xunzi, as well as by Zhuangzi), each person is pleased with his situation, since the particular tasks he performs "are neither too much nor . . . too little for him."¹⁰⁷ Thus are "the Hundred Pleasures born of a well-ordered court and capital."¹⁰⁸

But as Xunzi readily acknowledges, many people are not so fortunate as to live in that enviable situation where their work suits them and they can duly perform it in security. In an ill-governed realm, no one, however virtuous, "can necessarily earn honor . . . trust . . . or suitable employment."¹⁰⁹ And then, nothing guarantees that the good person will escape danger and opprobrium, even if "humane and dutiful acts are promoted as the means to long-term security."¹¹⁰ Good men and women sometime meet with misfortunes due to poor fates, natural disasters, social disorder, or the warped actions taken by petty people trying to conceal their own unworthiness. "The repulsively ugly may regard a beautiful woman as a calamity. To a petty-minded crowd, an official cognizant of his duty to the common good may be an ulcer. To the vile and perverse, a man upholding the Way may be a villainous traitor."[111] Slander and envy can do a great deal of damage.

But if exemplary conduct cannot guarantee conventional rewards to the decent person, why should he or she act virtuously at all, particularly when the inculcation process is so manifestly slow and strenuous? Xunzi answers with a stunning claim: that the person of integrity and insight lives the most pleasurable life, even when denied a suitable profession and rank commanding a comfortable livelihood, and, more astonishing still, even when denied an honorable death. For those intent upon doing good wrest from life an easy sense of self-worth that allows them to take profound pleasure in their actions, whatever their current status and situation.[112] Their bodies may be cruelly violated or destroyed, but no one can strip them of their astonishing achievement, their integral wholeness. Xunzi surely had in mind the examples of Yan Hui, who died an early death, and Confucius, who went to his death believing himself a failure, but Xunzi pushed this idea further, as we will see.

On the Ultimate and Inviolate Pleasure, Integrity

Important as Xunzi's premises and policy proposals are to a good understanding of the masterwork attributed to him, the most powerful of Xunzi's persuasions are found in the passages devoted to the topic of integrity and wholeness (both *cheng* 誠), passages deploying, quite uncharacteristically for Xunzi, the language of divinity that was common in his era.[113] The logic of his rhetorical appeals would surely have faltered, had it not rested on two allied concepts: the honor accorded the noble man or *junzi* who had no need to rely upon others for his pleasures, and the noble man's standing as an extraordinarily compelling work of art, the former being thoroughly rooted in classical traditions and the latter possibly Xunzi's inspired innovation.[114] As Xunzi tells it, under any circumstances, the *junzi* 君子 (a term for the aristocrat as well as the noble in spirit) takes pleasure in acting with grace in social situations, confident that the polish he has acquired by training and practice earns him a good name that will long outlive him. In adverse, even dire circumstances, the noble man does not panic, because he knows himself and also "what would constitute a *real* setback"—a lack of charismatic virtue.[115] Because he feels no panic, he looks for no one or no thing (for example, "fate") to blame. He remains unruffled, unhurried, and constructive to the end.

Perhaps the place to begin is Xunzi's lengthy description of the noble man's integrity and wholeness — a description that leaves the mundane world of Realpolitik far behind. The passage says:

> Nothing is better than integrity [*cheng* 誠] when nourishing the faculties of feeling and thinking in the heart [*xin* 心]. With the highest form of integrity, no other task exists [but that continual process of nourishing it]. Only humaneness is held fast; only duty is performed. Now when the *xin* 心 of integrity holds humaneness fast, it becomes manifest in the body. Being manifest, it assumes the wondrous qualities ascribed to the gods in Heaven. In this state, it is capable of exerting a transformative influence upon others [as anything divine is, by definition]. And when a heart of integrity performs a duty, then it appears as deep pattern. Being patterned, it assumes the brilliant qualities ascribed to the gods on earth. In so acting, it is capable of effecting change. When change and transformation follow one after the other in smooth succession, this we call "Heavenly virtue or grace."
>
> Heaven does not speak, and yet men deduce its [incomparable] height. Earth does not speak, and yet people deduce its [incomparable] breadth. The four seasons do not speak, and yet the people deduce the proper times for planting and harvesting from it. Each of these cosmic powers has a constant character, which constancy gives it integrity and wholeness. [Like them,] the noble man of consummate grace and virtue [*zhi de* 至德] need not speak to convey his ideas. Before he has reached out to others, they already hold him dear; before his anger erupts, they already credit him with awesome authority. They follow his orders because he has been careful to preserve what makes him singular.[116]
>
> What we can say without hesitation is this: without integrity and wholeness, nothing is singularly good, and without such singularity, nothing takes visible form, and it is crucial that singular goodness take visible form [in the whole person]. For even if it is ingrained in the heart, shown on the visage, and voiced in speeches, the people may still experience some reluctance to follow the person. Or, if they do follow him, they will always harbor doubts about him [and so fail to give their utmost allegiance to him].
>
> Heaven and earth are great, certainly. But if they were not whole and one, they could not transform the myriad sorts of things. The sage is wise, certainly. But if he is not whole and one in his person, then he cannot transform the myriad people.... Now, integrity is what the noble man holds fast to, and

it is also the basis of sociopolitical affairs. Being the only place where he abides, the noble man uses it to bring like-minded men [to his aid]. To clasp it is to obtain it.... As soon as he grasps it, then things become easy. And when things become easy, then he acts in a remarkable manner.[117]

Xunzi here portrays the gods of the universe as self-sustaining patterns of unalloyed brilliance and beneficence that powerfully affect the lives of those on earth who depend upon them.[118] He invokes the associations of the common religion of his time, which credited the unseen cosmic forces with effecting major changes miraculously, outside of the usual realms of cause and effect (they "do not speak," still less do they appear to intervene), without resort to compulsion or commands. Through parallel constructions, Xunzi establishes the analogy, suggesting the marvelous influence of the noble man's personal integrity, which need not speak, but must take form if he is to transform others by his singular goodness, ultimately rooted in and evinced by his absolute dedication to his commitments seen and unseen. Above all, Xunzi stresses the superb efficaciousness of the pleasurable and edifying spectacle presented by the noble man's activities in life: consistent in his initiatives and utterly compelling in his grace, the noble man displays (necessarily in brilliant forms visible to all) his inherent humaneness, so that his dutiful acts appear as strikingly patterned as the veins in jade. Because inner and outer, above and below, are perfectly correlated in his person, his comprehensive model of cultivated integration works its will on lesser human beings as easily as the gods are said to do. His compelling quality stems from the thoroughness with which the noble man develops his own potentials by working to elevate his feelings and thoughts. For him, no other task exists than preserving this integrity in its perfection. Hence, the distinction between the noble man and the petty person, who is a patchwork of diverse and unconsidered impulses.

> The learning of the noble man, having entered through the ear, then shines in the heart and spreads throughout the four limbs, taking its forms in action and in rest. All of his modulated words and subtle movements are exemplary, and so all can be taken as models. By contrast, the learning of the petty man, having entered through his ears, is blurted out through his mouth. As the space between the mouth and ears is a mere four inches, how could it ever be enough to make a grown body, six-feet tall, beautiful?[119]

Such a sarcastic estimate of the petty man, regardless of worldly status — sarcasm that would hardly have been lost on Xunzi's privileged listeners — contrasts sharply with Xunzi's frequent paeans of praise for the noble man's beauty — his existence as "perfect and ultimate pattern" for himself, his close associations, and the larger society. Xunzi stakes such claims about the noble man with high seriousness. Most important to Xunzi is the noble man's unique sway and commanding presence, which expresses itself throughout his bodily demeanor, not just in his words or in his heart of hearts. For that reason, "analyzing the heart is less effective than seizing the techniques"[120] — those artful ways of acting that gratify his person, imbue him with beauty, and lend geniality and grace to all his primary social relations.

As noted earlier, such techniques must be practiced, over and over again before their efficacy can ever be fully appreciated. One does good and through doing it slowly discovers its power. But to Xunzi, the best techniques are honed through ritual exchanges, even those conducted at the lowest level, with the village wine-drinking ceremony one fine illustration of ritual's transformative power. Performances at the village wine ceremony reiterate, confirm, and instill in the body the teachings that Xunzi has tried to advocate in his rhetoric, as is clear from his description of the ceremony:

> The host goes in person to greet the chief guests and their attendants. All the other guests follow after.... [After an exchange of ritual bows, the host] presents the wine cup in pledge.... Then the performers enter, ascend the stairs, and sing three pieces, following which the host presents them with the wine cup.... [After several sections of the orchestra have performed, the end of the musical performance has been formally announced, and the performers have left], two men are designated by the host to raise the horn tankard in a toast to the guest of honor, after which another is appointed master of ceremonies. *From this we learn that it is possible to be congenial and to enjoy oneself without disorder and dissipation.* The chief guest pledges the wine cup to the host; the host pledges it to his attendant; and the attendant pledges it to the other guests. Young and old quaff a drink from it, with the more senior going first.... *From this we learn that it is possible for junior and senior to drink together without anyone being left out.*... At the end of the formal ceremonies, [the participants], descending the stairs, remove their sandals. Ascending again, they resume their places. Now they "prolong" the wine cup, putting no limit on the number

of drinks.... *Moderate and patterned are they, to the end of the gathering. From this we learn that it is possible to take one's ease, yet in no way be disorderly.*[121]

Ritual, by this account, teaches many lessons at once: that one need not pit duty and pleasure against one another, for the noble man attains them equally; that the social hierarchies need not leave anyone feeling alienated and bereft of dignity; and that informality in community settings need not spell disorder. Noble men interact gracefully, guided as they are by the age-old institutions established by the sage-kings, so that their every gesture, howsoever casual, embodies the polite arts. That Xunzi registers such a belief is hardly startling, given the prevailing assumption that cultivation makes for charisma and harmonious accord. (Recall his third chapter, delineating the ideal ruler and his subjects.) Yet Xunzi never hesitates to excoriate those pompous types who disdain such simple social exchanges as occur during the village banquet. His vision of unforced and easy conviviality, wherein the entire assemblage, under their leaders' tutelage, finally drinks "long" from the cup, shows a humane imagination that transcends deft handling of social situations.[122] Men with the requisite grace perceive themselves and others performing as actors in meaningful dramas, and their whole lives are made up of such artful constructions, at once highly patterned, usefully prolonged, aesthetically pleasing, and free from regrets.

Before turning to a related passage in the *Xunzi*, it may be useful to review the usual associations of art, in premodern and postmodern analyses. At the fundamental level, "Art is art because it is not nature." (Nature makes no intentional selection, nor does human nature, as Xunzi insists.) A compelling piece of art, as an artifact of intention, must work as a whole and be perceived to have a kind of integrity; be consistently beautiful or arresting; move people to retain it in the mind's eye, the better to receive its message; have a recurrent impact independent of and beyond a one-time emotive transference or even a "sustained experience of mental synthesis" from artist to onlooker; exist as an end in itself, a *sub specie aeternitatis* that cannot be really "possessed" by its nominal owner; and yet thrill onlookers with intimations of a greater potential than they had hitherto known.[123] All of these well-known associations appear in Xunzi's descriptions of the noble man and his deliberate actions, for as Xunzi puts it, the noble man's goal in cultivation "is to beautify his person," not to please others, but to gratify himself.[124] Statements throughout the extant *Xunzi*, like the following, suggest that the noble man himself has become a sublime work of art whose final achievement of what is "beautiful and fine" surpasses the ordinary, the haphazardly fashioned, the fragile, and the fragmentary. Quietly ignoring the vexations of slander, opportunism, and mishap, the noble man's attention in extremis to form, cadence, and texture provides an avenue to heightened perception and thence to a sort of salvation:

> The noble man knows that whatever lacks completeness or wholeness and fineness does not deserve to be called "beautiful and fine" [*mei* 美]. Therefore, he intones his lessons over and over again, in order to make them penetrate [his entire being]. He mulls things over, in order to make them comprehensible and penetrating. He becomes the proper model of a man, so as to take his stand in it [integrity]. He excises damage, so as to support and nurture [integrity]. He banishes from his eyes any desire to see anything deleterious; from his ears, the desire to hear anything deleterious; from his mouth, the desire to say anything pernicious; and from his faculties located in the heart, the desire to think malicious thoughts. And when he has truly reached the point of preferring [integrity to anything else], his eyes then prefer it to the colors; his ears prefer it to the five notes; his mouth prefers it to the five flavors, and his faculties think it more beneficial than possession of an empire. For that reason, this unparalleled state of privilege and benefit cannot be overturned [by any external forces]. The crowd cannot force him to budge from it, nor can the court and empire disturb him. For he follows it in life and in death. This we call "grace resplendent," a holding onto virtue [*de zao* 德操/藻],[125] after which his heart is settled in such a way as to elicit responses from others.
>
> When his heart is settled in such a way as to exert influences, we call him a "model of perfection." In the heavens, there shines the sun, moon, and stars; on earth, too, there shine brilliant lights [that assume the form of the noble man]. The noble man prizes [above all] his wholeness and integrity.[126]

Here, Xunzi implicitly rejects two alternatives that have captivated many down through the ages: the allure of surface beauty and the fascination with the fragile and fragmentary or ruin. Unless cultivation pervades his entire person thoroughly, settling his heart, a man will seem flawed to others. Through this single passage Xunzi threads together with seeming artlessness a series of linked propositions about the ineffable beauty acquired by the man steeped in cultivation: that any object is deemed beautiful and hence desirable insofar as it is whole and fine; that the noble man endeavors to make his own conduct whole and fine, that is, without any admixture of coarseness in it, because he desires to achieve the honorable state of "grace resplendent" that will move others toward the good; and that these endeavors cloak him in a singular grace so robust that it cannot help but attract widespread admiration, producing a wonderful effect that "settles" men's hearts (his own, certainly), freeing them from anxiety, after which their responses accord with his own efforts. But as with other remarkable sources of beauty and light (for example, the starry heavens), the special quality of the noble man — akin to the mysterious divine — exists independently of its reception by observers. "Gracious and splendid / Like a jade baton / Of good fame and good aspect"[127] is he. The noble man is supremely aware of his own perfection, regardless of others' opinions of him. When he speaks of his own shining excellence, "he compares himself to Shun and Yu [the exemplary sage-rulers of the distant past], thinks himself a fit companion to heaven, and knows this is no exaggeration."[128]

Thanks to his own circumspection (in both senses of the word, "seeing from all sides" and "prudential action"), the noble man finds in himself, "in activity and at rest," a serene source of satisfaction, though he is duly mindful of the hazards of social isolation and self-absorption.[129] And since the very "thoughts of the sage give pleasure,"[130] the noble man need not rely on externals for his physical and psychic sustenance.

> So long as the faculties located in the heart feel serene and at ease, then even colors and beauties that are less pleasing than usual suffice to feed the eye, and tones that are less pleasing than usual suffice to feed the ear. A diet of vegetables and a broth of greens suffices to nurture the mouth. Robes of coarse cloth and shoes of rough hemp suffice to support the bone structure. And a cramped room, reed blinds, a bed of dried straw, plus a stool and mat, suffice to support the bodily form. Thus, even without the finest things of the world [associated with high rank and status], a person can nurture his own sense of pleasure. And even without high rank, he can nurture his good name.... This we call "recognizing his own due weight and *gravitas* while putting others to use."[131]

Ideally, this lessened reliance on externalities for a sense of self-worth extends to the sociopolitical sphere, though Xunzi more than once exhorts people, as social beings, to avail themselves of every opportunity to participate as fully as possible in that sphere. Whereas the petty man is enslaved by things, the noble man, by contrast, deploys every available thing he has to sustain him. He is not overthrown or undone by others' reactions. The lure of power and profit cannot disquiet him, nor the masses sway him, nor the court agitate him. The noble man has too strong a sense of his own honor and integrity.[132] He knows how to sustain himself in dire straits and he knows, too, that he has the courage to meet every exigency.[133] Hence the couplets "Seeing the benefit in the mastery of his dispositions, / Knowing his good name to be the real source of glory, / With hordes [of supporters], he may feel in harmony. / But alone and in isolation, he still has pleasures enough."[134]

And "since everybody who outwardly attaches great importance to other things is inwardly anxious," the noble man's relative indifference to externals serves him well as he pursues his own pleasures in daily life.[135] "This is why the noble man, even when he has not yet gotten [his rightful place in society, with all its perquisites], takes pleasure in what pervades his heart."[136] (As Yang Xiong, the late Western Han master, succinctly put it two centuries later, the "sage takes pleasure in being a sage.")[137] From this, Xunzi concludes, one may say of such a man that he "is without an [extra] moment's anxiety and concern [*you* 憂]."[138] Anxiety and concern, after all, generally stem from nothing more than deep-seated fears about a future deprivation of sources of pleasure. And the noble man is clear on this: there can be no real "poverty and misery where true humaneness is found, and no [long-lasting] riches and honor where it is absent."[139] For all of these reasons, the determination to acquire goodness and grace (*de* 德) holds out the fairest hope of obtaining success (*de* 得) in life,[140] for the ordinary desires for conventional goods depend too much on others' arrangements. That said, the noble man's wishes are complex, always: to make pleasure durable while attaining a degree of admirable autonomy, being "not divided against oneself," even in the face of disgrace or death. Such integrity and wholeness exerts so remarkable a mastery over one's physical form and inner states as to impart an air of grace and ease that color all one's chosen activities.[141] Because "the area of choice is small but real," I suspect that Xunzi would have concurred with E. M. Forster's summation: "It is not all gossamer, what we have delighted in, it has become part of our armour, and we can gird it on, though there is no armour against fate."[142] Xunzi, a master teacher whose pupils looked to him to "utter words at once pleasing and helpful to life,"[143] would have agreed, I warrant.

CHAPTER FIVE

Vital Matters:
The Pleasures of Clear Vision
in the *Zhuangzi* 莊子

> Pleasure is finally the consent of life in the body—the reconciliation, momentary as it may be—with the necessity of physical existence in a physical world. —Fredric Jameson, "Pleasure: A Political Issue"

There is, as Aristotle famously remarked, a great deal more to living than survival.[1] Pleasure in being alive, a capacity associated with feeling secure and at ease,[2] as well as freedom from humiliations and self-preoccupation[3]—all this depends upon developing a certain kind of clarity. Zhuangzi sets out to convey what that clarity might look like or entail. But neither the imagined author of the *Zhuangzi* compilation assembled over centuries nor the protagonist called "Zhuangzi" who appears therein are prone to arid theorizing.[4] (See the Appendix to Chapter 5.) To the contrary. The historical compilers and the fictional protagonist named Zhuangzi alike scoff at all types of thinking that employ hard-and-fast logical categories, as every novice reader learns. That said, the *Zhuangzi* is not an easy text to parse, because it uses every trick in the rhetorical book to shake and startle readers out of their most cherished biases and preconceptions. Nonetheless, the *Zhuangzi* advances a sophisticated argument about the enormous pleasures to be had from trying to see things whole and as they are while ably demonstrating that no person can ever see with perfect accuracy, given that he or she is part of the very things in continual flux. And yet, by the *Zhuangzi*'s reckoning, the person alive to the mysterious life pulsing through all things, most vividly experienced in complex interactions, feels intense pleasure from participation in such scenes, caught up in the delicious injunction, "Make it be spring with everything."[5]

The style of the *Zhuangzi* is unique. For Zhuangzi, there are no carefully organized treatises devoted to one topic, though there are spoofs of that genre. Zhuangzi works most often through parables and fables and fast-paced dialogues jam-packed with snappy punchlines traded back and forth between likely and unlikely partners in the animal kingdom, or the afterlife, or at court, in settings both familiar and not. Some of Zhuangzi's prose is deliberately circular. In other passages, the points and counterpoints erupt so furiously as to leave the reader dizzy. The stunning fictiveness of the *Zhuangzi* engages the imagination and redirects the gaze to finer prospects from the conventional world where rules, regulations, routines, and institutions play out. The fictions allow readers to try out different roles in some safety, seeing how others act and think and glimpsing the strong appeals lodged in novel ways of thinking. Friendships are performed, but often to different ends than envisaged in other texts, with a larger dollop of tolerance and generosity doing the preparatory work for satisfying relations. And while writing in the same high style that only potential and actual candidates for office and the members of their families could command, Zhuangzi, unlike Mencius, is not much interested in the rulers he encounters, though he is ready enough to serve family and court with intelligence and guts, when necessary. But why confine one's thoughts to such locations, with so many sights to see, he asks? In complete agreement with Xunzi's analysis of the typical human condition as frantic and fragmented, Zhuangzi sharply diverges in his prescriptions for healing the usual ills that beset humankind. For whereas Xunzi advises us to refine our desires and single-mindedly craft the self, until it appears to be a jade polished to a luminous sheen, with Zhuangzi, we tread new ground. Zhuangzi would have us "return to the basics" and "return to the source" (both *fan ben* 反本), learning to be cognizant of and comfortable with our human limitations while defusing our learned preferences for moral certainty and complacency before setting out on a leisurely journey to enjoy the curiosities of the cosmos.

Modern readers may find this fresh look at the *Zhuangzi* unfamiliar, even discomfiting, for several reasons, chief among them the overwhelming emphasis hitherto placed on the first seven chapters in the *Zhuangzi*, the so-called "Inner Chapters," out of laziness or in the mistaken belief that the writing there captures the original author's voice.[6] In addition, there is the thick overlay of "explanations" for the

Zhuangzi provided by members of religious groups, Buddhist and Christian, espousing truths not only extraneous, but downright contradictory to the *Zhuangzi*, not to mention the modern agendas reading "purity and freedom via reclusion" into the *Zhuangzi* compilation in more recent times.[7] These need not concern us for the moment. Far more tempting is the invitation extended by the *Zhuangzi*, as it gestures toward a path of several steps: from recognizing the skills that facilitate mutually transformative exchanges with other living things and people, to clearing away the underbrush impeding full appreciation of the world, to apprehending the ineffable beauty that surrounds us in the simplest interactions and mundane objects. Clumsier in its expositions than the original text, what follows in this chapter builds upon the "Supreme Pleasure" chapter in the *Zhuangzi* and two additional scenes of instruction in the vital arts as it tries to reconstruct that path. It is perforce plodding and additive, insofar as it tries to make explicit lessons that Zhuangzi subtly intimates. It aims nonetheless to recapture a whiff of the excitement generated by the *Zhuangzi* in early times, privileging close readings of the text over the wide range of interpretations that it has spawned. As Zhuangzi advises, "Let it be! Leave them alone!"[8]

To reveal one thread in the multivocal *Zhuangzi* while sidestepping most of the thorny issues that past commentators retroject into the text is itself a profound pleasure. As Lee Yearley once noted, in every tradition, the acknowledged masters tend to present far more radical visions than their followers can comfortably accept.[9] More conservative types, including opponents of pleasure, have nearly occluded the distinctive talk of the pleasures outlined in the *Zhuangzi*. This chapter therefore opens with two episodes drawn from the *Zhuangzi*, one each from the Inner Chapters and the Outer. These episodes feature Woodworker Qing and Butcher Ding, both of whom extrapolate insights into living well from the practices developed in their respective crafts. The *Zhuangzi* passages devoted to the heightened state of being "fully present" presuppose an understanding of the sacred dimensions of the world, which are invisible or simply unnoticed, as well as the prevailing resonance theories of the day. Naturally, then, those topics merit review in the context of Zhuangzi's sublime trust in the multifaceted dimensions of reality.

In classical Chinese, the single graph for "life" and "living" means three things, as in English: to survive, and not to die; to earn a livelihood; and to retain vitality, in part because "insecurity" and "worry," as well as the desires to control and interfere, have been given up. It is the third meaning that primarily concerns Zhuangzi, although Zhuangzi is mindful how much fear of death and anxieties about career diminish vitality. That the marvelous gift of life should elicit far more from recipients than a grudging assent to mere survival is a pervasive theme. Once that gift is seen (rightly) to be the source of supreme pleasure, other insights no less pleasurable enter the realm of vision. From pleasure to pleasure, existence opens.[10] From wonder to wonder, vistas widen. The world is so packed with mysteries and marvels that it makes little sense to rush in single-minded pursuit of a limited range of perceived "goods" (fame, career, wealth, long life) that will likely prove elusive in any case. Such heady dreams may entice, but they are apt to divert attention from the living out of our allotted days in flourishing ways.

Paired Scenes of Instruction:
Woodworker Qing and Butcher Ding

Two rather lengthy episodes in the *Zhuangzi* sketch the wonders inherent in a person's being "fully present" (*zhi ren* 至人) to *mutually* affecting and transformative experiences. It is hardly coincidental that the ideal protagonists of these two stories are skilled craftsmen of relatively low status. Evidently, the insights they derive from the practice of their craft do not depend upon hereditary privilege or remarkable erudition; instead, they rely upon a strong intuitive "feel" for other things and people and a patience described elsewhere as an ardent waiting upon the unfolding situation (see below).[11] Such experiences hardly qualify as "transcendent," there being no sign of the mystical, yet they do transport us to new insights. This quality of experience the *Zhuangzi* treats with awe, fully cognizant, on the one hand, of human limitations, but also of the human potential to partake of the sacred. So while people can never become truly spontaneous on the model of the cosmos, people can still manage, through the alternative applications of rigorous logic and of learning to let go, to untie a number of the knots that bind and chafe as they occupy their stations in life, high or low, and fulfill their social commitments. Welcome release ensues, but only through the radical deprogramming that permits the restoration of the vital *qi* 氣 flow.

The two stories bear testimony to Zhuangzi's serene faith in the divine, but ineffable presences exerted within ordinary experience. Woodworker Qing outlines the extraordinary pleasures gained from being fully present to the sights around him. The more familiar Butcher Ding story describes his approaches to objects of interest, which require, for lack of better terms, a kind of withdrawal of the senses from certain things and a dramatic opening of those same percepts to others. On such foundations did Zhuangzi fashion his own distinctive views of the world, as well as of the processes that typically shape or distort people's sense of reality. That ordinary habits and conventions prevent people from seeing clearly what is right before their very eyes is a continual theme in the *Zhuangzi*. It is in this context that I place Zhuangzi's references to "balancing the faculties of the heart" (*qixin* 齊心, later [?] *zhai xin* 齋心)[12] and "quieting the mind" (*jing xin* 靜心), that is, stepped withdrawals of attention from the perceptual overload that afflicts ordinary life and fragments awareness, then as now, so as to listen more attentively to the polyphonic structures of the world, perceiving the totality with a kind of clarity before responding with increased acuity and renewed purpose.

The less familiar story about Woodworker Qing, from the Outer Chapters, describes a protracted process, a deep form of engagement that might be alluded to, if not captured by the term "vision presence," a receptivity to the divine:[13]

> Woodworker Qing carved a piece of wood into a bell stand. When it was finished, all who saw it were awestruck, thinking it to be the work of the gods. When the Lord of Lu saw it, he asked him, "What technique did you use to make such a thing?" Qing replied, "Your servant is but a mere artisan. How could I claim to have any special art? Even so, there is one thing: Before making a bell stand, I never dare to let my energies flag. I always fast, in order to quiet my heart.
>
> Modulating [my body and spirit] for three days strips away any hopes I cherish for congratulations and rewards, high office and stipends. Modulating for five days more strips away any concerns I have about praise for ingenuity or blame for clumsiness. Modulating for seven days means that I even forget that I possess four limbs and a bodily frame. At that point, no ruler or court exists for me. My skills are concentrated and all outside distractions melt away.
>
> Only then do I enter the mountain forests, to contemplate the natural patterns the trees were given. With the bodily forms fully present,[14] then and only then do I see a bell stand with perfect clarity, after which I apply my

hand to it. Otherwise, I'm finished! In this way, I join something of Heaven to something of Heaven. That's why some suspect[15] the object to be a work of the gods. They are, in a way, right about that!"[16]

This passage attests the sheer length of the preparatory processes that led to final, albeit temporary mastery, involving the gradual surrender of self-preoccupation and the concomitant rise in attention to the absorbing task at hand. The lengthy process of "balancing" the faculties, a process with quasi-religious connotations (*qi* 齊 or *zhai* 齋), invariably meant more than the ritual observance of obvious taboos, abstinence from sex, and limitations on diet or dress. "Balancing" the physical form and vital spirit implied stilling the feelings and judgmental mind, an untying of knots and loosing of struggles, followed by a concentration of one's resources, sometimes through meditation, to settle the person, so that he or she becomes freer to notice the particularities and patterns of the present.[17] Eventually, the skilled artisan, with self-preoccupation behind, discerns the unique structures inhering in another's form, at which point the woodworker finally intuits the best approach to engage fully with them. The woodworker becomes the wood. (Compare Matisse's comment, when painting a bird, "I became a parakeet. And I found myself in the work.")[18] With the divine qualities of the tree conveyed to him, and his to it, the woodworker may attend wholly to the task and work toward it, in the full confidence that the singular vision of beauty already vouchsafed to him can be realized in a form perceptible to others less prepared than he. Note the recurrent element of *time* here—and not only in the lengthy preparation: the tree's patterns have grown up slowly over time, as the rings and knots attest; so, too, their counterparts in Woodworker Qing's makeup. Having mastered a domain of physical, emotional, and mental skills, Woodworker Qing, in all his complexity, is thoroughly prepared to contemplate another layered presence, with the sedimentary memories and experiences making each partner in the exchange utterly unique.[19]

Notably, Zhuangzi does not portray an artisan merely following the injunction to "Be aware!" ("How could I claim to have any special art?" Qing asks, not disingenuously.) Yes, an appreciation of the proclivities of others serves as precondition or trigger to alert the accomplished artisan to his own distinctive powers. But this woodworker is one who, in being fully present and so alive to the vital patterns in the tree, is irrevocably changed by the power of his resonant encounters with others. Crucially, then, Qing does not work *on* an object, though his carving knife is sharp; instead he works *with* it, "joining something of Heaven to something of Heaven," bearing witness to the essential unity of himself with the tree as living things. Thus, Woodworker Qing's account illustrates the mutuality of this exchange while casting it as divine. Evidently, an awareness of the singularly divine in human and nonhuman elevates each, leading to the "godlike" result. So while many recent books on the *Zhuangzi* repeat the old chestnut that Zhuangzi is "tranquil and inward,"[20] the text rather enjoins us to look outward and to engage.

That the wonder of it lies in looking to the task and working steadily, but cautiously toward it, a second anecdote in the *Zhuangzi* confirms:

> Butcher Ding was cutting up an ox for Lord Wenhui. With every touch of his hand, every leaning in of his shoulder, every step of his foot, every thrust of his knee — zip! zoop! He slid his knife along, and all was in perfect rhythm, as if he were performing the Mulberry Grove dance or keeping time to the Jingshou song. "Great! Fantastic! Your skills have reached such heights!"
>
> Butcher Ding laid down his knife and replied, "What I care about is the Way, and that goes further than skill, certainly. When I first began to butcher oxen, what I always saw was the form of the ox itself. After three years, I no longer saw the ox as a whole. And now — now I let my spirit encounter it, rather than using my eyes to perceive it. Understanding through my sense organs has come to a full stop, and that spirit goes where it will, relying on the qualities found in the whole ox, following the inherent patterns endowed by Heaven. I strike in the big hollows, channel through the great openings, cleaving to what is fixed already. As for my technique passing through the smallest ligaments and tendons — well, never once has that happened [meaning, I would not try *that*]; still less would I chop through the large joints.
>
> A good cook changes his blade once a year — that's because he cuts. A mediocre cook changes his blade once a month — because he hacks. I've had this knife of mine for what must be nineteen years, and I've cut thousands of oxen with it, but the blade is as new as if it had just come from the grindstone. There are spaces between those joints, and the knife blade really has no thickness to speak of. If you insert something with no thickness into a place where there is space, there turns out to be more than enough room for a knife to play around in. That explains why the blade of my knife really is just as sharp as the day it came off the grindstone.

That said, whenever I come to a complicated place, *where I see I will have trouble*, I am on my guard and most cautious. I keep my eyes fixed on what I'm doing, and I work quite slowly, nudging the knife the smallest little bit, until flop! The whole thing falls apart like a clod of dirt crumbling. I then raise my knife and stand up straight, looking all around me, somewhat reluctant to move on because of the immense satisfaction I experience from achieving my objectives. Then I wipe the knife clean and put it away."

"Excellent," said Lord Wenhui. "Having heard your words, Butcher Ding, I will now be able to use them in nourishing life and vitality."

Butcher Ding is one of the five best-known stories in readings of the *Zhuangzi*,[21] and deservedly so, for in that story, Zhuangzi lays out the ways to "nourish life and vitality" (*yang sheng* 養生) — ways that entail knowing when trusting to gut intuitions will serve and when it is better to step back to try to gather more information before proceeding slowly (being confronted with "a complicated place"), with due attention to the particularities of the situation. Such ways entail seeing, really seeing, something beyond oneself. Many write as if they believe that the Butcher Ding story is all about effortless flow and fluency. After all, they reason, the story likens Butcher Ding's maneuvers to "perfect rhythm," "performing the Mulberry Grove dance or keeping time to the Jingshou song," with even the final thud produced by flesh falling from bone hinting at the inspired attunement existing between the butcher and the ox. There are rhythms to the craft and his artistry. And Butcher Ding himself says that he usually lets his spirit (and inspiration?) lead him where it may, willing or just letting his conscious "perception and understanding" to come to a halt.

The story as written is considerably more demanding, however. As the *Zhuangzi* phrases it elsewhere, "The sage takes his ease only where it is suitable, and is uneasy and insecure where suitable."[22] On closer reading, it becomes more obvious that a person's spirit and his intuitions, howsoever fine, can carry him only so far in life. The butcher, albeit well practiced and insightful, admits to encountering places that indicate trouble ahead. At that point, he takes extra precautions: he looks more intently into the whole animal, applying his powers of perception, his undivided attention, to the specifics of the situation at hand. After that, he proceeds very slowly, tentatively edging the knife forward bit by bit, feeling his and its way around the hidden places, for knife and butcher are now essentially one. In other words, the foremost expert has recognized that his intuitions do not suffice, and neither an expert solution nor a heuristic rule will serve him in this tricky situation; instead, he must switch to a slower and more deliberate and effortful form of thinking.[23] The expert's modus operandi can be fast or slow, in other words, depending on how much effort is required for him to proceed with exemplary caution beyond his initial spontaneity. With prudence and foresight supplementing and modulating the impulses in play — such precautions being learned through long years of disciplined practice — he finally manages to dismember the ox without seeming to perpetrate any violence upon it or his knife. His spirit energies have discerned the energetic paths within the ox, and he and his knife are guided by them for one brief shining moment. Once the ox falls apart as easily as a clod of earth, Butcher Ding, registering the triumph, pauses for a moment to acknowledge the wondrous beauty of it all. And then he changes gears, attending to the next job of caring for his tool, to which he will also give his undivided attention.

Apparently, even the most accomplished people can never reach a transcendent, stable state where continual engagement with disparate things is entirely effortless action. A person schooled in "mastering life" can at best reach a sane point of balance where he operates, for a time, effectively and in thoroughly satisfying ways. As Zhuangzi notes, "In antiquity, what they called a 'noble recluse' was not the man who hid his person and refused to show himself, nor the man who stanched the flow of talk and was not forthcoming, nor the man who kept his knowledge and erudition in reserve. It is a great error that the current age speaks this way."[24] The inescapable conclusion: keeping oneself as open as possible to mutually transformative exchanges — rather than seclusion or withdrawal — is the single best way for human beings to nourish life and vitality.

Kindred insights can be culled from a number of other pleasing *Zhuangzi* stories. One of the most memorable appears, not coincidentally, in the "Attaining Life and Vitality" chapter, where Kongzi/Confucius is off on a sightseeing tour, during which he misconstrues a situation:

> Kongzi had gone to take in the sights at the Lüliang rocks, where the water falls from a height of thirty fathoms and rushes along for forty *li* with so swift a flow that no large turtles or water lizards or fishy creatures of any kind can swim in it. There he saw a tall man swimming in it, and he mistakenly assumed

that the man had found life so bitter that he wanted to die. So Kongzi ordered his disciples to line up in the direction of the current, so that they could haul him out before he drowned. But after several hundred paces, the man emerged from the water with his hair streaming down his back. He strolled along, singing a song, as he ambled at the base of the embankment. Kongzi followed him, and asked, "I first took you for a ghost, but upon further investigation I see you are a man. I beg to inquire whether you have any special way of making your way through the water?"

"No, no special way. I began from what I was used to; I developed my inherent nature; I then perfected the capacities conferred upon me. I go in together with the whirlpools, and come up with the rapids. I follow the way of water, rather than selfishly insisting on acting in my own way in it. That's how I make my way through water."

"What do you mean when you say that you began with what you were used to, developed your inclinations, and perfected the conferred?" The man replied, "I was born on dry land and feel safe on land; that's what I am used to. I grew up in the water and came to feel safe in it; that's what I identify as second nature. I don't understand how I do what I do; that's the destiny I was dealt."

Once again, a superb master of technique denies that he is special; to his way of thinking, he acts with no special art. He takes things near to hand (that is, familiarity with the dry land, the separate qualities of the water, and his own habits and talents), and he just keeps doing what he finds pleasurable until doing it well becomes second nature. "I don't understand how I do what I do," he confesses. He sees only that he has been given opportunities and talents by some unknown power, whether on land or in the water. For lack of a better explanation, he credits his miraculous way of operating in the world to the "destiny" he was given. He has trust in the water and trust in the process, and that trust helps to keep him afloat.

All three stories (and many others like them) employ ordinary language to convey Zhuangzi's extraordinary conviction that the inexplicable and ineffable within the most mundane actions gives us pleasure. The Zhuangzi persona claims to know a few things, in other words. Usually, we categorize Zhuangzi or the *Zhuangzi* compilation as skeptical; "agnostic" might be more accurate, insofar as the text prefers to reserve judgment on whether any unseen forces, including the deceased ancestors, the gods, and ghosts, exist or intervene in the realm of the living. Nonetheless, even the small selection of passages translated above — not to mention many others[25] — repeatedly affirm the ultimate value of life while directing attention to an unseen realm where monumental things happen, even if the living cannot possibly discover the motive force behind the happenings. Recall such lines as "thinking it the work of the gods"; "deem the object to be a work of the gods"; "I let my spirit (or indwelling god, a daemon?) encounter it" and "go where it will."[26] Elsewhere, the *Zhuangzi* muses,

> It would seem as if the myriad things have some True Master, and yet I find no trace of it. It can act — that much is certain. Yet I cannot discern its form. It has an identity, but no form.... There seems to be something that responds — so how can we say there are no spirits? There seems to be something that does not respond — so how can we say that spirits do exist?[27] ... I try to discover who or what is doing it, but I can't get the answer.... It must be my lot.[28]

Why would the *Zhuangzi* tarry over such a puzzle? Clearly, it is no easy matter to make the gods or the divine "objects of study" or "explanations." Since other early texts urging a kindred suspension of belief about the extrahuman deploy similar language, early thinkers evidently needed a catchphrase for the mysterious processes, if only because too many turns of events remained deeply puzzling. (Plato's *Protagoras* records comparable arguments.) For example, in the classical era, no ready answers existed for the onset of various afflictions, physical and mental, that left even the high-ranking feeling acutely vulnerable. Nor could facile reasons adequately account for most sudden reversals of fortune.[29] To refer to the gods or spirits (*shen* 神), for thinkers like Zhuangzi, was to provide a placeholder for "forces working in invisible ways that nonetheless exert dramatic, hence visible impacts upon the social world."[30] For to extend and adapt language once reserved for the unseen powers, the gods and ancestors, simultaneously acknowledged the confusions endemic to phenomenal existence and registered the futility of launching further inquiries. The "Supreme Pleasure" chapter shows that the *Zhuangzi* intends to signal and register these very points:

> Day and night the emotions replace one another right before our eyes, and no one knows the place where they have germinated. Enough! Enough! It is enough to know that we get these, morning and evening, and they are the means by which we live. Without them, we would not exist. Without us, they would have nothing to take hold of. *This* comes close to the heart of the matter, even if I do not know what drives them.[31]

Thinking people then, as now, continually confronted the awful prospect that they must somehow determine how to act within the context of situations exquisitely in flux, without adequate information to guide them in their deliberations, because time was running out. Trust in some intelligible order, however ill understood, as premise for concerted action has often served those loath to indulge in reckless behavior. Forethought would be pointless, because the situation so seldom matches anticipations,[32] even when an agent subscribes to the dubious belief that even partial understanding of larger patterns at work permits the wise to plan acts and gauge priorities. Nonetheless, as the modern philosopher Donald Davidson observes, "when we ask why someone acted, we want to be provided with an interpretation ... which fits into a familiar picture ... which *redescribes* the action [so as to give it] a place in a pattern" easily assimilated to our current world view.[33] Thus, any resort to explanations about the likely truth of an emerging "pattern" is probably circular, because the pattern has already been deemed to work in the past. Fully alive to the circularity behind most appeals to logic, Zhuangzi nonetheless has few options beyond pointing to prevailing patterns in the world that he and his readers think they see or know or feel, from experience, practically speaking.

No less germane to any competent reading of the *Zhuangzi* compilation is an appreciation of the resonance theory of its time.[34] As numerous passages assert, vital exchanges and operations occurring on unseen and seen levels sustain us ("Without them, we would not exist.... They are the means by which we live").[35] By implication, human beings endure grave, probably irreparable losses as soon as they fail to see the connections between life's riches and their own existence, for once the sorry paths we tread become thoroughly ingrained in us, real receptivity to life's riches is no longer possible. We become "closed off, as though set with seals."[36] Acknowledged or not, the interconnectedness of all the myriad things, tying microcosms to macrocosms, is one of the few assertions the *Zhuangzi* insists upon.

The Woodworker Qing story, in particular, illustrates resonance theory beautifully. Like the early Greeks, the early elites in China accepted that all sensation through eye, ear, mouth, or skin occurs after "direct contact" between the sensory organ and a thin "material effluence" or film (in Greek, *eidola* or *simulacra*, images or simulations) emitted from the object to the senses. By the time of the *Zhuangzi*, this effluence had come to be called *qi* 氣 ("breath," "air," "animating spirit," or "pneuma," for lack of better translations). By the prevailing resonance theory, every human body with perceptual equipment does two things: each continually sends something out, thereby giving an indication of its peculiar character, even as it continually "waits for some sensory effluvia to come to it in return."[37] (See Chapters 1 and 2.) Early music theories took it into account that a string always vibrates, producing the same note, in response to a sound made by a tuning fork or stringed instrument placed at some distance.[38] Here was "proof" that the animating impulses in people and things operate by laws of sympathy across time and space within the wider macrocosm, with like seeking like and dissonances repelling. Where reasoning typically requires separation and autonomy, the disjunction of subject and object, resonance entails adjacency, similitude, sympathy, and the collapse of the boundary between perceiver and perceived.[39] As the Woodworker Qing story puts it, "With the bodily forms [of the woodworker and the nascent bell stand in the wood grain] fully present, then and only then do I see a bell stand with perfect clarity, after which I apply my hand to it.... In this way, I join something of Heaven to something of Heaven."

Woodworker Qing's account of conjoining something of Heaven [the unique configurations in the tree patterns] with something of Heaven [the unique patterns within Woodworker Qing] obviously falls outside notions of mechanical cause and effect in important ways. The tree neither imposes itself upon the woodworker nor the woodworker upon the tree, yet neither remains the same. Nor is there a unilinear movement, as there would be when nudging a line of dominoes or trusting to contraptions.[40] Rather, Woodworker Qing's careful preparations render him sufficiently receptive to unseen currents that he can sense the tree's own readiness to convey something of its particularity to him, even as he gradually lays himself open to transmit something of himself to the tree. The result: this human being becomes palpably full of life, because of the connections he has forged with things outside himself that feed his vital energies.

Because the foregoing has sketched the attributes of the Fully Present Man (*zhi ren* 至人), a brief word is in order on the figure of the True Man (*zhen ren* 真人), aka the Holy Man of Old, as a familiar concept well before the time of the *Zhuangzi* compilation.⁴¹ Down through the ages, these models have been solemnly put forth as synonymous, worthy goals for people to aspire to, even if they are but Supreme Swindles.⁴² I would like to suggest a heresy: that a waggish Zhuangzi is having a bit of fun at the expense of the older, religious notions crediting the monumental figures of legendary antiquity with virtual divinity ("Are not such men as I've described gods in Heaven?" 所謂[真]人之非天乎 ... 至人神矣),⁴³ rather than promoting such exemplars in all situations with deadly seriousness. After all, the True Men of old, who are possibly immortals, supposedly come equipped with magical powers: they can "enter the water and not get wet, enter the fire and not get burned." They "have no love for other men" and "nothing much that they hold very dear."⁴⁴ One might have more confidence in the incredible stories told about the True Men in the *Zhuangzi* if the qualities ascribed to them tallied better with Zhuangzi's sketches of true friendships or if the marvelous qualities of the spontaneously generating "cosmos" were ever said to be mirrored in a human capacity for "spontaneity," innate or acquired. In general, one can say the same for most of the occurrences where the Fully Present Man (apparently Zhuangzi's invention) turns up — except for a few notable passages, such as Woodworker Qing, where the Fully Present Man seems to bespeak a lesser goal, something more down-to-earth and approachable: the goal of "being there," that is, alive to the full spectrum of realities because one is less preoccupied with one's own ego (*wu ji* 無己).⁴⁵ Occasionally, the *zhi ren* is interested in helping others, evading the usual giving and receiving of blame, and "roaming in supreme pleasure."⁴⁶ For contra the "common wisdom," the wise man, in Zhuangzi's telling, learns to live contentedly within the constraints of the hand he has been dealt in life. He does not waste his days yearning for impossible perfection; neither does he aim for full awareness, because that is no less attainable. Obviously, in some circumstances, the "fully present" person may briefly display a remarkable virtuosity, as do Butcher Ding and Woodworker Qing, but that condition never lasts, nor should anyone expect it to, despite the immense satisfaction derived from it. After all, a holdover would interfere with making adequate preparations for the next situation that is unfolding.

Supreme Pleasure: What It Means

In the opening passage of the chapter entitled "Supreme Pleasure," the *Zhuangzi* confronts us with the question of what it takes to feel more alive, on the assumption that we intuitively move toward whatever we take (or mistake) for pleasure-enhancing activities, objects, and people while eschewing whatever we calculate will diminish pleasure, now or in the future. Because the passage raises so many of the issues to which the *Zhuangzi* returns again and again, it seems best to quote it nearly in full:

> In our world, is there such a thing as supreme pleasure or is there not? Is there something that may be used to enliven our persons or not?[47] In today's world, how are we to act, and how are we to make a secure foundation? What are we to shun or to make our place? What are we to go for and what are we to eschew? What are we to regard as pleasure and what are we to detest?[48]
>
> Now, as we all know, the world esteems riches and high rank, long life and others' approval. Bodily ease, rich flavors, gorgeous robes, attractive sexual companions, and fine music — these it regards as pleasurable. What it denigrates are poverty, low station, early death, and social censure. What it holds to be bitterly hard is the body deprived of rest; the mouth deprived of rich flavors; the forms deprived of gorgeous clothes; the eyes deprived of sexual companions, and the ears deprived of fine music. Those unable to get these pleasures are greatly aggrieved and cowed. This is a stupid way to act on behalf of the body, however.
>
> Now, as we all know, the wealthy drive themselves hard and frantically engage in activities so as to accumulate more wealth than they can ever possibly use. This is to treat the body as something external or alien. And as we all know, those who want to be paid honors spend their nights and days calculating what may bring them social approbation or disapprobation. This likewise is to distance oneself from the realities of the physical body. Anxiety accompanies life itself, and the elderly grow dull and doddering. To endure long-standing worries and not die, what suffering is this! Such calculations on behalf of the outer bodily form stray far from its proper aim. Why are they so very hard on themselves? For this [daily struggle of the ambitious] likewise creates a gap between the person and the realities of the body.
>
> Stellar examples of courage [*lie shi* 列 (=?烈)士] act so that the world will approve them, certainly, but they fail to do enough to keep alive. I have never known whether others' approval is truly a good thing or not. But let us treat

it as a certain good for the sake of argument. Then we still find that approval does not suffice to enliven the body or keep it alive. But if, on the other hand, we treat it as an evil, that may suffice to keep some people alive.[49] Hence the proverb, "If your loyal remonstrance goes unheard, give way and do not fight it." In effect, it was Wu Zixu's stubborn resistance that destroyed his own person. But had he not fought, his reputation would not have been made. So is there truly some good to making a fine name for yourself or not?

How could I possibly know whether what the common run of men identify as "pleasure" is indeed pleasurable or not? Whenever I contemplate the conventional sorts of pleasures and the herd's pursuits, all I see is those people scrambling and dodging around, seemingly unable to get a grip on themselves. Still, they all, to a person, would identify this sort of pursuit as "pleasurable." I do not know whether to regard this as a pleasure or not. In the end, is there such a thing as pleasure or not? I regard activities without overt pleasures[50] as true pleasures, certainly, but this mode of action is precisely what conformists find a bitter pill to swallow. Hence my saying, "The greatest pleasure is no [conventional] pleasure, and the greatest fame, no fame."

In the world, rights and wrongs, what is so and not so cannot be determined, in point of fact. Even so, activities without fixed goals and polarizing effects (*wuwei* 無為)[51] prove of *use* when settling what is so or not so. When it comes to the greatest pleasure and enlivening one's physical person, one is *nearly there* if one acts without fixed goals and polarizing effects. For such activities are *nearly enough* to sustain or preserve a person.

Let me try putting this into words: Heaven's undirected activities are the reason for its limpid quality, and earth's undirected activities are the reason for its settled quality. Together, these two types of undirected activities produce and transform the myriad things in all their variety. Such a marvelous jumble, with no single origin. Such a marvelous jumble, taking on no specific shape [permanently]. The myriad things carry on, each attending to its own work.[52] All of them flourish by drawing upon the formless [the unseen patterned operations in the cosmos]. Hence the proverbial saying, "Heaven and earth do not toil, yet nothing remains undone." But what human being, pray tell, ever really manages to attain this same sort of activity without specific goals and impulses to intervene 得无為?[53]

No matter what the current shape of phenomenal existence, it will change, sooner or later. Of all the creatures in the universe, only people ask why this is so and whether the processes will come to an end. To formulate and try to answer the question distracts them from the tasks at hand. The cosmos will sustain us, as surely as the earth, until it does not, and the transformations will take place whether we heed them or not. Better to trust to the unseen order, then, for the time that remains.

As we will see, the *Zhuangzi*, in an expansive vision that incorporates the fact of human limitations and misapprehensions, offers two spectacular promises if we can eschew the fixed goals, with their polarizing effects, of the common herd: that the clarity of vision associated with a divine spark (*ming* 明, *jing* 精, or *shen* 神) can infuse daily life with a profound vivacity, vitality, and vividness of perception and that these positive infusions, in turn, can induce a pleasurable awareness of wholeness, integrity, and well-being (*quan* 全 or *cheng* 誠).[54] Far better to appear slow-witted ("being in place, like heavy and inert bodies," in company with Zhuangzi's famous clods) than to join the febrile rush after the latest fashions and lifestyles. Hence the *Zhuangzi*'s catalogue of signs and signposts for the vital spark and any directions or modes of activity likely to preserve it. As one passage states plainly, "The ordinary man puts great weight on profit, and the swordsman for hire on reputation. The worthy man of service upholds his ambitions. The sage, by contrast, places the highest value on vitality itself."[55] No relativism is in sight,[56] for in the hierarchy of values and motivations, the pursuit and maintenance of vitality in life is to supersede every other consideration. Unfortunately, the goals embraced by the common run of men place an unduly high priority on securing luxuries and sensual delights, ego-feeding triumphs, and afterlife reputations.[57] But even from the most selfish standpoint, the disadvantages of giving oneself over to these conventional pursuits should be obvious enough. For in their drives to secure an abundance of material and psychic goods deemed gratifying, people engage in a range of frantic endeavors so laden with anxieties and likely failures that they forget, in the end, to "know what is enough" when taking stock of life's pleasures.[58] (Here one thinks of Gilbert Ryles's "thrills, twinges, pangs, throbs, wrenches, itches, pricklings, and so on.") Thus they blindly consume their vital stores of energy, rather than keeping or replenishing them, while leaving themselves open to states of craven dependence upon the whims of others. No measure of success can provide long-lasting relief, since hedonic adaptation insures that people will not derive satisfaction from it for very long.

Not surprisingly, perhaps, Zhuangzi's portrayal of the typical human condition borders on the bleak:

> In their sleep, they make contact with other souls, and in their waking hours, their bodies hustle. With everything they meet they become entangled. Day after day, they use their minds in strife. Sometimes they are arrogant, and sometimes coarse or secretive.[59] Their petty fears are mean and trembly; their great fears leave them stunned.[60] They bound off like an arrow or a crossbow pellet, thinking themselves the arbiters of what is right and wrong. They cling to their positions as though they had sworn to uphold them before the gods, saying to themselves that they will grasp some victory through this. Their destruction comes like fall and winter: such is the way their days melt away. They drown in what they do. They cannot be made to turn back. They grow closed off,[61] as though sealed with seals: such are the deep channels cut by old age.[62] And when their minds draw near to death, nothing can restore them to the light.
>
> Once a man receives his complete bodily form, he keeps it, waiting for the end. Sometimes clashing with things and sometimes bending before them, he completes his course like a galloping steed, and nothing can stop him. Is he not pathetic? To the end of his days he sweats and toils, never seeing his own accomplishments. He utterly exhausts himself and never knows where to find a true refuge. Can you help pitying him? "I'm not dead yet!" he says, but what good is that? His body decays and his heart and mind follow. Can you deny that this is a great source of sorrow? Man's life has always been a muddle like this. How could I be the only one who is muddled, while other men are not muddled?

Here we discern our own likeness: desperately seeking approval, in life and in death, to get our rewards or make our reputations, and to leave something behind, driven perpetually by externalities. Elsewhere, Zhuangzi considers those who are unduly preoccupied with success or with stifling family ties. "They please other people, but do not have the pleasure of pleasing themselves."[63]

Nevertheless, on this grim assessment Zhuangzi grounds his talk of the supreme pleasures afforded by greater clarity. For acuity in the *Zhuangzi* means, above all, seeing how little we see and how little we know, given our limited perspectives and powers as human beings. By one rhetorical tour de force after another, the *Zhuangzi* circles around variations on a single theme: that we should exercise due caution when tempted to intervene or impose our views on others, given that we are in the dark about so much of phenomenal existence, including our own motivations. Temporary suspension of belief is ideal, but people usually have to act on the basis of their fallible assessments of a situation in order to earn a livelihood. To elite men of Zhuangzi's station (the first readers of the *Zhuangzi*), to delay decision-making or taking action was seldom an option; still less available would be leaving family and court behind. Such suspension as can be mustered, however, could prove to be the key to survival in high places and, more crucially, flourishing. "It is easy to keep from walking; the hard thing is to walk without touching the ground."[64] Since quiet balance and the light it affords are not that easy to embrace, due preparations must be made for the long journey.[65]

Sweet Release and Clarity via the Uncertainty Principle

People can be cured of at least some of the ills that plague them, so long as they can be persuaded to set aside some of their cherished convictions about value in the world, and it matters little whether they transform their views or merely suspend them for as long as useful. In either event, so long as humans "try on" Zhuangzi's vision, just for the time being, they become more indifferent to all that the world holds dear—in large part due to the sudden realization that, as Zhuangzi remarks in the "Supreme Pleasure" chapter, "it does not seem possible to live—let alone, live well—while behaving as they have been behaving."[66] Thus, *for strategic purposes* rooted in the contingencies and unknowns of each individual situation, the *Zhuangzi* urges an embrace of uncertainty as a method to conserve or even increase the vital energies while reducing needless risks.[67]

The *Zhuangzi* delivers five neat jabs to puncture our vast reserves of complacency by asserting that the human predicament mandates that no person is so powerful, politically, intellectually, or morally, as to be able to escape human limitations; that death is likely an improvement over the miserable existences eked out by the careerists (if not over a life of supreme pleasure), and since it is inevitable anyway, any reduction in the fear of it will likely mean a correspondent reduction in desperation or defiance with a concomitant potential for greater openness to the other personal, social, and cosmic realities; that experiences in life are like dream scenarios, hard to parse and quickly shifting, so continued trust in the wisdom of our judgments is sorely misplaced; that while language may be the best human tool for thinking and communicating, it also limits understanding, because it never fully captures the range or varieties of human experience; and finally that consequently, the wise learn to accept the fact that all the living wander about "in a daze," befuddled and muddled in the deep unknowns. After all, like a fish that finds itself out of water and gasping for breath, human beings who have left the elements where they can easily swim may, with luck, still manage to survive, if given just a dipperful of water.[68] And if revived, the promise is this: they can learn to wander in bemusement. With the "underbrush" cleared from the path, the journey becomes appreciably more enjoyable, especially when one gives up interfering with others (*wuwei*) in order to reach a specific goal—though Zhuangzi is not so silly as to promise surcease from pain and care.[69] As the *Zhuangzi* puts it, "Great understanding is broad and unhurried, in contrast to petty understanding, which is cramped and busy." To sketch these practices, this section first reviews passages that allude to or illustrate each of the first four teachings, then follows up with the fifth, the probable payoff for living strategically.

No Entity, However Exalted, Escapes Limitations

Repeated passages in the *Zhuangzi* "prove" that no person or thing escapes his lot and limitations, so the best possible outcome is to accede to this sumptuous reality and then get on with it. The *Zhuangzi* compilation opens with an image of incredible power: a gigantic Kun fish whose body spans thousands of miles in space is transformed in a flash into a Peng bird of equal size. When that bird rises up and flies off, his wings appear to be like clouds stretching across the heavens. "Beating the whirlwind," his wings roil the waters below and the skies above, setting off "wavering heat, bits of dust, and living things blown about by the wind." Nonetheless, the sky continues to look very blue. "Is that its real color," the *Zhuangzi* innocently asks, or does it merely look like that because "it is so far away and without any ends"? Zhuangzi notes that "when the bird looks down, all it sees is blue, too."[70] Apparently, the Peng bird—despite its extraordinary size and strength, despite its godlike ability and the superior vantage point from which it surveys all phenomena—can see no more and no less than the mere mortals below.[71] No living thing, it seems fair to conclude, can wrest an accurate picture of anything. How ludicrous to expect an absolute or lasting perception of things whole and intact, seen from every conceivable angle, while passing swiftly over or through the land as one of the continual things in flux. No one can align their senses, as is evident from this fabled creature streaking across skies and deeps.⁷² Nor can any living thing ascertain precisely the origins of its perceptual problems (Figure 5.1). The limited cannot limn the limitless nor see the wind piled high, as the sage Lao Dan avers: "Day after day these things go forward, yet no one has ever seen anything bringing them about."⁷³ And let us not forget that the Peng bird changes, almost certainly without intending to, into the Kun as the seasons turn, just as humans grow old and die, whether they will this or not. So the Peng bird, incredible and majestic as it appears, reminds us of home truths that ground many of Zhuangzi's astute observations.

Having established this picture, the *Zhuangzi* lodges three related claims: that for Master Zhuang, "seeing" for human beings implies noticing plus evaluating, and since all evaluations rely upon seeing things from a particular vantage point, all evaluations and thus all forms of seeing are skewed, or worse.

> If you look at things from the point of view of their differences, they will be as liver and gall, or Chu and Yue [completely different sites]. But if you look at them from the point of view of their similarities, all the myriad sorts of things are one. Now, a [worthy] man ... does not know what his ears and eyes should deem appropriate. He lets his mind wander in the harmony of what things are [*de* 德].... Seeing things as one, he does not notice overmuch their [individual] losses and deaths [*sang* 喪].⁷⁴
>
> He who knows he is a fool is not the biggest fool, nor is he who knows he is confused in the worst confusion. The man in the worst confusion will end his life without ever getting it straightened out, and the biggest fool will end his life without ever seeing the light.... And if I know something does no good and I still make myself do it, this, too, is a type of confusion. So it is best to leave things alone and not force them or me. If I don't force things, at least I won't cause anyone any worry and trouble.⁷⁵

In Zhuangzi's estimation, there is always too much to know and too little time, energy, and talent for perfection, so no worthy person feels compelled to show off (*shi* 示) his opinions in public debates and quarrels.⁷⁶ Equally obviously, no worthy person adheres to a deluded belief in the reliability of his senses and powers. However, his chances of seeing more of the world increase in proportion to the steadfastness of his refusals to be puffed up by or preoccupied with fruitless self-regard (*zishi* 自視).⁷⁷ True friendships, for instance, require consistent

Figure 5.1. Adrian Gordon, untitled photograph, 2017. Reproduced by permission of the artist.

Gordon uses no special filters or lenses in her camera work; what viewers see is what her naked eye is seeing. The complex interplays of layered fragments of motifs recalls Zhuangzi's remarks about our inability to capture the totality of any phenomena and make sense of it. Nonethless, the viewer's percepts seem no less real for being phantoms or projections.

looks to others (*xiang shi* 相視).⁷⁸ In stark contrast to the usual run of men (many in high places), the worthy man neither invites others to see him nor sees others from the vantage point of convention; he does not objectify others or welcome being the object of another's intrusive gaze.⁷⁹ As a result, while he may have something of the look and demeanor (*mao* 貌) of an ordinary man about him, he leaves others "no way" to physiognomize" (*xiang* 相) him, that is, to measure his surface likeness in order to predict what his future will hold.⁸⁰ As we might say today, he travels under the radar, "leaving no traces."

That no percept escapes distortion⁸¹ does not mean that the worthy man feels free of ordinary commitments; he can hardly afford to be *that* deluded.⁸² Practically speaking, in Zhuangzi's world, the literate population could at best hope to live without undue impediments or burdens, contra the many readers (not only Buddhists and libertarians) apt to prefer a Zhuangzi hawking a form of lofty detachment easier for moderns to swallow from their gated communities.⁸³ In Zhuangzi's time, such men and women could not have freed themselves from the holds of their families or their states upon their thinking, perceiving, and acting, nor did they necessarily long to do so.

> In the world, there are two Great Decrees: one we name destiny and the second we name duty. That a child should love his parents is fated. It is not possible for a child to be released from those ties in the heart. That a subject should serve his ruler is duty, we say. There is no place where a subject can go where there will be no ruler. There is no place in the cosmos to escape from that. Hence the name "Great Decrees."⁸⁴

In antiquity, members of the small governing elite would have been less apt to parade their "autonomy" than moderns, with autonomous action often viewed as selfish and dangerous. Certain situations might call for independence of thought and action, but the glory invested today in the idea of the "self-made man" was unthinkable, even to a Robber Zhi sporting noble birth. From childhood men and women grew up among dependent relations among kith and kin, which patterned professional lives, patron-client bonds, and marriage alliances, an occasional irksomeness notwithstanding, and, in any case, relative freedom or autonomy was granted or earned within a hierarchy of powers.⁸⁵ No early thinker in China yearned, *pace* some of Zhuangzi's recent translators and interpreters, to transcend the world to become an *Ubermensch* "unbound" to others, this being a thoroughly modern construction.[86] (The most famous "recluses" in pre-Han and Han times boasted hundreds, if not thousands of disciples and hangers-on.) And while, in the post-Freudian age, the language of everyone loving his parents may trip up the unwary reader, the phrase translates, with little effort, into the line that no child gets to choose his or her parents, whose images will nonetheless remain constantly in the mind's eye. And beyond those proximate duties to family and state, each person participates in the larger cosmic workings as but one of the myriad things, an insight that undercuts the main impetus for erecting strict hierarchies ranking faces, forms, voices, and activities, not to mention the great transformation that is dying.[87] Zhuangzi goes "six degrees of separation" one better, then. On that insight much can be built.

Fears of Death as Limitations on Life
The *Zhuangzi* insists that death may not be worse than living, because it is probably only a different state of being. Besides, it comes to all uninvited. People are "now alive, then dead, now dead, then alive."

> Here is a man of the Central States, neither yin nor yang, living between Heaven and earth. For a brief time only, he will be a man, and then he will return to the primal ancestors. Look at him from the standpoint of the Source, and his life is a mere gathering together of breath/*qi*. And whether he dies young or lives to a great old age, the two fates will scarcely differ—a matter of a few moments, you might say. So how, then, could such distinctions suffice to substantiate any rights ascribed to a sage-king like Yao or any wrongs ascribed to an archetypically bad ruler like Jie?... Man's life between Heaven and earth is like the gallop of a white colt as glimpsed through a crack in the wall—whoosh, and that's the end.[88]

The bottom line: "How do I know that loving life is not a delusion? How do I know that in hating death I am not like a man who, having left home in his youth, has forgotten the way back?"[89] The *Zhuangzi*'s advice is fairly modest: in small, homeopathic doses, we are to examine all aspects of daily life, especially death itself, so as to live in full consciousness of what we cannot ever explain or hide from.[90] Besides, loss reminds us to notice, transience to cherish, fragility to remain humble.[91] A man's stops and starts, his life and death, his rises and falls, "none of these can he do anything about, yet he thinks he may hold the key to mastery over them!"[92] "Attach greater value to life and living well!" is Zhuangzi's advice. "Surely doing this has greater value than the largest and most luminous of pearls."⁹³ (Guo Xiang's commentary to the "Supreme Pleasure" chapter seems helpful here: "Only when worry is forgotten does life become pleasurable, and only when life has become pleasurable is this bodily form truly something that belongs to me.")⁹⁴ To live with full vitality is to have the emotions and thought processes "never taste death" prior to death's own arrival, since the anticipation of death tends to cast a pall over one's appreciation of life's blessings.⁹⁵ That explains the statement, "There is no grief greater than the death of the heart and mind. Beside it, the death of the body is a minor matter."⁹⁶ And "only if one truly understands how to live *and* die does one prevent greed for life and fear of death."⁹⁷

Mortality remains the single biggest hurdle to seeing that major transformations will happen all the time, regardless of individual desires. For most, the strategic decision to "live less in your head and more in your body," to try to be more unfazed by change, serves as the first and most difficult step in the process of moving beyond fear of death. The tale of Zhuangzi's mourning for his wife—typically misread as a sign of Zhuangzi's disinterestedness and detachment from the mundane—shows that adults are generally less aware of the need to take proper care of their bodies than newborns, for even newborn piglets are quicker to abandon a dead mother than Zhuangzi is to leave off bemoaning the loss of his companion. The piglets, far more in tune with their bodies' needs, soon recognize they must turn elsewhere if they are to find the nourishment they need to continue living. Again, *time* and timing are essential factors in deciding what to do, for it is critical not to nurse emotions, prolonging them after they would otherwise have dwindled. As Zhuangzi says elsewhere, "Beasts that feed on grass do not fret over a change of pasture, nor do creatures that live in water fret over a change of stream."⁹⁸ Not coincidentally, the "Supreme Pleasure" chapter relates story after story of legendary heroes, including Zhuangzi himself, forced by circumstances beyond their control to cope with their own mortality and that of their loved ones.

It is not that people can or should become inured to death. The Zhuangzi persona mourns his wife, although scholars are apt, like Huizi, to mistake his informality as sign of his lack of emotion.

> When Huizi went to convey his condolences after Zhuangzi's wife died, he found Zhuangzi sitting with his legs sprawled out, pounding on a tub and

singing. "You lived with her, she brought up your children, and grew old with you," Huizi chided. "It should be enough not to weep at her death. But pounding on a tub and singing — that's going way too far, surely!"

Zhuangzi replied, "You're wrong. When she first died, do you think I didn't grieve like anyone else? But I came to look back to her beginning and the time before she was born; not only the time before she was born, but the time before she had a body or even a spirit. In the midst of that jumble of wonder and mystery, some change took place, and she had a spirit. With another change, she had a body; with another, she was born. Now there's been another change, and she's dead. It is just like the progression of the seasons. Now she's going to lie down peacefully in a vast room. Were I to follow after her, bawling and sobbing, it would show that I haven't a clue about what is ordained. So with that, I stopped [my public lamentations]."[99]

To everything there is a season. In doling out such advice, the *Zhuangzi* usefully distinguishes between emotional suffering, on the one hand, and death and chronic disabilities, on the other, since it is anguish that mainly grinds people down, even though it is largely avoidable. Whereas death, certainly, and some bodily pain are inevitable (no good opiates or antibiotics in Zhuangzi's time), a heap of misery stems from "your own mind making it so."[100] (Recall that "the problem of life's meaning cannot arise for an animal.")[101] Thwarted in her desires for largely or wholly unattainable goals (for example, immortality, omniscience, or permanent glory, achieved without exciting malice or envy), the person is apt to be outraged or depressed by the irreconcilability of the realities of the larger situation and inappropriate personal cravings. Self-inflicted injuries, no less attacks from an enemy, deplete a limited store of vitality. As Zhuangzi says, "when you treat your vital energies like a stranger or less central matter, you wear out your vital energies."[102]

Unfortunately, self-inflicted injuries come in all forms and shapes, great and small, as Duke Huan 桓 of Qi discovered when he fell seriously ill just after spotting a marsh ghost while out hunting, only to recover as soon as a court official persuaded him that the sighting was a good omen.[103] Paradoxically, holding onto the bodily form so tightly results in a looser grip on the business at hand, with the person "never knowing where or to what he can return."[104] That said, men born to the governing elite, the *Zhuangzi*'s intended readers, had to confront a great deal of pressing business, including family affairs and matters of state,[105] or else risk great danger for dereliction of duty. For better or worse, Zhuangzi's readers are in the thick of things ("in the world of men," as he puts it).[106] "Howsoever irksome, public affairs must be attended to.... Yes, ritual constrains, but it must be repeatedly practiced. Therefore, the sage ... responds to the demands of ritual and never shuns them; he disposes of affairs without making excuses."[107] "Can you afford to be careless?"[108] the *Zhuangzi* inquires. This need to attend carefully to duty while avoiding too much mental and emotional investment in a particular outcome — that is what the *Zhuangzi* enjoins. And it can be done, says Zhuangzi (here channeling a fictive Confucius of his making):

> Make your will one! Do not listen with your ears, listen with your heart and mind. No, better yet: do not listen with your evaluating faculties of thinking and feeling, listen with your vital energies (*qi*). Listening stops at the ears, and the mind stops at the point where it thinks it has found a match for reality. But *qi* is empty, in the sense that it waits upon other things and people. The Way gathers in such emptiness, which restores balance to the faculties housed in the heart.[109]

Many of the *Zhuangzi* commentaries, influenced by Buddhist asceticism, perhaps, speak of the highly ritualized purification of the mind, but enough early evidence exists to show that Zhuangzi calls for something less rigorous and less rooted in self-denial: a balancing of one's faculties, a refusal to stick unthinkingly by fixed rules and stances, these being required preparations for an attunement with less than obvious realities, not to mention pleasurable roaming beyond conventions.[110] Rational calculation isn't likely to elucidate much. In hopes of more illumination, better to trust to the animating spirit that imbues the *qi* and rebalances the faculties.

Experience Is Like Dreams
The *Zhuangzi* finds dreams fascinating, and not only because ancient theories held a person dreaming to be "a [wandering] soul coming into contact with others."[111] In Zhuangzi's day, as in modern English, "dreams" functioned as near synonyms for "ambitions." One justly famous tale (not coincidentally in the "Supreme Pleasure" chapter) plays upon both these ideas. While traveling, Zhuangzi happens to come upon a skeleton lying beside the road, at which point he muses about the possible ambitions that may have led to the death. That night, the skeleton appears to Zhuangzi in a dream, and after chiding

him for his stone stupidity, informs him that any person equally experienced with both phases would surely choose a peaceful death over life, with its endless succession of troubles.

That memorable episode notwithstanding, I used to hurry over the famous dream of Zhuangzi as butterfly, unsure why I should care whether it's Zhuangzi dreaming he is a butterfly or a butterfly dreaming that he is Zhuangzi. Quite belatedly, I see that the famous butterfly passage explodes the very categories we typically deploy for what we think we see. With the butterfly, the simulacra of the daily world in dreams are no pale and decidedly lesser manifestations of reality, as they would be with the Greeks or possibly the Mohists. Zhuangzi's butterfly is substantial, one prop of our complexly related multiple experiential worlds, since people's percepts, impulses, and desires seem no less real for being phantoms or projections. Thus, "he who dreams of drinking wine may weep when morning comes."[112] This upending of the ontological hierarchy of percepts and phantasms as one state gives over to another "we call the transformation of things." (Death, then, is but a dream, by such reasoning.) People always respond to external stimuli with genuine concern, no matter what the prompt, no matter how limited their capacities to address it. Dreaming or waking, it's all the same. To each world they perceive, people may respond with dim awareness, depending upon the time and inclination. Nonetheless, each encounter and response leaves palpable traces in the bodily habits brought to thinking and acting before yielding to the next sensation. (Fast-forwarding to the neurosciences, we learn that dreams, no less than waking experiences, carve our neural pathways, with an estimated 80 percent of our dreams nightmares, products of an overactive amygdala rehearsing potential situations that leave us tossing and turning.) To correlate the partial glimpses — the white colts racing by cracks in the wall — is a slow process, at best. So if Zhuangzi now sees himself as a butterfly, he cannot see himself as Zhuangzi, as well, and vice versa.[113] Moreover, honest people readily admit that things happen in dreams that they cannot begin to fathom, though they turn sheepish when making such admissions about scenes they take to be "real." And so, caught up in a tangle of overinterpreted and underinterpreted scenes, people dream their lives away,[114] and it matters not one whit to their well-being in the end whether or how well they separate the worlds they dream from their waking worlds. As the *Zhuangzi* muses,

What's more, we go around telling one another, I do this, I do that, but how do we know that this "I" we talk about has any substantive "I" to it? You dream you're a bird soaring up into the sky. You dream you're a fish diving down into the pool. But now when you tell me about it, I don't know whether you are awake or whether you are dreaming.[115]

The Limits of Language and Logic
In the awesome presence of a truly great person, those nearby, if they are astute enough, are at a loss for words.[116] No words can convey the awesome mystery of the Dao itself (that is, the totality of the cosmos, rather than an anthropomorphic force or perfect order),[117] and the highest pleasures resist our best efforts to describe them;[118] any person's memories and experiences are largely uncommunicable. At the same time, "Words are not just wind. Words have something to say. But if what they have to say is not fixed, then do they really say something or not? People suppose that their words differ from the peeps of baby birds, but is there really any difference or not?" This query comes from the chapter "Seeing All as Equal," where Zhuangzi grounds one of his most compelling discussions by conjuring the cosmic *qi* that breathes life and expressive form into each of the distinctive myriad things in succession before passing on. By likening human language there to the chirping of birds, Zhuangzi suggests that people cannot help but talk all the livelong day to those of their kind, feeling the urge to communicate *something, anything*, even if the messages they send out and take in vary mildly or wildly from person to person, given different memories and experiences, no less than from time to time and from place to place. As Zhuangzi says,

> Out of the murk, things come to life. With all your cleverness, you declare, "We have to analyze this!" You try putting your analysis into words, though it is not something to be put into words. You cannot, however, attain understanding.... Let me try to describe this analysis of yours. It takes life as its basis and knowledge as its teacher, but from there proceeds to assign "rights" and "wrongs." So the end result is "names" and "realities," with each and every man thinking himself the best final arbiter of abstractions.[119]

When we speak to others (and sometimes even to ourselves?), we might as well be speaking a foreign language badly, unaware of the possibly grave consequences of misspeaking or piping up out of turn. As Zhuangzi wryly remarks, most of the words we use to "explain ourselves" 自說 ("nine-tenths" of our babble, by his estimate) originated with others anyway.[120] Zhuangzi, one suspects, would have concurred with Wittgenstein's witty remark, "Say what you like, so long as it doesn't stop you from seeing how things are. (And once you see that, there are plenty of things you won't say!)"[121] The main problem is not inauthenticity, however, but rather that every single phrase or sentence forces protean reality to assume a definite shape, when it is not so easily captured: "Small bags won't hold big things." Meanwhile, people are apt to deem "right" only what conforms with their preexisting views; we might call them "presuppositional evidentialists," insofar as they accept only such evidence about the world as props up the beliefs in their comfort zone.[122] More damaging still, definitions imply identities of things and speakers, so words can spark dissension and tribalism. And no one worth their salt likes the "lip service" of the voluble hypocrite or sycophant; as a Sanghu advises, "The noble man's friendships are as flavorless as water; the petty man's, as cloyingly sweet as rich wine. The first leads to affection, while the second leads to a sense of revulsion. Those with no particular reason for joining together will, for no particular reason, part company." In crafting a life, as in artisanal work, gesture often excels language in communication, its import less liable to fray. And yet birds cheeping and chirping are not sheep bleating or cows lowing or horses neighing; they are more hopeful harbingers of spring. So while using fewer words is sometimes advisable, since too many words leave little room for speaker and hearer for face saving, we are dissuaded from becoming mute and "sealed."

Purely as a practical matter, the worthy man, in office or out, can hardly abandon language and reasoning (the latter reliant upon language), simply because the human equipment is less than perfectly suited to the purpose at hand. The *Zhuangzi* considers ordinary ways of talking not only as major contributors to the human muddle, but also as the sole entry points through which to assess the muddle.[123] Hence the continual play with the expectations we bring to language to spur an improved awareness of people's shifting frames of reference. If one learns to give shorter shrift to inherited values and societal imperatives (often distorted guides to sensible living), one can arrive at a more accurate reflection of the totality of things. In

Zhuangzi's own words, the wise person "uses anything he knows to get at his *xin* 心 [organ of the evaluating mind and the emotions], and then he uses his [ordinary] *xin* to try to reach the constant *xin*."[124] While an ultimate vision of the "constant *xin*" is bound to prove elusive, the attempt alone teaches many useful lessons[125] and can impart greater clear-sightedness.

One of the best indicators of the inability of logic (a special brand of language) to capture the full panoply of sensory experiences and intuitions comes in the form of two exchanges between the logician Huizi and the more perceptive Zhuangzi. In the first, Huizi queries one of Zhuangzi's central claims.

> Huizi said to Zhuangzi, "Can a man really 'have no feelings or inclinations,' as you are wont to claim?"
>
> "Yes."
>
> Huizi retorted, "But how can you call a human a human if he has no feelings?"
>
> Zhuangzi replied, "The Way gave him a look and a bearing; Heaven gave him a form. So what's wrong with calling him a human?"
>
> Huizi: "But once you call him a human, you imply things. So how can a human be without feelings?"[126]
>
> Zhuangzi: "Clearly this is not what I meant when I brought up the subject of feelings. My phrase 'having no feelings or inclinations' describes the person who refuses to allow likes or dislikes to worm their way in and inflict harm on his physical person. Such a person habitually cleaves to what is so by itself, letting things be the way they are.[127] He doesn't try to help life along or increase his lifespan."
>
> Huizi argued, "Well, if he doesn't improve his lifespan or livelihood, then what will he use to sustain his living body?"
>
> Zhuangzi: "The Way gave him a look and a bearing. Heaven gave him a form. He doesn't allow likes or dislikes to get at and into him, where they can do him harm. You, now — you treat your spirit like a total stranger, and so you deplete your vital energies. You lean on a tree and moan or slump at your writing table and doze. Heaven picked out a bodily form for you and you use it to prate on about 'hard and white' puzzles in logic."

Right before this exchange, an unnamed voice in the *Zhuangzi* — not identical with that of the Zhuangzi persona — claims that the sage has "a human form, but no human feelings." By ordinary logic, then, to lack "human feelings" or "human inclinations" should guarantee that "rights and wrongs cannot get at" the person.[128] That sounds like a sensible course, given how easily "right and wrong" slide into the sort of "likes and dislikes" (that is, values attached to phenomena) that create unruly feelings, if not outright discord.[129] And to Western ears, this advice sounds comfortably like the *apatheia* or *ataraxia* espoused by Cynics, Atomists, and Stoics, although closer examination points to dissonances.[130] But for Zhuangzi, emotions and feelings spring up like mushrooms—they are fully natural, as his second chapter shows ("Day and night, the emotions replace one another right before our eyes, and no one knows the place where they have germinated.")[131] To be emotionless is to be virtually dead, something Zhuangzi does not advocate. Notably, Huizi's cocky way of parsing the world requires stark logical categories, A and not-A, categories that ignore a linguistic tic common to many languages, including classical Chinese, where a phrase couched in terms of a negative absolute ("be without X") often means something scalar ("do not do or have too much of X.")[132] Clearly, people's propensity to speak in categorical imperatives is unhelpful at best and mystifying or dangerous at worst. Zhuangzi's impatience with abstract rules and overconfident assertions readily translates a few exhortations to "abandon the world" to "let the world's opinions slip away from your focal attention," for solemn vows to "take oneself off to the hills to live in strict reclusion" do no one much good.[133]

So when Zhuangzi is lucky enough to be strolling with his favorite sparring partner, the logician Huizi, he cannot resist the fun of pointing out that language, and logic itself, far from trafficking in absolutes, assume meaning through context, so Huizi would do well to pause and reflect before pouncing on "logical contradictions" (in this case, wrongly). Granted, Zhuangzi may have provisionally borrowed a well-worn slogan, but he never counted on Huizi to throw the baby out with the bathwater.[134] The injunction "Have no feelings" is no better than the slogan "Eliminate desires," which Zhuangzi denounces elsewhere,[135] for no living person can be free of such inclinations.[136] Feelings and desires, like language, well up from unknown sources, and no amount of effort can root them out.[137] But to overestimate their importance or duration impedes simple well-being.[138] The point for Zhuangzi is mainly not to hold onto emotions once they are spent or to hold onto language or logic once it has been delivered or served its purpose. A gradual release is generally safer than a buildup or outpouring. Once the fish trap catches a fish, it's time to throw it away. Meanwhile, the Zhuangzi of the story, unlike his friend Huizi, who is full of objections and recriminations, takes care not to mandate a particular way of being or acting in the world, lest he contradict his own warnings against "driven activities." Instead, he figures out how to avoid harm, a full-time job that involves not clinging overlong to any emotion or stoking it. It is Zhuangzi, then, rather than the dealer in fabulous paradoxes and piled-up negations, who becomes expert in traveling "two roads at once," tacking back and forth between common sense and rigorous logic, between trust and doubt, employing whatever will serve for the moment.[139] At points appearing to be one of the most systematic of thinkers, he sooner or later laughs at all human follies produced from systems and certainty, in the firm belief that grandiose and totalizing talk is at odds with the genuinely human. Look how hard it is to for two best friends to settle upon a good definition of the human.[140]

Readers trust the *Zhuangzi*, I suspect, because it accepts, nay, celebrates human limitations, but urges some salutary mending of fences, too. Most people ("nine-tenths," it says again) will feel a rush of pleasure, a form of emotional expansiveness, when "catching sight of the old homeland and dwelling,"[141] and the most enlightened will beat the pot and wail when a spouse or partner dies, "grieving at first just like anyone else."[142] As the short exchange between Huizi and Zhuangzi alerts us, albeit quietly: only machines function totally without any negative feelings of dread, loss, or mourning. Just to live out one's days requires complex interactions that tax the vital human functions and capacities, not the least of which is figuring how to earn a livelihood. If only for this reason, no human being has ever succeeded in privileging efficiency and the avoidance of missteps above human decency without doing lasting harm to herself or others.[143] So what if negative feelings arise!? They will. The question is how the person is to prevent them doing significant and permanent harm. One remedy is to recall that human flourishing implies profound human dependence upon other things, people, and events, no matter what the situation and personal assessment. Thus, a refusal to take on unnecessary distortions and malformations goes hand in hand with a firm decision to stay connected with those who sustain. Essentially, "less I means more world,"[144] as well as an opening up to what I need not become.

This axiom the *Zhuangzi* supplements with a reminder that

people have already been given more than enough from which to fashion good lives full of compelling interest. Blessed is the person who "has all the ordinary apertures [ears, eyes, nose, mouth, anus, urethra] and hasn't been struck down midway by blindness or deafness, lameness or deformity"; the good career, high social standing, and immense wealth are extras that typically bring trouble in their wake.[145] Zhuangzi denies any imperative to expand, reinvent, or "improve upon" the basic endowment of form and bearing given each person. "Heaven picked out a bodily form" for each, but most then put it to uses for which it was never designated. In place of Huizi's hyperactivity and avidity for novelty, Zhuangzi calls for appreciation of the "gift of life," which hardly needs to imply strict conformity with all sociopolitical arrangements, especially if the person manages to fly "below the radar," drawing no particular attention to herself. Nonetheless, for excellent reasons, Zhuangzi finds the logic chopper Huizi a stimulating companion when he wants a chat and is ready to enjoy challenges given or received. (Zhuangzi confesses himself bereft when Huizi dies, believing he may never again find such an engaging friend.)[146] Yet Zhuangzi is no less alive to the severe constraints Huizi's habitual modes of activity impose on his thinking and acting. Improbable as it may seem, Huizi's form of logic chopping never comes as close to the mark as Zhuangzi's vague bundle of emotions, feelings, and intuitions:

> Zhuangzi and Huizi were strolling along the dam by the Hao River when Zhuangzi said, "See how the minnows come out and dart around where they please! That's what fish really take pleasure in!"
> "You're not a fish, so how do you know what fish take pleasure in?"
> "You're not me, so how do you know that I don't know what they enjoy?"
> Huizi responded, "I concede the points that I am not you and so I certainly don't know what you know. At the same time, you're not a fish, so it's equally certain that you don't know what they take pleasure in!"
> Zhuangzi said, "Please, let's go back to your original question. You asked me *how* I know what fish take pleasure in, so you already knew that I knew it when you asked me the question. I know it by being here on the Hao."[147]

Perhaps Zhuang's reflection, as he pores over the water, has merged with the scene of the fish, so that they have become one. Regardless, Zhuangzi once again teases the logician, using the tricks of Huizi's game while hinting at the limits of language, despite all

Huizi's vaunted cleverness. "People who do not observe cannot converse,"[148] and the good listener, like the good watcher, is on the lookout for common ground, sometimes in the unspoken ellipses.[149] There are more things in Heaven and earth than Huizi can account for in his games,[150] for in the end, robust experience is never entirely reducible or amenable to logic, and communication can happen mysteriously. Far more is at stake here than mere winning and losing a word game, however.[151] Again, the fundamental questions are: What is a human being? How different are people from other animals? And how does a person seem to know certain things without formal instruction via language or the allied processes of acculturation? One simply cannot know. It suffices to close some gaps through lively exchanges and to narrow the distances created and validated by conventions. Had Zhuangzi been so inclined, he could have said,

> "You big fool! You never get it. If you think I cannot know what fish are thinking, what else must I explain to you about my experiential world? You've misconstrued my sense of things and their potential, not to mention the basis for truly knowing things. Because you know only logic, I'll give you a logic-chopping answer. You don't leave me, your dear friend, another method to communicate with you. Why constrain our conversation so?"[152]

To throw off encumbrances, to untie knots, to undo constraints, to consent to live in the body — surely that is a signal achievement to be relished.

Learning to Wander in Bemusement:
Clarity and Illumination, Plenitude and Pleasure
So what are we to make of *Zhuangzi*'s claim that it is possible to live life with greater vitality when so much of the *Zhuangzi* catalogues the countless losses and limitations to which each human being is always prey? Mutilation, rejection, ridicule, fear, the specter of death stalking intimate friends and family members, not to mention one's own looming mortality — these pains the *Zhuangzi* details, yet styles as part and parcel of the supremely great gift of life. And how can *Zhuangzi* posit vitality as the ultimate concern, when the occasional anecdote in it offers portraits of exemplary figures who seem dead to the world, "like ashes," men who appear to be "stupid and blockish"? The exemplary Ziqi of South Wall, for instance, sits leaning on his armrest, staring up at the sky and breathing — vacant and far away,

"as though he'd lost his companion."¹⁵³ The feverish grab for significance that once gave life its meaning has cooled, retarding the foolish expenditures of emotion and desire, in the process called "balancing" of the faculties located in the heart (*qi* 齊/*zhai* 齋 *xin* 心).¹⁵⁴ With "the eye / fixed & almost / averted," that's "how to see."¹⁵⁵

> [The wise man] will leave gold hidden in the mountain and pearls hidden in the deep. He will see no profit in money and goods, no enticement in eminence and wealth, no particular joy in long life, no particular grief in early death, no honor in affluence, no shame in poverty. He will not snatch all the profits for his private hoard. Nor will he lord it over world and think that dwelling in glory. What he glories in is light and clarity, [because he knows that] the myriad things constitute a single repository and life and death are of one and the same body.¹⁵⁶

Clearly, seeking light and clarity may require certain acts of temporary withdrawal, such as meditation, that leave us changed, as by a trance; by such methods are the most obvious encumbrances left behind. The search also entails a refusal to salve our wounds with treacly sentiments and Happy Faces. Honest men and women must acknowledge that no ready answers exist for any fundamental questions, also that neither logic nor erudition nor withdrawal get people very far.¹⁵⁷ Still, the freshness, the radical nature of the *Zhuangzi*'s images cannot be gainsaid; they emerge with "a sudden salience on the surface of the psyche."¹⁵⁸ Illumination beckons.

Soon we discover that the *Zhuangzi* is itself a virtual treasure house of viable methods to achieve greater vitality. Presumably because people are too diverse in their habits and inclinations for any thinker to mandate a single "path" or set of instructions that will help every person,¹⁵⁹ the *Zhuangzi*'s anecdotes, parables, and puzzles in logic — all strikingly visual — offer a dazzling array of possible leads and signposts. And so the *Zhuangzi*'s characters burst forth, the grotesque and rapacious jostling the most refined, with their eccentricities and contrary accounts leaving alert readers stunned at the sheer multiplicity of ways that people have found to make sense of the worlds they inhabit. Many kinds of wind blow through men and women.¹⁶⁰ When all is said and done, this intriguing awareness of diversity as a source of pleasures puts deadening conformity in the light,¹⁶¹ and so we see Zhuangzi positioning himself always deliciously somewhere "in between" the usual dichotomies¹⁶² and arrogant denunciation of beliefs. For the *Zhuangzi*, every person is a mélange of sentiments, affiliations, and behaviors that fail to fit neatly together,[163] and this the *Zhuangzi* underscores. Circumspection in both its senses makes for survival, nay, flourishing.

However, none of the strategies enjoined in the *Zhuangzi* will "make sense," let alone conduce to deep satisfaction, absent serious and sustained consideration of what it means to enliven our lives and to quicken our sense of the world around us. The goal is clear enough: to be fully present and receptive to exchanges, duly enamored with something inexpressibly good lying within and beyond the mundane. Then vision upon vision of plenitude opens to anyone willing to replace an easy acquiescence in truisms for the true ease of mind and heart and body that rests upon clarity. But to do that, we must return to the "basics" of human existence (*fan ben*), and their relation to illumination and clarity (*ming* 明).[164]

What are those "basics" in the *Zhuangzi*'s compilation? Because words connoting surprise and downright bewilderment pepper the entire *Zhuangzi*, consideration of those essential features of the human condition[165] seems a good starting point, for those who would experience more vital lives. Whereas a great deal of crass behavior is motivated by a desire to avoid confusion, or any appearance of it, while squashing our dread of the unfamiliar, real cultivation of the art of living demands people's awareness that they have "never once managed to hear a perfect teaching," which suggests that "realizing one knows nothing is the precondition for true understanding."[166] With limited powers of perception and attention, we find there is too much contingency and mystery to cope with unless we discern the rich potential in the basic fact of the human condition[167] — the infinite curves of human response that come unaided into view.[168] Thus, to embark on the journey to perceive anything about the world, it is best to acknowledge the unknowability not only of the world outside, but more importantly, of one's own person.

And yet there is light, undeniably. The muddle need not eclipse the pluriform and polyphonic experiential worlds, for, Zhuangzi says, "the most dazzling clarity and light are born of the blackest murk."[169] *Ming* 明, the word for "light," "clarity," and "illumination" in classical Chinese, also indicates the gods on earth and the sacred more generally. It is hardly divorced from politics, because the best rulers erected their worship halls as "Halls of Light." But the graph, in showing sun and moon conjoined, says more: their radiance by turns shines equally on all below, benefitting all without discrimination, though the light assumes different forms and different strengths, fulfilling different functions by day and by night, but in steady relation to one another, as when the new moon appears with the setting sun. One is tempted to say that light and enlightenment cannot exist absent this reflexive will to benefit all. And yet few human beings are prepared to see themselves as sufficiently powerful to become a beacon of light and illumination to others (Figure 5.2).

Yet all people, in Zhuangzi's view, regardless of their breeding or talent or training, have enough to learn what they need to know, so that they may "nourish what is within" and "fulfill their mission":

> If a man follows the faculties given him and makes them his teacher, then who can possibly lack a teacher? Why do you think you must comprehend the process of change, forming your mind on that basis, before you can have a teacher? Even the idiot has his teacher. Contrarily, to refuse to dwell in comfort with those faculties, preferring to insist upon your favorite rights and wrongs — this is like saying that you set off for Yue, far to the south, and got there yesterday. This is to claim that what doesn't exist exists.... Resign yourself to the inevitable and nourish what is within — this is best. What more do you have to do to fulfill your mission?[170]

That acknowledgment, that people from the beginning have the capacities that they need to flourish, sits well with a belief in the inherent goodness of the gift of life.

> You have been bold enough to take on human form, and your take delight in it. But the bodily form experiences an infinite number of changes. Your pleasures, then, must be too numerous to count. Therefore, the sage ... delights in early death, as in old age; he delights in the beginning and in the end. If he can serve as model for others, how much more is this true of whatever phenomenal existence is tied to and all changes wait upon [the unseen sacred orders]?[171]

> If I think well of my life, for the same reason, I must think well of my death. When a skilled smith is casting metal, if the metal were to leap up and insist on being made into the most ideal sword of all time, the smith would surely deem the metal inauspicious. Now, having had the audacity to take on human form once, were I to say, "I don't want to be anything but a man in his prime!" surely the Fashioner would regard me as a most unpropitious sort of person. So now I think of the cosmos as the great furnace, and the Fashioner as a skilled smith. Where could he send me that would not be alright?[172]

Figure 5.2. Karen McLean, untitled photograph, southern Arizona, 1980s. Reproduced by permission of the artist.

The foreground of this photograph suggests the benightedness and muddle we often experience, but sources of light always exist in our lives (as in the shaft of sunlight in the middleground), which in turn leads to the lush verdant spaces lying just beyond, in the distance.

People have been given pleasures in life and ease in death. As we have seen, the only truly moral response to these undeniable facts of life, known to the smallest child, is a capacious embrace of things as they are, combined with an ardent waiting for things to unfold, producing a form of moral pluralism, if you will.[173] (As someone else said, "we are here to keep watch, not to keep" or keep stock.)[174] So despite the propensity of modern academics to forge comparisons between Zhuangzi and the Stoics, Zhuangzi does not demand that readers give their allegiance to self-rule by reason, because reason cannot make out the cosmic or individual workings.[175] Zhuangzi, on the other hand, opposes moral absolutism and moral universalism, but not morality itself. Tacit or tactical knowledge — the sort of knowledge that serves in most cases to ground a provisional insight and tentative action — *that* lies within one's grasp, so while "no authority announces the *right* interpretation, competing voices open up various perspectives and angles."[176] And Zhuangzi throughout entertains thoughts of supreme pleasure: "When the men of old talked of fulfilling their ambitions, they did not refer to fine carriages and caps. They meant simply that their sense of pleasure [in their lives] was so complete that it could not possibly be increased."[177]

What specific sorts of pleasures might a person of sufficient clarity enjoy? Quite a few, for if too much of life is reductionist (for example, reducing another to her status or wealth; reducing a glowing sunset to "nice pic!" and so on), Zhuangzi repeatedly conjures visions of plenitude.[178] Ordinary events become "happenings" (notable occasions worth celebrating, if not idealizing), with heightened emotions bringing to greater awareness the sense that "life is good." Each beautiful form is distinctive in its construction and powers of attraction, each taste, each sound, each touch, and each dream. And while phenomena in the cosmos exhibit complicated tendencies, some in seeming contradiction with others, somewhat unaccountably, as "contradictions accumulate," "everything comes alive."[179] All prompt eager curiosity, if only we attend to them with the wild and unruly curiosity of the toddler or the very young at heart, who don't bother to claim the smallest semblance of knowledge and understanding.

Thriving connotes both "no unnecessary interference" and "beneficent interventions." Small wonder, then, that in reading the *Zhuangzi*, two reoccurring commands leap off the page: "Let it be!" (usually 已乎 or 已矣 in classical Chinese) and "Let it be spring!"

(*chun zhi* 春之 in classical Chinese). "Let it be!" is Zhuangzi's warning against strong human propensities to set up a serious impediments to enlivening ourselves. After all, the world, occasionally bruising, affords spaces for a person of sufficient skill to locate some "big openings" that allow the spirit "to move where it wills," like the keen knife of Butcher Ding. What Zhuangzi seeks in the world is wiggle room, space for maneuvers that are comparatively free of hindrance.[180] Ergo, the many stories taking wandering for their theme. But even more is possible, again, to those happy people who can let other things be themselves and thrive. Such people adopt the generous model of the cosmic powers, which provide sun and light, water and earth, for all in due measure and in due time. After all, the barely glimpsed Dao, we are told, "blows on the ten thousand things, each in a different way, so that each can be itself."[181] To understand that there is no common denominator, except that each being longs to thrive, is to intuit enough. Indeed, the single standard by which to gauge the success of one's actions may be, "Does it work well for others?" for empowering others imparts profound pleasures. Fully present, with generosity toward all.[182] That's the goal. To pursue any other means "a great deal of work, no [real] success, and certain danger to the person."[183]

Sure enough, "Let it be spring" is Zhuangzi's teaching, but the binomial phrase can mean at least three different things: that the wise person allows things to proceed merrily in their own courses, which turn out to be as good and as welcome as a spring day; that she nudges her peers in such positive directions as to quietly induce substantial improvements in various situations, so that all thrive together; that she senses the rhythms of her own life partaking of the eternal round of the seasons amidst impermanence. For each of these interpretations, the *Zhuangzi* offers support, and they are hardly mutually exclusive. All three readings elude the linguistic traps that designate passivity versus action. Ardent engagement assumes a spiritual dimension.

Forgetting and Laughing

> Running around accusing others is not as good as laughing, and enjoying a good laugh less fine than going along with things. Be content to go along and forget [to dread] change, and then you can enter the mysterious Oneness of Heaven.... Just go along with things and let your faculties for thinking and

feeling in the heart move freely. Best of all, resign yourself to what cannot be avoided and nourish whatever is within you.[184]

But how, pray tell, is the person to wait and "go along with things"? And how much can be forgotten, before forgetting itself becomes an impediment to survival and living well? Zhuangzi's heroes are continually found interacting with the things, people, and situations they happen to meet. Many lines in the *Zhuangzi* describe such lingering, dwelling, staying, and sojourning as pleasurable exercises, exercises that are paired with openness and quickened faculties, rather than static closure or sullen passivity. Forgetting some things (for example, the last unpleasant encounter with your difficult neighbor) allows less obstruction to others (the possibility that the neighbor may some day improve and that you must, regardless, muster politeness at all times, if there is to be any hope of improvement in future). And forgetting some things in the picture can result in a better view of the whole, and hence better reasoning.[185]

Musing about such passages, the idea of "reverie" may come to mind for readers who know Gaston Bachelard. After all, Bachelard's reverie is the fuguelike state that roughly corresponds to Zhuangzi's "stillness" or "resting,"[186] underscoring the strong connection between not letting energies flag and responding fully, between "balancing the heart and mind" and levelheadedness ("quieting the heart and mind"), as prelude to constructive action. In Bachelard's telling, reverie could hardly differ more from nostalgia, the retention and inevitable falsification of one's own emotions in memories likely to prove to be cold comfort to some.[187] For the moments spent in reverie, like those in *zuowang* 坐望 (literally, "sitting and forgetting," usually construed as meditation), neither strive to redo the past nor project into the future. Exploring the contours of the present is gratifying enough. Yet the single mention of *zuowang* that appears in the lengthy *Zhuangzi* characterizes that exploration as utterly transformative:

> Yan Hui said, "I can now sit down and forget everything!"
> Kongzi looked startled: "What do you mean by that?"
> Yan Hui said, "I let my limbs and body go loose. I dim my powers of perception. I leave my bodily form and jettison erudition. Then I can be like the Great Thoroughfares. That I call "sitting and forgetting.""

Kongzi replied, "If you are like that, then you now have no more likes, nor will you cling to constants and regularities as you are transforming. You are, in fact, such a worthy man that I beg to become your follower."[188]

Granted, even a person's "spirit and breath" are best forgotten, if an overestimation of their value will impede the elementary exercises in figuring out "how to look after the body."[189] Still, in the last line, it is hard to say whether Kongzi is genuinely admiring or gently mocking. To stay for long in transcendent fugue states is impossible. Yes, a temporary retreat from conceptual overload is welcome, insofar as it helps lift the layer upon layer of illusion that deter seeing and insight, intrusions of our own devising or fond adoption. In that light, the catchphrase "sitting and forgetting" limns a soul free of undue tension, calm, yet poised to plunge back in. Crucially, reverie, in or out of meditation, never precludes decisive action; it can even foster it, because it represents, at a minimum, a simultaneous disentangling from ordinary holds (chief among them, pride in one's own cleverness) and a setting of life's trajectory *for now*. A willingness to accept pain and vulnerability as necessary, even valuable components of the human condition, plus a lightheartedness — this is precisely what is lost in most of the pious academic accounts of Zhuangzi. After all, the *Zhuangzi's* goal is "healing," rather "curing" a person of the human condition.[190] Such an approach may reliably sustain us when most others fail. So will an ounce of self-knowledge, a refusal to take oneself too seriously (not to be often confused with nonchalance), whence Zhuangzi's talk of lightening up and laughter. For Zhuangzi, a self-deprecating modesty is *spirituel*, as is dallying with irony and paradox, less for immediate practical ends than because small jests can impart mutual pleasure (shades of Zhuangzi and Hui Shi).[191] (As Henri Bergson noted, laughter is an inherently social activity, the ultimate "social signal.")[192] So in an era valuing dramatic oratory and rhetoric, Zhuangzi would persuade himself (and others like him in positions of power) to steer clear of virtuoso performances and forget the slights that daily life inflicts, the *ressentiments* that gnaw away at one's core.

But there is laughter and laughter, of course.[193] The first chapter of *Zhuangzi* introduces the mocking laughter exemplified by the complacent cicada, turtle dove, and quail. We know these types well: thinking they know everything, they heap ridicule on anyone or anything unlike them, when, in actuality, they have not left themselves open to new experiences.¹⁹⁴ This sort of dissociating self-satisfaction exacts too high a price, as does the superficially affable laughter designed to trap the unwary.¹⁹⁵ But there is also diversion in the face of human follies, one's own or others' ("How peculiar—how silly!").¹⁹⁶ Rooted in a developed sense of multiple incongruities, diversion can often bring relief, relaxation, and release.¹⁹⁷ For how is one to ascertain whether a person has mastered a set of limitations like a frog in a caved-in well, adjusting too well to slippery sides and deep confinements, or whether she has grasped the subtle art of living through ever more capacious views? Better, perhaps, to assume a position somewhere in between or recognize that since "method has no place in art, folly is better."¹⁹⁸ And finally, there is belly laughter rumbling up from a delicious fit, whether a quirky juxtaposition of the odd and outlandish, a *bon mot*, an immersion in the task, or plain comfort in one's own skin.¹⁹⁹

Zhuangzi simply would remind us that he knows it makes sense to prefer whatever works to foster vitality, as the following tale from the "Old Fisherman" chapter puts it:

> A shamefaced Confucius asked, "Please, tell me what you mean by 'true.'" His guest replied, "The term means vitality and integrity that is developed to the highest degree. Whoever lacks this cannot move others. It follows that anyone who forces himself to wail cannot arouse grief in others, no matter how sad the sound, just as anyone who forces himself to be angry cannot arouse feelings of awe in others, no matter how fierce the sound. And anyone who forces himself to display or feel affection cannot ever elicit sympathy in others, no matter how genial his smile. True sadness doesn't need to make a sound to arouse grief. True anger does not need to show itself to incite awe. True affection needs no smile before it elicits a sense of harmonious accord. When a person is genuine inside, an unseen spirit moves the outside, the visible realm. This explains why we value the true and genuine [its transformative powers].²⁰⁰

Zhuangzi would have a person keep to whatever is true for that person (*zhen* 貞) and genuinely felt at the moment,²⁰¹ for absent true feeling, no exchanges have the power to move, transform, and inspire others. Besides, only the "truth of the [present] matter" exists, no absolute Truth with a capital "T."

That said, no specter of the modern and postmodern cults of authenticity lurks in this passage.²⁰² It is just that to operate from a still center of some awareness ("*this* is who I am" *for the moment*, and *that* is how the cosmos and current social relations look *now*) can accomplish a great deal all at once: the person tends to be more patient with the flaws of herself and others, being more mindful of present physical and emotional needs; this increases her effectiveness in nudging others to recall or realize something of their better selves. As Lao Dan advises Zigong, who visits him after Kongzi, the ideal sage-rulers of antiquity were alike in only one respect: they managed to "make the hearts of the people one," because they reached out to others and refused to indulge in amour propre.[203] For most purposes, it suffices to "ramble within the social constraints with nary a concern for one's title and good name."[204] The result is, often as not, a sense of joint pulling together (that harmony), as this passage mentions, and the lasting pleasure that is magnified by successful cooperations and collaborations, for true affection, initiated, grasped, and returned, is most moving. That awareness of pulling together, while knowing that this feeling cannot ever fully ever resolve large questions about life's meaning, may be the only thing worth having, for a "mastery of the art of living well" and the allied "restoration of the vital energies" derive from a keen sense of belonging to the cosmos and the social realm with others.[205] We are told that when Kongzi/Confucius finally learns "to abandon his studies and give away his writings, his disciples no longer bowed low before him, but their love of him grew day by day."

Many passages in the *Zhuangzi* forge links between slow retooling of the inclinations, which may then "return to the basics," and the potential restoration of the best (that is, most vital) version of the person, as in the following:

> Out of the flow and flux, things were born, and as they grew, they developed distinctive shapes called "forms." The forms and bodies held spirits within them, each with its own characteristics and limitations. That was called "inborn or inherent nature." If the nature is trained 性脩 [into a second nature], so that it returns to its Virtue and Power 反德, then that Virtue and Power, at its highest peak, is identical with the Origin, as it is empty (i.e., receptive) and thus great. You may then join in the cheeping and chirping [of birds], and be in harmony with the cosmos."[206]

It is on this note of joyous togetherness heralding dawn's early light and the promise of humanity's return to concord with the cosmos that this chapter should end, with the birds chirping and the visions of warm fellowship that the *Zhuangzi* offers for the reader's delectation:

"To walk among the myriad things in companionship is the Way.... Whoever comprehends the unity of all things will find the most important things have been prepared for. The spirits themselves will defer to such a person, though he or she lacks self-conscious desires for gain," knowing that such desires are apt to disturb and sap the vital energies.[207] As the border-pass guard Yin tells Liezi, "In general, anything with a distinct demeanor and form and voice is a thing, and why would things think it right to use something to distance themselves from each other?"[208] In most human functions, we perform singularly: we are born and we die alone; usually we don't require others to breathe, walk, eat, or sleep, and words do not always serve us particularly well in communication. But the gap between speaking and chirping is not wide, with human speech likewise anticipating responses from others,[209] even with those who operate on such starkly different principles as Zhuangzi and Huizi. People need never presume that their interests and needs will not coincide, since the roots of the human potential for individuation depend in crucial ways upon a willingness to see the need for cooperation. And if we are lucky, participation in the most demanding forms of cooperation yields insights into ourselves and others,[210] fostering a larger, shared expressive realm of greater ontological security while remaining vulnerable to change. Thus, people accustomed to make war, competition, and strife their primary reasons for living can learn to imbue their lives with meaning by looking outward, developing a vocation for listening and a taste for solidarity and its sustained and sustaining pleasures.

More than any Chinese thinker, Zhuangzi emphasizes the sustaining vital pleasures attached to curiosity as a counterpart to clarity. He analogizes life to a gift, unexpected and unrequested, but packed with potential delights. One cannot fathom much about anyone's person and motivations, even one's own. But one can playfully engage with things as they are (a category in which the personal is subsumed), like the traveler who notices things that ordinarily strike the person at home as of no interest. Unlike many thinkers in the Western literary and philosophical traditions, Zhuangzi evinces no interest in the reality/appearance dichotomy, itself an outer/inner conflict. While conceding the idea that a person's sensory equipment does not allow him or her to see the world in its totality, Zhuangzi does not linger there overlong. For Zhuangzi, the real damage comes from ignoring how much we never know and proceeding to make distinctions, assign values and priorities, and then make words reify our crazy categories. Clear-sightedness with respect to the unknowing and unknowable becomes the key to pleasure and a precondition for astute observation.[211] No incapacity prevents us from living fully, in other words. The goal for Zhuangzi is not to produce new knowledge, but to produce a transfigured person open to new experiences and prepared to live life fully.

If surprise is the most sensitive indicator of how we understand our world,[212] curiosity evokes "the care one takes for what exists now and what *could* exist in future; a readiness to find the strange and singular" in our surroundings.[213] Strange, singular, and singularly intriguing is how Zhuangzi finds the world, and endless curiosity is the main attitude that he would have us bring to it. Zhuangzi is no relativist hawking cool "going with the flow,"[214] nor is he an escapist longing to evade the inevitable. As a teacher, he is hardly oblivious to, acquiescent in, or complicit with the sorry forms of dehumanization that some have so proudly devised. A keener sense of plunging into the world with a heightened capacity to shape it, so as to leave it a gentler place — that might bring a loving smile to Zhuangzi's face. For Zhuangzi knows ordinary human beings in their wondrous humanity, wielding a battery of techniques and strategies to minimize their fears and maximize their pleasures in illumination and in life.

APPENDIX TO CHAPTER FIVE

On the Dating and Composition of the *Zhuangzi*

We know that Xunzi, shortly before unification in 221 BCE, knew a *Zhuangzi* manuscript, although that manuscript cannot possibly be the text we hold in our hands today. The chapters where Xunzi borrows heavily from the *Zhuangzi* seem to be the *Xunzi*'s early chapters, for example, the "Jie bi" ("Letting Go of One-Sidedness") chapter. Much of the text we hold in our hands today must date from the second or even first century CE, or still later, as Chinese scholars have painstakingly shown. But centuries after the Western Han, which ended in 8 CE, the first major commentator Guo Xiang (d. 312) did a number on the *Zhuangzi* text, trimming it, according to his liking, from a fifty-two-*pian* (bamboo bundles, probably one per chapter) to the current version of thirty-three chapters.

There is no doubt, then, that the *Zhuangzi* is a composite text, authored by several unknown people. The name "Master Zhuang" serves an author function while leaving many puzzles unresolved.[1] That being the case, we should hardly assume consistency or single authorship or editorship within or across pericopes. Still less should we expect to locate an "essence" of Zhuangzi's teachings, though there are themes. At the same time, for millennia, readers of the *Zhuangzi* have constructed intriguing views and a lovable eccentric character behind or beneath the text. And recent scholarly efforts to stratify layers within the text, though more or less plausible, are all clearly flawed, principally because they are undergirded by modern presuppositions about what the text *ought* to say.[2] Since I cannot come up with a better solution myself, I follow the lead of Esther Klein, who contends that we have put far too much emphasis on the Inner Chapters, because those to many expert ears sound more complex, more demanding, and more interesting than the contents of the Outer or Mixed Chapters. (However, perhaps the greatest *Zhuangzi* expert, Wang Shumin 王叔岷, demurs, thinking the Outer and Mixed Chapters equally good.)³ Following Klein, I decided to spend time reading the current *Zhuangzi* backwards (that is, reading chapter 33 first), rather than start at the beginning only to peter out, as has been my habit, somewhere around "Autumn Floods" (*Qiu shui* 秋水), chapter 17. It's many of these latest chapters—chapters 17, 23, 26, 28, 29, 31, 32, and 33, to be precise—that drew early attention, judging from the extant sources.⁴ Logical arguments, needless to say, can posit either a later or earlier date for the Inner Chapters as compared with the remaining chapters.⁵ But significantly, these Outer and Mixed Chapters are the chapters most consistently preoccupied with the problem of how to live well within sociopolitical constraints, and the Inner Chapters' concern with this same problem becomes more evident after one has read more of the *Zhuangzi* compilation.

The readings I give are consistent with many early readings of the text, because my research has been guided by parallels and explicit borrowings in the Western and Eastern Han. That said, my readings exclude one dominant tradition that insists that the *Zhuangzi* attests the superiority and special status of the sage. This tradition ignores the first story in the *Zhuangzi*, which explicitly says that the great Peng bird can't see any better than the tiny bird that looks up at him in the sky. That the chapter is entitled "Xiaoyao you" 逍遙遊 (usually translated as "Free and Easy Wandering," but more likely meaning "Wandering in a Daze")⁶ has also prompted my consideration of an allied set of suppositions. "Perspectivism" is a clunky academic term for a cheeky approach to knowledge, one that usually results in skepticism. This chapter posits a Zhuangzi who takes pleasure in seeing and seeing well while insisting, wisely and with forbearance, that we as human beings cannot ever see with perfect clarity.

For those who would like to know more about the compilation and editing processes of manuscript culture, I have written several pieces about this, including "Academic Silos, or 'What I Wish Philosophers Knew about Early History in China'" (2016), as in the Bibliography.

CHAPTER SIX

Yang Xiong 揚雄 on the Allure of Words Well Chosen

> What thou lovest well remains,
>   the rest is dross
> What thou lov'st well shall not be reft from thee
> What thou lov'st well is thy true heritage
>   —Pound, *The Pisan Cantos*

> Do not the archaic script forms and writings seem better to you than the nonstandard? (*guwen buyou yu ye hu* 古文不猶愈於野乎)
>   —Liu Xin, Yang's peer and rival

Yang Xiong 揚雄 (53 BCE –18 CE), the Han philosophical master, remarks at one point in his *Exemplary Figures* (*Fayan* 法言), "Books are as alluring as women."[1] Modern readers may frown at a comparison they regard as less than apt. Certainly, Yang knew that there are different kinds of sexiness, meretricious sexiness versus the indefinable appeal associated with classic elegance, with each exerting powerful effects. No doubt he did not mean us to succumb equally to both, for Yang became famous later in life for his diatribes against the very *fu* poetic form in which he excelled, on the grounds that it was showy, with a surface beauty and no substance.[2] In this, Yang's writings drew upon long-standing traditions, ascribed to the sages and the Classics, contrasting the unusual strength of the basic drives for food and sex with the general weakness of the acquired inclinations toward moral behavior (noted here already in Chapter 3).

To defy those commonplaces by insisting that books have the same appeal as beautiful women was to play with readers' expectations while plunging boldly into the business of reformulating traditions as a latter-day sage. It was also to elevate the value of *certain* texts, the

Figure 6.1. Alleged site of the Tianlu ge palace library in late Western Han times. Digital photograph by Michael Nylan, in the suburbs of Xi'an, Shaanxi Province.

This site marks the spot where tradition puts one of the most important palace libraries in which Yang Xiong and Liu Xiang labored. Essentially nothing remains, aside from a modern construction to mark the site.

Classics and the neoclassics styled on them, to the level of morality itself on the grounds that they encapsulate acquired tastes of supreme insight indicating the most desirable aspects of the civilized, deliberate life. Yang's statement is all the more intriguing when placed in the Western Han context, for in his era, as his *Exemplary Figures* shows, writings were frequently dismissed as the "mere dregs" of the sages' teachings, to be resorted to in the absence of living masters while training for careers in government.³ One early figure spoke for many when he complained with a sigh, "Reading makes me sleepy."⁴ Another equated the composition of original texts with "torturing oneself."⁵ Those conventions notwithstanding, Yang thought exquisitely refined writing — defined as spare compositions preserved in pre-Qin script or the latter-day productions written in the spirit of the true pre-Qin masters — as alluring in its forms as any natural beauty: "With women, one hates it when paints interfere with their feminine graces. And with writings, one hates it when overelaborate phrasing sullies or confounds the model and measure."⁶

This chapter's principal aim is to recapture the delicious sense of playfulness that superb stylists of the Han, beginning with Yang Xiong and his peers at the late Western Han court, brought to morally serious ideas about the edifying properties of reading and classical learning (Figure 6.1). To some degree, it focuses on the banter found in the *Exemplary Figures* dialogues, insofar as these reveal Yang's clever conflation of himself as writer with the character he constructed as Master Yang the classical master, once we plumb the biographical information that we have for Yang. Nevertheless, full appreciation of Yang's own enduring appeal rests upon situating his work within its proper historical context, for Yang's achievements are thrown into higher relief when seen against their cultural backdrop: the state of manuscript culture in Yang's world in relation to authors and authority.

Certainly, others before Yang and his peers had registered a serene belief in the capacity of some texts to impart infinite pleasure — and not merely because they improved one's chances of office-holding or promotion. This shines through an anecdote told about an older contemporary of Yang named Kuang Heng 匡衡 (active ca. 31 BCE). As a young man, Kuang dedicated himself to book learning, but he could ill afford to pay for a fire to read by at night. Since his neighbor's house was well lit, Kuang bored a small hole in the common wall so as to "borrow" a bit of light. (Note the pun on "light" and "enlightenment.") Inevitably, the hole in the wall was discovered, but the master of the adjacent house, far from being angry, was so delighted with Kuang's diligence in learning that he lent him a great many manuscripts to read.[7] The tale carries an implicit twist: in most pre-Han and Han stories, the motive for boring a hole through a neighbor's wall is the kind of voyeurism that more typically ends in an assignation with the neighbor's daughter, so the anecdote about Kuang Heng alludes to his sensual longings for book learning, which set men like Kuang apart from the common run.[8]

We see from Kuang Heng's story that a belief in the capacity of *some* texts to give pleasure was hardly novel in Yang's time. Many readers before Yang Xiong must have also *experienced* the pleasures of reading, as an idle comment in the *Huainanzi* suggests.[9] Still, to believe in or experience a pleasure is hardly the same as to *theorize* about it,[10] and it was Yang Xiong, in company with several of his companions at the late Western Han court of Chengdi (r. 33–7 BCE), especially Liu Xiang 劉向 (79–7 BCE) and Liu Xin 劉歆 (53 BCE – 23 CE), who constructed the first serious, sustained, and systematic case for the keen pleasures to be had from reading the Classics.[11] Building a solid case for his compatriots' passion for antiquity, Yang with the Lius, father and son, started a trend, fashioning parts of preexisting arguments into a distinctively new mix, which they then traced back, on tenuous grounds, as we now know, to Zhougong 周公 (fl. 1050 BCE) and Confucius (551–479 BCE), centuries earlier (Figure 6.2).[12] According to these advocates, reading classical texts — the older, the better — was critically important to the cultivation of the refined "taste" that underpins all well-informed and alluringly well-crafted writing,[13] and the court-sponsored institutions in which cultivated men served should follow suit, taking a "classical turn" to instantiate the values of the Duke of Zhou, to the degree that this was feasible in contemporary times. Within their own lifetimes, the "loving antiquity" (*haogu* 好古) movement, whose ideological underpinnings were constructed by Yang and the two Lius, took off among many men at court who commanded high cultural literacy and enjoyed superior access to the emperor, ushering in a host of policy changes, foreign and domestic, in addition to new forms of rhetoric and new objects of serious study.

Remote antiquity certainly held Yang's gaze. How fitting, then, that Yang, more than any other writer, defined what a classic of

Figure 6.2. "The Duke of Zhou." Song dynasty (?), album leaf in "Lidai shengxian banshen xiang" 歷代聖賢半身像 (Portraits from the waist up of sages and worthies through the ages), ink and color on silk, 88 x 59.4 cm, National Palace Museum, Taipei, Taiwan, R.O.C.

The colophon in running-regular script (*xing kaishu* 行楷書) reads: The Duke of Zhou (Zhougong), whose personal name was Dan, was the son of King Wen of Zhou and the younger brother of King Wu (r. ca. 1048–1043 BCE). He was uncle to King Cheng (r. 1042/35–1006 BCE). He acted as virtual king for seven years. His brothers Guanshu and Caishu circulated a slander about him saying that he was not acting in the best interests of the young King Cheng, after which on this pretext they fomented rebellion. Zhougong went east to carry out his punitive campaign against his brothers, and he executed both of them. Only when King Cheng was sufficiently mature did Zhougong "return the government" (*gui zheng* 歸政) to him, in the seventh year of King Cheng's reign. Before this, shortly after the Zhou conquest, when King Wu was gravely ill, Zhougong begged the gods to take him instead to the afterlife, as the Zhou ruling line still needed King Wu in place to secure its power. As Zhougong had transcribed his heartfelt prayer to the gods and stored it in a metal coffer, King Cheng eventually came to learn of Zhougong's selfless act, not long after he had banished Zhougong from his court, due to a second round of slanders. It was at that point that Zhougong composed for King Cheng the "Be Not Idle" chapter of the *Documents* classic, to further instruct his king.

enduring value was: a piece of writing that struck the perfect balance between saying too much and saying too little in the course of offering valuable interpretive keys to weigh diverse, prolix, or diffuse arguments.[14] Fitting, too, that Yang epitomized the ability to engage in play (*wan* 玩), a word denoting the tactile pleasure experienced when rolling something smooth around in the hand or the mental pleasure to be had from rolling something over in the mind or communicating one's thoughts to like-minded people of comparable sensibilities.[15] Arguably, it was Yang who made one source of play — classical learning acquired through reading — the very definition of refinement.

To realize his ends, Yang aimed to develop a new style of writing drenched in early allusions and classical turns of phrase in the hope that his immersive manner of living and writing approached the hallowed antique models, most of whose precise details had been lost over time. Today, we read Yang's poetry and prose as words on the page, sweating and cursing, perhaps, over the difficulties we encounter. By contrast, Yang expected his sumptuous compositions, inscribed into memory, bodily gesture, and language through repeated recitations and the painstaking copying of manuscripts, to thrill readers while confirming his standing as sage. (Invariably, his writings take for granted the conflation of classical master and influential text.) Through his distinctive writings, Yang communicated his passion for classical learning and neoclassical works so well that the shorthand phrase *haogu* soon came to characterize the fresh import and sheer stylishness of Yang's literary concoctions, not just the movement that he and the group of late Western Han reformers did so much to spawn. His own prose and poetry — however difficult to read, interpret, and imitate — became the standard for composition that was brilliant, lucid, and enlightened, even sacred (*ming* 明 embraces all these connotations, as we saw in Chapter 5), a standard against which authors compared themselves and were compared at least through the eleventh century.[16]

Throughout, Yang imparted the excitement he derived from playing the classical master, playing with words and scripts, roaming through manuscripts, and competing with his preferred Ancients.[17] He derived immense pleasure from deliberately deconstructing and reconstructing his court persona, as his autobiography, his grand display *fu*, and his philosophical masterworks attest. The early sources catch this propensity of Yang's to roam between the diverting play on others and self-valorization. It was unmistakable, since Yang went so far as to supply one of his philosophical tours de force, the *Great Mystery*, with autocommentaries written in two distinct styles, one of which he had famously scoffed at.[18] In his *Exemplary Figures*, some sixteen thousand characters in length, Yang strung together a series of snappy dialogues slyly mocking all manner of puerile propositions that passed for common wisdom. Several cross-cultural comparisons spring to mind—for example, Yang Xiong as counterpart to Callimachus (ca. 305–240 BCE), the famous poet-librarian who expressed in similarly strong terms his distaste for the meretricious lushness of Hellenistic poetry, or to Aulus Gellius (d. ca. 180 CE), "whose nightly joy lay in sucking out the meanings of words and the origins of customs."[19] So skillfully executed was each of Yang's turns at play that for far longer than they, he remained *the* master par excellence to emulate when basing one's life on the allied pleasures of reading and classical learning.

Once Yang's passion took hold, the entire curriculum of the late second-century CE Jingzhou 荊州 Academy came to be organized around Yang's masterworks and parsing his pronouncements on the Five Classics.[20] Few details have survived, although early records suggest that Yang's writings most likely inspired Wang Bi 王弼 (226–249 CE) and the Mystery Learning (*xuanxue* 玄學) movement of the Six Dynasties.[21] No less importantly, Yang's passion for antiquity was championed by the most famous writers in the centuries after the Eastern Han, so much so that later literary convention virtually required learned members of the governing elite to express their unbounded delight in reading about antiquity.[22] Echoes of Yang, then, resound in Tao Yuanming's 陶淵明 (365–427 CE) "There's pleasure to be had in it [the text], / Pleasure reaching unto the utmost limits," as in Ge Hong's 葛洪 (280–ca. 343 CE) line about the Ancients: "They... diverted themselves with the [Five] Classics."[23] Despite massive losses, the extant sources provide abundant confirmation of Yang's uninterrupted influence in the centuries spanning the Eastern Han through the Northern Song. The modern scholar Yan Lingfeng 嚴靈峰 (1904–1999) has found enough commentaries upon Yang's philosophical works for brief notices on those works to occupy some eighty printed pages.[24] Ironically, the very ubiquity of some of Yang's claims in later literature has had an unintended consequence:

some modern readers groan when they come upon such overworked phrases and regard Yang—not his imitators—as the boring hack.

At the same time, neither Master Yang nor his contemporaries in the *haogu* movement could have foreseen the book culture of the Wei-Jin through Northern Song periods that they helped to foster. Paper and woodblock printing did not exist yet. Writing paper, invented a century after Yang's death, came to rival silk in quality only in the fourth century CE, and woodblock printing became more common only in the late tenth century, judging from the available evidence. No less importantly, no one in Yang's day could have conceived the modern notion of an authoritative edition whose content was fixed by an author, a publisher, or both, an outgrowth of the early modern period and the development of Western copyright law.[25] In Yang Xiong's world, manuscripts worth saving and passing on were mainly written on bamboo strips or on silk scrolls, the former being bulky and cumbersome to prepare and store, and the latter quite expensive (see Figure 4.1).[26] Neither format facilitated "roaming through a manuscript" in the sense of reading at leisure and dipping back into earlier passages to savor them anew. Physical format alone made it difficult to produce tables or locate the contents and themes of a manuscript once read.[27] Conventions of punctuation, script, and format had yet to be regularized.[28] Reference works, as reading aids, were either unknown or in their infancy.[29] A few word lists were available, but no etymological dictionaries or lexicons of regional expressions.[30] (Yang dedicated nearly three decades to remedying this lack, eventually producing the first authoritative glosses for regional languages and obsolete expressions.) So few private libraries existed that if Yang Xiong had not been fortunate enough to gain access to the imperial libraries, then closed to the vast majority of Han officials and aristocrats,[31] it is doubtful whether even he, a man of genius, could have come up with the "old finery" that adorns his most famous compilations. Yang himself conceded as much.[32] In other words, between the late Western Han, when Yang flourished, and the Northern Song, when expertise in Five Classics learning finally became the primary qualification for office-holding, major shifts in the social practices of the text had to occur (see below).[33] One can hardly follow Yang Xiong's writings about pleasure without appreciating the particularities of the milieu in which he wrote so feverishly.

Since the manuscript culture that Yang knew well and shaped to his ends could hardly differ more from today's print and Internet culture, this chapter addresses a number of issues that are not as self-evident as it may appear. The first concerns what authority was invested in the Classics and in antiquity as models, and how the authority of texts came to be constructed and construed as the product of individual authors, culminating in the studied authorial persona of Yang Xiong. The second involves Yang's understanding of his actual and ideal readers—the practices and expectations involved in acts of reading and reciting, as well as the changes in reading habits that came with developments in manuscript culture. The third examines Yang at "play with the Ancients" as a historian and a philologist of ancient writings whose activities contributed to the passion for antiquity in subsequent eras.

Good reasons exist to welcome a review of these issues as prelude to considering the close ties between Yang's distinctive mode of writing and Yang's celebrated role as textual authority at a classicizing court supremely mindful of earlier pleasure discourses. In his writings, he is never thinking about mere words or phrases on a page, the content we might read in a printed edition of his works. He is talking about pleasing arrangements of well-chosen words that thrill the erudite reader to the very marrow and about what such words can accomplish in the world beyond the page.

Yang's Place in Early Manuscript Culture as Author

Yang Xiong's first taste of court life came late in the Western Han, during the reign of Chengdi (r. 33–7 BCE), a classically trained ruler then in thrall to two seductive beauties, Flying Swallow Zhao 趙飛燕 and her younger sister.[34] Perhaps the sex-drenched atmosphere at court, plus the emperor's considerable interest in book learning, prompted Yang's little joke about the allure of books.[35] (That Yang deplored the emperor's sexual license is evident from a *fu* dated to 11 BCE, however.)[36] Years before his arrival at court, Yang had immersed himself in the study of arcane vocabulary and phrasing, and his lifelong passion for rarified and refined language led him to cast the potentials of writing and classical learning in novel, yet pleasing ways. Yang's writing, in contrast to the "subtle wording" ascribed in Han times to Confucius, supposed editor of the *Annals* (*Chunqiu* 春秋) classic, does not lay claim to a systematic adherence to specific rules of composition. Indeed, Yang, as will become clear below, was far more inclined to demonstrate that he could play with the rules.[37]

Fortunately for those who would study him, Yang Xiong and his associates and admirers left an impressive paper trail. For Yang alone there exist two early accounts: the biography/autobiography preserved in Ban Gu's 班固 *Han Histories* (*Hanshu* 漢書) completed ca. 92 CE) and a second biographical notice recorded in Chang Qu's 常璩 (act. 265–316 CE) *Record of the Lands South of Mount Hua* (*Huayang guozhi* 華陽國志), which strings together anecdotes collected by Yang's friends, acquaintances, and critics, including his disciple Huan Tan 桓譚 (43 BCE–28 CE). There are the two philosophical masterworks, the *Exemplary Figures* and the *Canon of Supreme Mystery* (*Taixuan jing* 太玄經), designed to update and expand the *Analects* and the *Changes* respectively, not to mention Yang's corpus of long epideictic *fu* and two spirited defenses by Yang of his decision late in life to compose paracanonical masterworks after he grew disgusted with the propensity of the display *fu*, ostensibly stern reproofs, actually to encourage bad behavior on the emperor's part. The *Correct Words* (*Fangyan* 方言),[38] Yang's third masterwork, is the sole survivor among his several lexicographical works,[39] but to this already impressive list can be added Yang's reminiscences about his original intentions when composing court *fu* and a set of letters supposedly exchanged between Yang and his rival Liu Xin in connection with the compilation of an early version of the *Fangyan*,[40] as well as the semihagiographic sketches offered by Yang's later admirers, including Tao Yuanming and Liu Xie 劉勰 (ca. 465–ca. 522 CE).[41] Almost certainly, there survived in the first centuries CE many more pieces of Yang's writing, including some formal admonitions on the court hierarchy that are ascribed to him, not to mention stories about him.

Within the corpus of early writings, Yang is one of the few early masters to receive a substantive biography that adequately conveys a sense of psychological development.[42] (Confucius's biography is perhaps the only comparably rich treatment, until long after Yang Xiong's death in 18 CE.) Already in Yang's biography, purportedly based on his autobiography, two competing portraits of Yang contend for supremacy, just as they would later in history.[43] The first depicts the poor, earnest scribbler, abused or ignored by inattentive or obtuse rulers, who is nonetheless hailed as a great poet and thinker by a small band of dedicated followers who recognize his genius.

(Yang's writings express his profound dissatisfaction that some at court thought him little better than a hired entertainer.) The second depicts Yang the traitor, the slavish supporter of the usurper Wang Mang 王莽 (d. 23 CE), the consort clan (*waiqi* 外戚) regent who seized the throne in 9 CE while posing as a second Duke of Zhou.[44] But more likely to be true is a third portrayal that can be pieced together from Yang's biography. That third portrait has Yang enjoying considerable prestige as an authority regularly consulted at court on important affairs of state, including omen interpretation and foreign policy, despite the elegant, yet forthright stances he used to skewer the Han emperors and his fellow courtiers, annoying powerful factions at the late Western Han court.[45]

Yang conceivably was catapulted into prominence by the pervasive uneasiness that many of the best classicists expressed at the likelihood of a severe break from earlier traditions — a break occasioned not only by the momentous change from ornate seal script to simpler clerical script (see below), but also by new settings, occasions, and practices for reading, writing, and thinking about the guiding role of the distant past in the present.[46] Astute thinkers among Yang's peers had noted the confusion engendered by innovations in writing technologies and practices and in the preferential treatment accorded certain types of texts.[47] Along the way, then, my account will bring to modern readers some basic information needed to gauge the particularities of manuscript culture as it evolved during the first half of the classical era (323 BCE –316 CE), the better to understand what denotations and connotations the "right people" at court would have assigned to authorial authority in terms of ideal readers and classical learning. As it happens, the outline given below of the development of authorship in early manuscript culture lets us view Yang Xiong in a proper light, as the first indisputably self-conscious author within the entire extant corpus — for a self-conscious author, by definition, must create original works, rather than splice together passages from earlier works, with or without attribution; moreover, he must be cognizant of his own creative powers, as Yang indisputably was.[48]

Of course, in taking up such subjects, the usual caveats must be duly registered. Many scholars believe that a fully developed culture of the written word cannot exist absent three preconditions: the circulation of multiple copies of key works; the expansion of access to books via the growth of private libraries; and the development of public repositories for books. By those measures, a sophisticated culture of the written word did not emerge before the Song (960–1279), which witnessed the phenomenal growth of private academies and libraries, as well as court-sponsored and private printings of the Classics, some commentaries, and a few technical manuals.[49] But a thirst for a classicizing form of writing was created by Yang Xiong and company, so arguably, the seeds of Northern Song development should be traced to late Western Han.

In addition, because every sentence below rests on an analysis of the surviving texts (received and excavated), a single archaeological discovery could dramatically alter the chronology laid out here, though the prospect of such a major reversal diminishes with each new manuscript find.[50] Obviously, too, any chronological outline tends to overstandardize practices and attitudes within discrete eras and across many separate textual communities located in the North China Plains and nearby territories—lands whose cultures were far from unified in any modern sense. Older ways of thinking about texts and textual authority did not simply disappear overnight, for older texts in circulation served to keep earlier views and conventions alive before the mind's eye, allowing novel ideas to be grafted onto the old and contradictory assertions to coexist happily within the same manuscript, as Yang himself acknowledged.[51]

Nevertheless, surveying these early manuscript cultures, we can now imagine a gradual evolution in ideas about literary authority unfolding as a culture that was based primarily on oral traditions attributed to legendary figures developed into a literary culture devoted to the written words attributed to those who actually produced them, leading to the advent of new types of texts and new locations and opportunities for reading and writing superb literature. Accordingly, what follows traces a trajectory that begins in an era without authors as we understand them and that, with the transition from archives to libraries, leads to the emergence of practices increasingly consonant with a sense of authorship as a self-conscious act and ultimately the emergence of a literary culture of the written word and of individual authorship.

Prior to unification by Qin in 221 BCE, there were no true "authors" in the modern sense of the word. The very ease with which those in the pre-Qin era moved between oral and written transmission precluded the possibility of authorship in a strict sense. As Dirk Meyer notes,

"The interplay of the spoken and the written, but also the fusion of various traditions within one text, exclude the existence of a single identifiable author for the different texts."⁵² For if the word "author" is to have meaning at all, three criteria must be met. First, there must be an original text, a determinate product attributable to a producer, and as Jack Goody argues in many of his works, the concept of one fixed "original" version is inappropriate when discussing oral cultures. A culture of the written word therefore must develop before a concept of self-conscious authorship is both possible and valued. "Authorship" also must "claim that the particular words belong to a specific person and are not merely one possible articulation of a general truth,"⁵³ let alone a compilation of many earlier passages spliced together with newer bridging materials. Finally, as in the modern word, "authorship" implies intentional composition to achieve a particular end within a particular context; it must denote more than the mere transcription of spoken or imagined words into writing.⁵⁴

During the Zhanguo period, attribution of a piece or set of writings to a legendary figure tended to be purely notional, in that it signified a belief that the writing(s) somehow reflected the teachings or actions associated with a politically prominent person who nearly always was a sage-minister or sage-ruler, Mozi and Zhuangzi being the only conceivable outliers by this rule.⁵⁵ By most modern accounts, the first person to put a personal stamp on his compilations was Kongzi (that is, Confucius), but he, of course, was initially cast not as "author" but as a highly activist "editor" of the *Annals* (*Chunqiu*) compilation, just as he was supposedly editor of the *Odes* and *Documents* compilations. (The latter supposition is demonstrably false.)⁵⁶ Only in the Eastern Han, after Yang's death, once Yang had trumpeted the inestimable worth of true authors, was the figure of Kongzi reconfigured as an author, instead of an editor or compiler. Until Yang, members of the governing elite (the main body of those who could read and write) generally deemed prominent figures' exemplary decorum, as expressed in heroic deeds performed during the court's service and in formal rhetorical speeches on policy matters, to be the most potent way to pass on important lessons.⁵⁷ To take one example, a century before Yang Xiong, Senior Archivist Sima Qian had Kongzi apologizing for his editorial labors in the sincere belief that a distinguished career instead meant serving as minister for a decent prince. Lacking a good opportunity, Kongzi purportedly made do with illustrating the efficacy of moral action not through "empty phrasing" (*kong yan* 空言), but through historical examples showing extraordinary men and women making decisions and taking action.[58] The supposed limitations of "empty" words, hypothetical abstractions, was a theme Yang Xiong and his peers would take up, but in Yang's view, the accounts of noble houses and the forms of professional expertise (*jiayan* 家言) that survived the wars leading to unification under Qin and Han had been preserved at the expense of the crucial ethical writings associated with the early sages of hoary antiquity.[59] Hence Yang's preoccupation with the Five Classics (not the Four Books that would come to dominate after Zhu Xi's 朱熹 late twelfth-century construction).

However, in the early to mid Western Han, conditions propitious for the development of a culture of the written word and of self-conscious authorship began to emerge as compilers and editors came to assemble authoritative written texts traditionally ascribed to famous people in high places. Yet even in that era, serious thinkers were still arguing the superiority of spoken rhetoric over written (as in the saying "those who understand how to speak do not hoard books").[60] Moreover, many assumed that the finest writing demonstrating high cultural literacy would always be composed of many excerpts from previous compilations known mainly from oral traditions, so that authorial intent was said to have less to do with the expression of personal, possibly idiosyncratic feelings cherished by a self-conscious author addressing a specific audience than with "the historically situated, *public* responses" credited to a fictional or heavily fictionalized minister or ruler.[61] Quite possibly the first emerging authors, like Sima Qian, whose monumental *Historical Records* (*Shiji* 史記) compiled before 86 BCE, were motivated to establish the authority of teachings tied to their own families or localities.[62] Yet by Sima Qian's own account — reiterated several times for emphasis — the truly great masterworks of earlier ages were compiled after famous political actors had been forced off of the political stage and could find no better arena for a display of their talents. No literary figures could win renown purely on the basis of their writings, with service in governing and in teaching esteemed much more highly.

Then, sometime during mid to late Western Han (206 BCE – 8 CE) there occurred a series of changes in the social practices and attitudes associated with textual learning. Chief among these changes, which eventually led, centuries later, to a culture where the written word could explicitly challenge the superiority of oral teachings and transmissions, was the transition from archives to libraries. When archives began we do not know, but we hear of pre-unification local lords "preserving" their lineages' genealogies in their temples, along with signed copies of sworn oaths and various charters. And certainly rulers and advisers from the fourth through the second centuries BCE prized administrative documents, maps, and registers of land and population — the "charts and registers" (*tuji* 圖籍) or "charts and writings" (*tushu* 圖書) of the time — if by a single standard, that of each document's utility to the centralizing projects of the powerful states contending for supreme power.[63] But naturally enough, items in archives tended to be pitched when no longer of immediate use. Behind the Qin and early Western Han palace walls, then, we should imagine enormous refuse heaps of bundled bamboo accounts, maps, and lists.[64] (This explains the discarded Shang oracle bones found tossed together in waste heaps at Anyang.) By contrast, items in full-fledged library collections acquire value in roughly inverse proportion to their practical utility. Texts became prestige items, cultural capital for demanding connoisseurs, only when specific aspects of the manuscripts had acquired value from their age, fragility, lack of mundane utility, relative rarity, place within the entire library collection, and the number and range of ritual activities entailed in their textual production and transmission.[65] It is libraries, not archives, then, that sparked real passions, even pathologies,[66] insofar as they were more than the sum of their individual parts. It is they that became the brainchildren of supreme intellectual ambition, since libraries combined aspects of a palace, museum, and shrine to the hallowed past.[67]

Legend credits Han Wudi's uncle, Liu De, King of Hejian 河間 (active 155–129? BCE), with taking one major step to further the transition from archives to libraries. But whether Hejian's prince enjoyed gathering together a group of recognized experts in a wide range of fields[68] (a well-attested practice accruing social prestige for princes and kings) or collecting the best editions of older texts — on that score, the sources do not agree. The earliest traditions have Liu De collecting not antique manuscripts, but classicists of immense learning, as seems more probable, given that it was nearly a century before a late Western Han emperor, Chengdi, ordered the creation of new palace libraries replete with fine texts, ably demonstrating his appreciation of antique writings carefully transcribed in new recensions. His illustrious reign saw the birth of the first self-conscious "antiquity-loving" (*haogu* 好古) movement promoted in the court circle that included Yang Xiong, Liu Xiang (Yang Xiong's mentor and a prominent member of the imperial line), and Liu Xin (Yang's rival and Liu Xiang's son). Under the leadership of these three towering classical masters, this countercultural movement ridiculed many long-standing traditions cherished by the court Academicians, the reputed experts,[69] while enshrining a new vision of the Five Classics' role in civilizing processes and a considerably more robust vision of authorship.

Let us return to Yang and the two Lius at Chengdi's court. By Liu Xiang's telling, a century before Yang's time, the palaces had begun filling up with manuscripts: "In the outer court were the collections [*cang* 藏] of the Superintendent of Ceremonials [*taichang* 太常], the Senior Archivist [*taishi* 太史], and the writings of the Academicians. Inside the palace residences were the manuscript repositories of Yan'ge 延閣, Guangnei 廣內, and Mishi 秘室."[70] (See Figure 6.1.)

Some of Wudi's successors continued to have copies of new writings deposited in the imperial palaces, with the result that after a century, a veritable mountain of books had been gathered, which "filled the palace storehouses."[71] By 26 BCE, then, Wudi's fourth-generation descendant had decided that the separate palace collections needed to be reorganized. Accordingly Chengdi named the members of a commission who were to identify lacunae in the palace collections, locate copies of any missing texts, produce new and better recensions through collation and editing, and classify all the versions assembled for the new imperial libraries.

An imperial envoy, Chen Nong 陳農, was dispatched from the capital to scour the suburbs and countryside for lost books. Palace Superintendent Liu Xiang, once the official supervisor of the imperial clan affairs, was duly appointed to oversee the all-important task of revising the Classics and their commentaries, as well as the philosophical and the poetic works. He was well chosen. A distant member of the imperial family, descended from a king of Chu who was likewise a brilliant classical scholar and patron of scholars, Liu held the rank of minister in charge of the palace residences and their contents. In addition, before Chengdi came to the throne, Liu had submitted a strongly worded memorial urging Chengdi's father, known to be sympathetic to the classicists at court, to restore the antique values espoused in the old texts as a way of shoring up the waning dynastic fortunes.⁷² Ren Hong 任宏, an infantry colonel, was to preside over the acquisition and editing of military texts for the imperial collections; Senior Archivist Yin Xian 尹咸, the technical manuals (*shushu* 數術) largely devoted to divination and calendrical computations;⁷³ and Li Zhuguo 李柱國, a court physician, the medical texts. Meanwhile, Liu Xiang, as head of the entire commission, set out to compile an annotated catalogue of the new redactions produced for the imperial collections that would serve as reference guide for the emperor and his staff. (This library catalogue, so far as we know, was the first of its kind.) An elegant persuasion piece dating a century later tries to capture the realities of Chengdi's era, when men of "great talent and vast erudition" were put to work "collating palace writings" in the four major libraries within the palaces (Figure 6.3).⁷⁴ And several early sources place Yang Xiong in those palace libraries, working under Liu Xiang's direction, which confirms the picture of Liu as Yang's patron.

Supervising the collation, editing, and cataloguing of the imperial collections, including illustrated books, charts, and maps, occupied Liu Xiang for nearly twenty years, until his death. Shortly after Chengdi and Liu Xiang died in 7 BCE, Liu Xin, Xiang's son, presented to the new boy emperor Aidi a catalogue entitled *Seven Summaries* (*Qi lue* 七略),⁷⁵ long presumed to reproduce his father's work.⁷⁶ That work we only glimpse from a third work, the bibliographic treatise of the *Hanshu* (compiled ca. 82 CE), supposedly based on Liu Xin's work. At second or third remove from the first library catalogue devised by Liu Xiang,⁷⁷ we find in the *Hanshu* bibliographic treatise a total of 13,296 scrolls in the imperial library collections grouped under seven discrete subject headings: 1. the "Six Arts" or Six Classics (the Five Classics plus a lost Music Classic); 2. philosophical masterworks; 3. verse; 4. military works; 5. treatises on the technical and quantitative arts, including divination; 6. medicine; and 7. miscellaneous works.⁷⁸

Here, for the first time, so far as we know, texts — not just masters — were treated as experts in their own right, which marks a crucial change in the transition to a culture of the written word and a more rigorous conception of authorship.⁷⁹ Yang Xiong's own philosophical works might resist easy classification under his patron's

Figure 6.3. The tomb, in Jinpenling 金盆岭 near Changsha, which dates to Yongning 2 (302 CE), was excavated in 1958, height 17.2 cm, Hunan Provincial Museum.

Tombs in China often include figurines designed to illustrate those in the entourage of the dead. Here we see a sculpture where two figures are engaged in the process of collating texts (probably bamboo rolls). In this important stage of textual transmission in manuscript culture, the one collator recited the text that had been copied while the second checked that version against the original. Verbatim transcription was needed, of course, with certain kinds of documents, including imperial rescripts, and, increasingly, the Classics, as authoritative guides to cultivated behavior and improved policy-making.

bibliographic categories,⁸⁰ as erudite neoclassics drawing upon several disciplines, but Yang's writings suggest his wholehearted support for Liu's decision — surely approved by their ruler Chengdi — to give pride of place to the Classics, above the masterworks, verse, and manuals on discrete subjects.⁸¹ In retrospect, however, that decision may have been the first step on a road that led finally to later scholars' easy conflation of library catalogues with "national history" and thence to serious distortions of the distant past.⁸²

Two millennia after the compilations of palace library catalogues, it is easy to overlook the sheer magnitude of the editorial changes wrought by the activist editors working under Liu Xiang's direction, massive changes that ended not only in the compilation of "new texts" (Liu Xiang's term) but also in the heightened attention to textual authority that would prove central to the existence of a literary culture and the emergence of self-conscious authorship.

To take just one example, the new edition of the *Liezi* 列子 in eight *juan* (silk scrolls or chapters) was produced after comparing, collating, and extracting passages from five short works totaling twenty *juan* that originally circulated under separate titles, only one of which was credited to a Master Lie in a shorter *Liezi* redaction. By Liu Xiang's own account, he found many incorrect characters and some duplication among the various recensions. Assuming the old palace editions to be more reliable than those circulating "among the people," Liu made a new edition, which he hoped would show more internal consistency within chapters, despite Liu Xiang's doubts whether the original chapters "were by the same hand."⁸³ Similarly, Liu Xiang compiled a "revised and reduced" *Zhanguo ce* 戰國冊 in thirty-three bamboo bundles from six different manuscripts whose content he described as a total mishmash.⁸⁴ More astonishing still, Liu Xiang rejected as "duplicates" all but 32 of 320 *pian* of the *Xunzi* 荀子, excising nine-tenths of the originals he had at his disposal, when making his new version of that masterwork; that tenth he duly had transcribed in a fair hand on strips of properly seasoned bamboo in order to minimize future damage.⁸⁵ Like father, like son. Liu Xin's laboring in the imperial library reduced a *Shanhai jing* 山海經 in thirty-two *pian* (bundles of bamboo strips) to a mere eighteen.⁸⁶ Judging from such examples, many, if not all of the received masterworks we routinely dub "pre-Qin" should rightly be dated to the last years of Western Han.

Perhaps it was an implicit faith in the authentic text's potential for perfect communication across great gaps of time and space (across "clouds and seas," as the poet Su Shi 蘇軾 would later put it)[87] that allowed Liu Xiang, Yang Xiong, and the other activist editors to believe that the authentic (or "true" or "accurate") content of manuscripts would become evident after careful perusal, comparing, and judging *by them*, as devout lovers of antiquity. Yang Xiong knew that his beloved texts of the Five Classics had sustained "additions and subtractions"—interpolations and losses—however regrettable such alterations were in the case of such supremely authoritative works.[88] At the same time, Han legends countenanced, even celebrated activist editing, styling the sage Kongzi as the first editor to take a hatchet to antique writings in order to distill and refine their moral message: by Han legends, Kongzi chose less than a tenth of the early source materials at his disposal for inclusion in his *Odes* and *Documents* compilations.[89] Plenty of later examples of drastic editing took place within the circles of classical scholars, Han through Song.[90] Evidently, manuscripts were like natural organisms in being subject to cycles of florescence and decay; hence the occasional need for sages or worthies to engage in therapeutic interventions so as to restore balance.[91] Sometimes to preserve the authenticity and authority of a text, one had to emend it: "put it in order," "master it," or even "heal" or "cure" it (*zhi shu* 治書).

As students of modern literary history know, the distance between activist editing and authorship can be short—the role played by an editor such as Maxwell Perkins in the careers of not just F. Scott Fitzgerald and Thomas Wolfe is only one example—and the notions of a person mastering and intentionally imposing order on a text are essential components of a conception of authorship. Such massive efforts directed toward remedying (also *zhi* 治) the flaws in hand-copied manuscripts would have led editors in the late Western Han imperial libraries to try to devise more reliable methods for evaluating competing variants and editions, with the result that those same editors insisted that others show greater deference to textual origins, greater awareness of the conditions of text production, and equally great selectivity when choosing the models for their own compositions. For instance, Yang Xiong's *Exemplary Figures* openly speaks of "borrowing" or "rejecting" the models and slogans associated with the Han and pre-Han philosophical masters. In tandem with this editing and cataloguing project came the studied use of relatively new textual forms, such as the addition of the editors' own postfaces or prefaces (many in verse), which served as early tables of contents (as does, for example, the last chapter of Yang's *Exemplary Figures*).[92]

Even more importantly, over time the origins of authoritative texts took on added social and political importance. Students, disciples, officials, and clients often put down in writing what they had just heard from their social or intellectual betters.[93] Broadly speaking, early manuscripts served as a visible locus and medium of exchange in the promotion of social solidarity within relatively small communities. By contrast, broader and more casual access to this textual culture over the centuries after Yang Xiong threatened to strip away the pleasurable cachet attached to manuscripts as site and symbol of complex human relations[94] — hence the aesthetes' steady sniping about texts available for purchase.[95] One Tang writer, Li Deyu 李德裕 (787–850), complained loudly in 845, "Everywhere you look, you see men of breeding and cultivation, some already occupying important positions and some already at an advanced age, pounding away at composing texts, looking only for compensation." Until manuscript culture came to an end, this strong association of manuscripts with formative social interchanges made them into highly authoritative guides to deliberation and conduct in a culture avidly searching for exemplary models in life or in writing.[96] The singular authority of such ethical examples necessitated an identifiable author as their source, in place of a messy process whereby oral traditions disseminated general truths for elite consumption.

Texts more transparently could become authoritative when a single person or family was responsible for a particular composition or set of compositions that circulated within small groups of likeminded people we call "textual communities." Yang several times conflated text and person, whether creator or transmitter,[97] and scholastics of his day tended to tie every act of composition or compilation to a particular legendary figure confronting well-defined circumstances. For the emperor and his officials hoped to discern authorial intent through biographical details, before weighing the relevance of a text to policy making or establishing plausible precedents for new initiatives.

The more immediate result of the expansion of textual authority tied to new methods in text criticism was the production of several of the first reference works in classical Chinese (see below). Accordingly, attention to pre-Qin epigraphy grew dramatically in the wake of the library organization. Imperial collecting would henceforth be propelled by the notion that putting old writings in order (*zhishu*) — the older, the better — would aid government and might even "restore antiquity" (*fugu* 復古) to the halcyon days of yore.⁹⁸ Small wonder, then, that scholars such as Yang Xiong who were lucky enough to be appointed to the imperial library staff felt that they had died and gone to heaven or, at the very least, to one of the Han paradises at Mount Kunlun or on Penglai Island.⁹⁹

Strikingly, Yang himself never aimed to deploy his writings to secure a higher political office. Instead, he strove for eternal fame¹⁰⁰ — this in itself marks the emergence of a culture of the written word. (Supposedly, Yang in his dreams saw his compositions as "phoenixes" miraculously emitted that would rise and rise again.)¹⁰¹ His writings consistently styled Yang as an anomaly in his day: a classical master who intended to live most fully inside his manuscript rolls, rather than in the royal courts, the ritual centers, or the schoolrooms.¹⁰² By such self-fashioning, he parted company with Confucius, Mencius, and Xunzi, who wanted, above all, to advise the ruling powers of their day. Let us recall that because poets and thinkers during the Zhanguo period sought positions as clients of the powerful local lords, their compositions frequently alluded to what they perceived to be the all-important relationship of host and guest, ruler and client.¹⁰³ Once the empire became highly centralized under Emperors Jing and Wu (r. 157–87 BCE), the single court in the Han capital superseded all others as the center of prestigious appointments and patronage, yet Yang's poetic predecessors continued to behave and to compose their *fu* as if they were still the "honored guests" of one or another local lord, on the old Zhanguo model. Yang came to reject the primary formula used in the traditional display *fu*, this pretense of delivering a "guest's reprimands" to his host and ruler for the latter's benefit. Instead, Yang placed himself, rather than the emperor, at the center of his later writings, entertaining a gaggle of interlocutors in ways that allowed his mature writings to divulge his own experiences as author and classical master. (Even Yang's *Fayan*, this author would add, seldom discusses the Han emperors as moral models, aside from its final chapter of thirteen.)¹⁰⁴

Yang Xiong, in his mature years, introduced further innovations in his writings — innovations that built on and extended prior developments in literary culture. These relegated the work of activist editing to a status secondary to that of the full-fledged author capable of expressing both his character and commitments through a culture of the written word. To do this, he had his own prefaces to various writings imitate the most successful appraisals of historical personages that he found in the *Shiji* by Sima Qian. He drew attention to the chronological order of his productions, claiming that he had grown wiser in the course of time's passing and his ever-more refined notions had, in turn, altered the intention, content, and style of his writing (a point *Exemplary Figures* reiterates).[105] And he insisted that he lived for the time when he could work on the compositions he himself felt committed to, rather than produce the confections required by patrons or employers.[106] Thus did Yang manage to bring together in a seamless narrative, evidently for the first time, a compelling authorial voice, a corpus of memorable works, and a volley of statements detailing his fond hopes for the positive reception of his best writings, whether among contemporaries or posterity.[107] For all these reasons, one may cast Yang Xiong as the first fully self-conscious author in Chinese history.

Few authors in Chinese history have been as good as Yang at fashioning a distinctive persona that, despite a goodly dollop of eccentricity and gruffness, still exerts a powerful appeal upon discriminating readers who might not share his particular viewpoints (though he can be very persuasive). Quite rightly, even those who disapproved of Yang's politics regarded his construction of the authorial voice as a model for all time — at least for more than a millennium after Yang's death in 18 CE,[108] at which late date aspersions cast on Yang Xiong's character made it harder for those with political ambitions to admit his superiority.

Unfortunately, it is next to impossible to convey to those who cannot read Yang Xiong's prose and poetry in classical Chinese how delicious those writings are and how artful. I will try nonetheless. Yang Xiong first won fame and favor at the Western Han court for producing *fu* poetry that celebrated the imperial hunts in exuberant language that was simultaneously graphically and semantically lush. A few lines extracted from one of those early *fu* may suggest not only Yang's love of exotic vocabulary, but also his extraordinary versatility in fashioning visual puns, for once readers spy the repeated variations on the graphic components for "trees" 木 and "birds" 鳥 in the passage, they find themselves easily transported to the dense forests where birds of every type abound:

木則樿松楔㮨。㮨柏杻橿。楓枰櫨櫪。帝女之桑。楈枒栟櫚。柍柘檍檀。結根竦本....
鸑鷟鵉鷋翔其上....

Its trees 木 include:
Tamarisk, pine, cherry, water pine,
Vitex, arborvitae, fragrant cedar, sweet oak,
The mulberries of the Celestial Lord's girl,
Coconut and windmill palm,
Black plum, silkworm thorn, linden, dalbergia...
Birds 鳥, marvelous phoenixes of all types, whirl aloft....

Yang Xiong's philosophical writings are no less clever, if more profound. I offer just a few extracts from his *Exemplary Sayings* in thirteen chapters to impart a sense of Yang's deftness in constructing his series of dialogues. There, he casts himself as the supreme classical master, a latter-day Confucius, to whom lesser men pose their questions about the ultimate value of Five Classics learning in the mistaken belief that they will be able to stump him, since they are more worldly wise:[109]

Someone asked me, "Learning confers no possible advantage. What does it have to do with matters of substance?"

"You certainly have not thought it through yet! Now surely those with knives grind them, and those with jades polish them. Unground and unpolished, of what possible use are knives and jades? Certainly, there is something substantial in the grinding and polishing processes. Were that not so, then one would stop doing it." *(Fayan* 1/4)

Someone asked me, "In today's world, people are always talking about casting alchemical gold. Can gold really be produced in this way?"

"From what I have heard, those who see a noble man ask about casting men, not about casting gold."

Someone asked me, "Can men really be cast like metal?"

"Surely Kongzi in a sense cast [his best disciple] Yan Hui."

Startled, the interlocutor responds: "Well done! I asked about creating metal and I get an answer about transmuting a man's mettle." *(Fayan* 1/8)

Someone asked me, "If a given piece of writing were to agree with the Classics in form, style, and quality, but the age did not value it, would it be right to study it?"

"It would be right to do so."

An interlocutor remarks with a laugh, "But one had better wait for the topics to be set in the exams for office!"

"For the great man, the Way is the goal of learning, whereas the object for the petty person is profit. Are you for the Way or for profit?"

Someone asked me, "Yet to plough, but not to harvest, or to hunt, but not to feast — surely *that* is not ploughing and hunting!"

[Yang replies:] "A person who ploughs the Way gets it. A person who hunts for virtue gets it. That is quite a harvest and a feast!" (*Fayan* 1/19)

Someone objected: "But if you took Yidun's fabulous wealth and used it to fulfill your filial duty, would that not be the very best kind of filiality? Surely Yan Hui [the beloved disciple of Confucius] often went hungry!"

"The former used his own crude methods, while Yan Hui used his refined sensibilities. The former employed a bent and partial way, in contrast to Yan, who used the most reliable, upright, and true method. In what way was Yan Hui inferior? In what way, pray tell?"

Someone said to me, "Were I to have crimson sashes and stores of gold — the pleasure would be infinite!"

"Crimson sashes and stores of gold — the pleasures that come with those are inferior to those known by Yan Hui. For Yan Hui's pleasures were internal, while high rank and wealth are external."

Someone asked me, "I beg to ask about the inner pleasures that you imagine coming from 'repeatedly going hungry.'"

"If Yan Hui could not become another Kongzi, he would not have thought even possession of the empire sufficed to give him pleasure."

"But did he not, for all that, have his bitter sorrows?"

"Yan Hui was deeply troubled by the perfection of Kongzi."

The interlocutor is startled.

"Yet that particular form of trouble — was it not the very means by which Yan gave himself immense pleasure?" (*Fayan* 1/22–23)

Someone asked me, "When you, sir, were young, you liked composing *fu*."

"That is true. Young lads *will* carve insects and seal scripts." A moment later, I added, "But grown men do not engage in such activities."

Someone asked me, "Can the *fu* be used to deliver indirect remonstrance?

"Remonstrance! If it would serve for such, that would be all well and good. But it does not stop there, with reproofs. My fear is that it cannot avoid encouraging bad behavior."

Somebody asked me, "But the *fu* are as lovely as the filmiest of sheer gauzes."

"Surely it is rather more like wood moths in women's work."

"The Swordsman's Treatise says: 'Swords are the means we use to protect our persons.'"

"By that logic, you would say that feints and parries make men all the more courteous!" (*Fayan* 2/1)

Reading through the various writings is like contemplating mountains and rivers. After an ascent up the sacred Eastern Peak, one grasps how meandering are the slopes of the lesser mountains, not to mention the small hillocks. And after floating on the azure seas, one understands the turgid flow of the Yangtze and Yellow Rivers, and even the dried-up marshes. One can never set aside the boat if one intends to cross the river. Likewise, one can never set aside the Five Classics if one intends to cross over to the Way. To discard the time-honored fare while developing a craving for unusual delicacies — how could such a person be deemed "a connoisseur of tastes"? To abandon the great sages and prefer the Masters — how could such a person be deemed "a connoisseur of the Way"? (*Fayan* 2/9)

"As the saying goes, 'A trail through the mountains is too narrow a path to follow,' and 'a door facing the wall' leaves too little space to enter."

"How is a person to go in, then?"

"Kongzi. Kongzi is the door."

"Are you, master, also a door?"

"A door! What a door! How could I possibly fail to be a door!" (*Fayan* 2/10)

Someone asked me why the sages' deeds cannot shine forth as brilliantly as the sun and moon. "Why allow later generations to be mired in endless debates [over their meaning]?"

"The legendary blind Music Master Kuang could maintain silence [to mark the changing rhythms], but even he could not elicit the right response from a tin ear. The legendary chef Yiya knew how to season things, but even he could not impose a connoisseur's sense of taste on an unrefined palate." (*Fayan* 5/10)

The brilliant light of day — that is of use to everyone's eyes. And the vast mystery that is the sages' Way — that is of use to everyone's hearts and minds!

(*Fayan* 8/11)

In the *Exemplary Figures*, as in his second neoclassic, the *Canon of Supreme Mystery*, Yang Xiong skillfully builds the case that of all pursuits, only devotion to the antique writings of exemplary figures who embody the loftiest and most capacious ideals will refine a man's tastes and hone his powers of perception. Those refinements in turn, at a minimum, will induce a sense of deep satisfaction, as once experienced by Kongzi's disciple Yan Hui. (In a civilized age, they may procure the person other rewards, in terms of high rank or reputation.) Yang considered this sort of rewarding engagement to yield a singular and valuable experience of human existence.

Thus, Yang Xiong's ideal readers, in his own and future generations, were those sufficiently learned and cultivated to engage in serious reading of such works as the *Odes* and *Documents*, the two antique Classics in which the legendary Confucius schooled his pupils.[110] During the Han, learners such as these were rare birds perched on high branches, inescapably aware of their superior erudition—or so Yang said.[111] Rulers, the ultimate patrons for classical scholars, often themselves lacked the requisite discernment.[112]

So "Why should words be a treasure to hold on to alone?" asked one early author.[113] Why indeed? How did Yang imagine the select group of his ideal readers, current and future, and their practices and expectations?[114] And why would Yang constantly find solace in the notion that like-minded people of sufficient taste might form a close-knit group defined by their love of classical learning?[115] These sorts of questions come flooding in as soon as we notice Yang's serene faith in later generations of discriminating readers who would appreciate and transmit his classicizing writings—and this in an era, the late Western Han, when but 5 percent of the population might have aspired to or commanded the level of high cultural literacy that Yang's writings demand. Because we have no proof that literacy rates rose much in Yang's time or in the centuries immediately after him, despite the throne's boast that it patronized classical learning,[116] further analysis of manuscript culture in Yang's time may bring us closer to answers for such questions.

Literacy and Claims for Writing in Yang's Era

Several features of the early manuscript culture of Han should compel our attention, the first being that "reading" in most cases meant "reciting." Given the paucity of inexpensive writing materials in Yang

Xiong's time, to learn the contents of a manuscript, one often had little choice but to commit those contents to memory by reciting them again and again. Then, too, the labor involved in copying a manuscript was considerable. After all, rough estimates put the pace of copying by seasoned professionals at no more than one manuscript of average length per month of full-time employment.[117] Even so, the gist of legal, administrative, or religious notices, if not the precise wording, needed to be orally conveyed by literate to semiliterate and illiterate populations. Less discriminating readers/reciters generally sought from manuscripts the requisite knowledge to follow in their parents' professions. In addition, they consulted technical manuals that promised to improve their daily lives by providing the proper formulae for prayers, useable calendars of lucky and unlucky days, or sound advice on medical treatments, irrigation, or rhetoric.[118] Whereas many professionals—artisans, merchants, doctors, soldiers, scribes, and functionaries—would have realized the advantages accruing from even a rudimentary grasp of a limited number of graphs,[119] the barriers to advanced cultural literacy were formidable. For in the estimation of the confirmed lovers of antiquity, high cultural literacy meant producing as well as reading manuscripts that wove together elegant phrasing redolent with classical allusions (*zhu wen* 屬文, literally "connecting" or "putting together" earlier passages) via complex prosodic and semantic rules.[120] This may explain why in Yang's era, judging from the available sources, the act of reading was not exalted as a virtue per se (only a sign of the diligence that could contribute to virtuous habits), nor were manuscripts fully commodified (reduced to a set price), even in the large metropolitan areas.[121]

Writings by Yang Xiong and his famous peers consistently claim that the exquisite pleasures engendered by classical learning derived chiefly from the manuscripts' associations with intensely gratifying relations binding master to pupil, father to child, patron to client, friend to friend, or ruler to subject. Manuscripts often circulated within the context of gift exchanges, and attitudes toward manuscripts copied by oneself or by kin and kindred spirits could differ markedly from attitudes toward documents produced by scriveners and clerks.[122] Hand-copying texts could confer moral and intellectual benefits upon the copyist, but only when undertaken as a vital part of the learning process, designed to steep the copyist in the virtue of fidelity to the authority who conferred the gift of knowledge and experience so as to lead to faithful transcription and transmission.[123] These were the preconditions for inscribing the text's contents onto the copyist's heart and mind, for internalizing the manuscript's underlying message, and for instantiating the authorial intent. Yang joined other authorities in telling novices that if they could "read and recite it [a manuscript] a hundred times, the sense of it would become apparent"[124] enough for them to reproduce its merits.

There was a profound faith in the transformative potential of hand-copying and memorizing model manuscripts, so long as the text and its author were thoroughly good. But by the same token, if the manuscript so carefully copied by hand and memorized was inherently misguided, its errors might well color the learner's experience of the world and skew his judgments. Hence Yang Xiong's frequent warnings against reading literature that was "mixed" (that is, not strictly focused on the related tasks of cultivating refinement and governing others).[125] Because writings carried such enormous transformative potential for good and for ill, it was incumbent upon the wise person to find a reliable standard by which to judge their contents. Yang wrote, "Just as countless numbers of ideas may exist in a single alleyway marketplace, so, too, innumerable arguments may coexist in any one scroll of writing. Just as a balance must be set up for each little marketplace in a narrow alley, so, too, a teacher must be set up even for a single text only one scroll in length."[126]

Yang's insistence on the necessity of finding a good teacher made supremely good sense in his own era. First, before the regularization of formats, genres, and script types and the development of scholarly apparatuses in Yang's lifetime and beyond, most readers could not reasonably expect to understand a text from the distant past without the aid of a teacher.[127] Second, uninformed perusal of a manuscript *on one's own* still smacked of superficiality, if not indolence — a negative association strengthened through Yang's disparaging comments about the *fu* and caught by such sensuous phrases as a "floating of the glance" and the "drifting of a look."[128] But once the moral reader, with the guidance of a good teacher, had ascertained the superiority of those texts that prompted reflection, that reader, in Yang's view, would be fully prepared to maximize his own potentials through repeated "immersion" (*qian* 潛) in the properly copied and memorized manuscripts whose contents could be traced to the finest classical masters. According to Yang, the written traces of such exemplars — in whose number he included himself — trained the avid learner to emulate, achieve, and communicate true greatness. After all, it was solely through the medium of old texts that "Confucius had inculcated in himself the qualities of the Duke of Zhou" in a manner as fully complete as Yan Hui emulating the qualities of Confucius, his master in life.[129] Thus, to discern the extraordinary value of the classical and classicizing texts was to be well on the Way to becoming a consummate authority oneself by a process Yang deemed "natural," if "not spontaneous"[130] — "natural" because the best people wanted to emulate the best in others, yet hardly "spontaneous," because immersion in the profundities was predicated on informed choices and hard work. (In Yang's formulation, we sense echoes of Mencius and Xunzi.) Yang consistently portrayed himself in his writings in this light, as an outstanding classical master who had duly acquired consummate authority through his painstaking study of the antique writings. As the culmination of a long line of distinguished teachers, Yang promised to guide the befuddled to reach the same glorious state of connoisseurship as he enjoyed, so long as they evinced a comparable taste for classical reading.

 Recall that in Yang's view, there existed two categories of texts. Some writings conveyed merely the dicta of one or more authorities, including contracts, edicts, legal documents, coins, seals, tallies, censuses, written orders, and so on. Such documents neither wanted nor invited reflection, but rather were specifically designed, by content and format, to forestall reinterpretation. By contrast, a second category of writings positively invited imaginative responses,[131] so that devoted readers, transcending time and space,[132] might bring to life their dearly beloved texts by ongoing engagements with and frequent resorts to them. In this connection, Yang had foremost in his mind the Five Classics and Confucius's injunction to "reanimate [that is, invest new life in] the old" or *wengu* 溫故, which encouraged the writing of neoclassics such as his own. Only the second class of writings — those that required a series of ardent commitments and communications — struck Yang as "alluring," in that they aroused Yang's own edifying attachments (now virtually second nature to him) to the most exalted of long-ago authors. Manuscripts full of resonant patterns and inspired juxtapositions of graphs and images simultaneously appealed to the imagination and mirrored the cosmos, thereby propelling people to interact and conceive of cultural exchanges well beyond their own limited experiential worlds. So when he could please himself and was not on palace assignment, Yang wrote texts to form rather than to inform; he spoke of "casting men."[133] That sort of manuscript, he maintained, would always find devotees eager to hand-copy and transmit it down through the ages.

As Yang Xiong himself remarked, writing can preserve long and intricate lines of argumentation across time and space, while oral discourse cannot. Writing in the most admired Han mode — weaving passages together in artful compositions — facilitated a merging of disparate ideas from multiple authorities; it was a gathering of sorts, to which one could invite the most stimulating companions.[134] Witness Yang's own selectivity when harnessing the pre-Qin persuasions to his larger vision. Good writing, for Yang, allowed authors and readers to try out various ways of sorting and categorizing knowledge, as is evident from the tales of Yang's shuffling and reshuffling his set of file cards on flat wooden boards in his search for the most revealing order.[135] And writing and composition could summon more satisfying reflections than oral discourse, if only because the writing process was frequently more drawn out, with slow ruminations about repeated drafts preceding the painstaking production of the final fair copy.[136]

In the class of texts inviting reflection, Yang believed the Classics and neoclassics — whether composed by the Ancients or by latter-day sages like himself — afforded readers the closest possible analogues to sustained relations with the ancient sages and worthies,[137] the best possible medium for communicating with those "far away" and remote in time, and the most gratifying to those of refined taste. (Hence Yang's decision to linger over the Classics in all his masterworks, whether in prose or in poetry.) But because verbatim transcription was almost an oxymoron in the era before printing, Yang saw that "engaging" texts also meant encouraging the copying and recopying that increased the chances that the original wording of any given text would be traduced, especially when so many of the badly trained in his day viewed manuscripts as mere aides to memory or pretexts for adaptation. Playful acts of literary impersonation, or *prosopopeia,* must be reserved for acknowledged masters such as he. With no apparent irony, then, Yang praised the "Redactors of the Canon of Yao, / Correctors of the *Odes* with their Hymns."[138]

Both the permanence and the impermanence of the written word could pose dangers when coupled with the attribution of personal authorship. Access to or ownership of each manuscript testified to a valued commitment — whether the manuscript came from a father, patron, ruler, or friend or was acquired through a gift exchange or some other means. Therefore, those in power generally chose to restrict access to their most treasured manuscripts, lest these rare and valuable objects fall into hands that would use them to the discredit of the original owners. Even a prince of the Liu ruling house could be denied permission to read a manuscript in the imperial collection.[139] Xunzi, a writer whose works Yang knew well, once observed that silence is often the better part of valor,[140] since any piece or form of writing is liable to *post-facto* consultation, and writings placed in changed or changing contexts tend to acquire unintended meanings and associations. Hence Yang's remark to an unnamed critic in one of his imaginary dialogues: that a single ill-judged word could occasion a fall from high office to bloody extinction for an entire clan.[141] Yang apparently on this basis decided to deny Liu Xin, his social superior, access to draft writings that he deemed too risky to circulate.

At least twice in Yang's career, he was threatened with jail or worse because of the ideas and images that he had entrusted to writing. He had composed a letter of recommendation on behalf of a fellow countryman from Shu who later fell afoul of the law.[142] Worse, he had tutored a Liu clan member in the arts of "strange graphs" (*qi zi* 奇字), the rare or obsolete characters sometimes used in divinations about dynastic fortunes, in magical medicine, or in apocryphal writings.[143] After an aristocratic pupil of his was accused of high treason, Yang jumped out from the upper storey of the library where he worked, fearful that he, too, would be implicated in the capital crime.[144] Given the Han legal system, which charged the intimates of criminals with guilt by association, Yang's suicide attempt can hardly be diagnosed as paranoia. But it is nonetheless striking that Yang's compositions often failed to grovel sufficiently before the powerful, risking charges of lèse majesté, as careful comparison of one *Fayan* passage with its parallel in the *Hanshu* shows.[145] In one *fu*, Yang Xiong had the temerity to implicitly compare his own emperor, Chengdi, to Jie and Zhou, the last tyrants of the Xia and Shang dynasties.[146]

Many more passages in the *Fayan* show Yang Xiong in his element, however, playing with words in artful ways modeled on the Classics, while driving one or another of his points home. One passage, for example, averred that a person's moral worth is easily assessed, simply by seeing whether that person becomes the object of others' admiring gazes or a truly repulsive sight in their eyes. Odes 295 and 297, two temple hymns, describe members of the governing elite "yearning" for Kings Wen and Wu of Zhou, paragons in their day and ever afterward. With those odes in mind, Yang composed the following exchange, setting out the basic qualities required of charismatic rulers:

> Someone asked me whether there are any keys to good rule.
> Reply: "Yearning and repulsion."
> "What does *that* mean?"
> Reply: "It refers to that time long ago, when the Duke of Zhou launched a punitive attack in the east, and 'Him the Four States took as king'; also when the Lord of Shao faithfully carried out his duties, and he was said to be 'Lush and verdant, the sweet pear.' Those certainly had the effect of making others yearn for them, did they not? But when Duke Huan of Qi wanted his army to cross through Chen, and Chen refused, in point of fact, to let him enter, the duke detained Chen's envoy, Yuan Taotu. This certainly caused others to feel revulsion, did it not? Ah! Those in power had better look to yearning and repulsion — that and nothing more!"
> Someone asks what creates yearning or repulsion.
> Reply: "We speak of a ruler being the object of 'longing' if he treats the aged and the orphans as they should be treated, if he tends the sick and buries the dead, if he puts the males to fieldwork and the women to silk production. But we label the person in power 'repulsive' if he mistreats the aged, humiliates orphans, neglects the sick, exposes the corpses of the dead, or lets the fields become wasteland while the looms lie idle."[147]

By continual resort to classical models, Yang lays out in admirably succinct and clear terms his fundamental challenges to those in power, whose actions can attract or repel their subordinates, leaving no doubt about (a) the basic human capacity to evaluate people, things, and events correctly, (b) the inevitability that wise judgments will be made by those whose classical learning puts them in an excellent position to judge, and the consequences for human motivation that (a) and (b) must entail for those in high places. To the written word, then, Yang entrusts his final evaluations of historical actors, in the belief that his beloved Classics have faithfully preserved the tenor of earlier examples drawn from history. Like Thucydides, Yang writes his own neoclassics not "to win the applause of the moment, but as a possession for all time." The principal subject of the antique writings—the playing out of human needs, desires, and predilections in society and politics—would change remarkably little over the ages. Hence the plausible contention that "Those who truly understand people are fit to assess the different ages."[148]

At Play with the Ancients' Writings: Passionate Devotion

In promoting the Classics and neoclassics, Yang clearly saw himself undertaking an uphill battle. True, many court officials professed to love antiquity, and they might even recite a line from a classical text when it suited their purposes. Long before Yang, the benefits of dressing up new arguments in old finery had become apparent to all aspirants to prominence at the court, since cultivation then, as in all premodern civilizations, was synonymous with knowledge of the past.[149] Still, in Yang's view, most who claimed to be erudite were in the thrall of hoary traditions or crass ambitions and all too inclined to mistake memorizing a heap of facts for attaining true wisdom. Hence Yang's sharp critique of the court-appointed Academicians who, deluded or self-seeking, misinterpreted the teachings of the Ancients.[150] Yang lambasted them as the typical careerists who expend time and effort in the pursuit of fame, fortune, and long life, goals that depend far more upon luck and timing than upon worth. The unreflective character of their "judgments" about value made a mockery of the serious content of the pre-unification teachings:

> The average person nowadays holds the ancient in high esteem, but looks askance at the new. Those who work out methods must attribute them to the antique sage-kings Shennong and the Yellow Emperor, and only then will they be admitted into the debate.... Today, if the writings of new sages were to be taken and labeled as writings by Kongzi or Mozi, there would certainly be many disciples who would wag with their fingers excitedly [in a gesture signifying warm approval] and accept them uncritically.[151]

Arguably, however, Yang Xiong and his peers in the late Western Han, even as they relished more recent narratives like the *Zuozhuan* and *Shiji*, were the first generation truly to feel passionate about antiquity (*haogu*) in the specific sense of wanting to *restore* a distant past they believed to be significantly better than their own time and organized along very different principles.[152] The early to mid Western Han courts had prided themselves on the unprecedented achievements of their strongly centralizing rule, and the prominent thinkers of the time had uniformly celebrated the newly unified empire and presumed its benefits so long as it was guided by the right sort of men. But the pitfalls of overcentralized rule had become apparent during Han Wudi's reign in mid Western Han, when Wudi's foreign wars of aggression, angry rejection of criticism, and preoccupation with his own immortality had left many of his right-hand men dead and the imperial treasury bankrupt. In the view of Yang Xiong, his followers, and some of his famous peers, it was necessary to revive ancient institutions and practices, and a viable plan for such a restoration became feasible only through accurate reading in the most ancient texts. Thus, when Yang spoke of one remote sagely figure, the Duke of Zhou, he portrayed him less often in his usual guise as capable Zhou regent than as author and influential model for a successful life lived through his writings, a portrayal distinctly at odds with that offered by more conventional advisors at the Han court.[153] But given the slow pace of copying and memorizing in a manuscript culture largely dependent upon oral instruction, the final authority of even the most renowned model and his writings could be trusted only insofar as circles of admirers committed themselves to preserving and transmitting their favorite texts intact. At the same time, as we have seen, every act of copying in early China was seen as an invitation to reinterpret the text. Thus, the sincere admiration of successive generations did not prevent emendations to the Classics and masterworks during centuries of transmission, as Yang readily acknowledged.[154]

Deeply troubled by the carelessness with which his classicist peers had handled past traditions, evaluated early texts, and promoted the exemplary figures they knew from texts, Yang set out to change the very way that manuscripts of the Classics and neoclassics would be handled. In company with Liu Xiang and Liu Xin,[155] Yang became the leading advocate of a countercultural "return" to classical learning that really represented — as such "revivals" generally do — a reinvention of the revered past accomplished through a careful selection of useable features. (That does not mean that Yang and the Lius invariably saw eye to eye. At points, Yang and Liu Xin traded insults.)[156] Given the fame of Confucius as a "praise-and-blame" historian, Yang Xiong's self-conscious claims to being a good historian speaking truth to power were a profession of his ardent love for the antique sages, just as his innovative proposal to use pre-Qin inscriptions to correct the texts of the Classics (see below) was part of a concerted campaign to restore integrity to the court-sponsored traditions associated with Confucius. Apparently the *haogu* campaign was effective in part because a cash-strapped throne was forced to come up with new ways to shore up its legitimacy, and linkage to the honored past, largely constructed in the imagination, was far cheaper to promote than public or palace spectacles or wars. In any case, frequent allusions to pre-Qin sources in authoritative arguments became commonplace in Yang's time, where before they had been comparatively rare, even perfunctory.[157] Meanwhile, Yang took considerable pains to challenge the supremacy of the literary genre that he excelled in, which he deemed a diversion from more edifying works, though *fu* had dominated the literary scene at court for at least a century, since the time of Han Wudi (r. 141–87 BCE).[158] In his estimation, not even the best *fu* could ever hold a candle to the Classics, masterworks, and neoclassical works in beauty or in the moral and practical effects they encouraged.[159]

Yang Xiong did not hesitate to excoriate members of the Han court for abandoning the model of the sages and preferring the methods and institutions associated with those fabled tyrants Jie and Zhou of old and the First Emperor of Qin. Indeed, Yang's contemporaries would have found some of the very passages in Yang's writings most likely to strike today's readers as insufferably insipid and eminently forgettable as harsh, if suitably veiled criticisms of the reigning emperor, his immediate predecessors, and their inner circles of advisors, all of whom he charged with falling far short of the sublime model of the Ancients. For instance, an innocent-sounding query in the *Fayan* about the advisability of courtiers amusing an emperor by playing chess with him was in fact an allusion to Xuandi's (r. 74–48 BCE) purported inattention to the matters of state and his unconscionable slyness in twisting a phrase ascribed to Kongzi to defend his weakness for games of chance.[160] Few of Yang Xiong's peers would have missed this.

Episodes in Yang's life and passages in his writings reflect his fervent desire to "play with the Ancients" in more demanding ways more consciously faithful to the Master, Kongzi. First, Yang labored long and hard as a historian composing chronicles (no longer extant, but evidently incorporated into Ban Gu's *Hanshu*), while delivering "praise and blame" (that is, pointed historical judgments on major historical figures) in no fewer than four chapters, or one-third, of the *Exemplary Figures*. Like Kongzi himself, he asserted the profound play of history in the present by such endeavors.¹⁶¹ As a result, Yang was thought of as "a historian's historian" by chroniclers of later ages, including the influential critic Liu Zhiji 劉知幾 (661–721).¹⁶² And second, Yang spent nearly three decades collecting many examples of pre-Qin script and obsolete regional expressions in the hope that their methodical application to textual studies would breathe new life into the manuscripts of the Classics and make them more reliable guides to the past.

For those in power, the chief subject of epistemological inquiry in pre-Han and Han times was not "knowing facts," but "knowing men": knowing how to gather and weigh evidence about the character and proclivities of candidates for office. Accordingly, aspirants to greatness sought in history a useable past and a secure professional status. Deliberation on exemplary historical and contemporary figures gave historians the material from which to construct impressive arguments about controversial topics before the court that were "worthy of careful consideration."¹⁶³ Pointing to contrasts between the time of the sages and the present day afforded authoritative authors such as Yang numerous opportunities to take stances and provide rationales for them.¹⁶⁴ Adopting the role of impartial judge over historical figures and events could elevate an author in the eyes of his contemporaries, given the widespread assumption that only the most astute of men, the sages, understood how to evaluate men and their activities properly.¹⁶⁵ That the root word for "investigate" was cognate with the word for "the aged" or archaic (*kao* 考 = *lao* 老) was hardly lost on men such as Yang.¹⁶⁶

Yang as Historian, Conversing with the Ancients

In line with his own strong historicizing impulses and the increasing interest among his contemporaries in the distant past, Yang Xiong sought to define himself as a "praise-and-blame" historian on the model of Confucius. For example, Yang's *Hanshu* biography, based on his autobiography, gives this account of the circumstances leading to the compilation of the *Fayan*: "Whenever people came to pose questions of Yang, he would invariably use a method [of consulting the histories and Classics] to answer them, and thus he came to compose the thirteen *juan* of the book made in the image of the *Analects* that is entitled *Exemplary Figures*."¹⁶⁷ While almost no one today thinks to place Yang Xiong in a long line of eminent "praise-and-blame" historians, that great Tang historian Liu Zhiji 劉知幾 (661–721) compared his own project to Yang's writings on history, which for Liu included the *Taixuan* and the *Fayan*; this was the highest commendation that Liu could confer.[168] Much closer to Yang's era, Wang Chong 王充 (27–97) testified to the existence of a chronicle compiled by Yang for the reigns of the Han emperors from Xuandi to Pingdi (r. 1 BCE–5 CE). Future courts, Yang reckoned, would benefit from his recounting of the follies of the late Western Han courts that he had heard of or known, so long as he showed the same degree of courage as Sima Qian in his monumental compilation,[169] and we know that the Ban family in the Eastern Han consulted copies of these chronicles (now lost) when compiling their own history, the *Hanshu*.[170] *Hanshu* cites Yang explicitly in several chapter appraisals, in language that echoes Yang's approach in his *Fayan* evaluations.[171] When describing the state of historical studies in their own day, the Bans likened Sima Qian and Yang Xiong to "the Yellow and Han Rivers"—that is, the two great reservoirs into which all lesser streams flowed.[172]

The continued coupling of the names of Sima Qian and Yang Xiong in the post-Han period almost certainly owes a great deal to Yang's decision to champion Sima Qian's work as a "true record" of antiquity at a time when it was barely in circulation, even at the Han court.[173] That said, the concision of Yang's historical evaluations in the *Fayan* and elsewhere presents a stark contrast to the lengthy judgments recorded in Sima Qian's *Shiji*. Yang felt Sima Qian's full-length treatments of exemplary figures, good and bad, far too sympathetic to the morally dubious types among them. This probably spurred Yang to compose for a host of historical figures the pithiest of final judgments, only one or two Chinese characters in length—judgments that in many cases reversed those offered in the *Shiji*.[174] Many of Yang Xiong's historical evaluations anticipate Ban Gu's curious "Tables of Men, Ancient and Recent," or the "Pure Talk" (*qingtan* 清談) epigrams favored by the leading figures populating the fifth-century *Shishuo xinyu* 世說新語. But, far more importantly, his forthright, even caustic pronouncements show Yang's unwavering support for dovish policies, periodic land-redistribution schemes, and low taxation rates on farming households.

Yang Xiong's mature writing was married to his delight in beauty and stylistic rigor, despite his propensity to focus on deliberate narratives of actual events; like Confucius before him, he was worried lest his followers be misled by "empty words" detached from human realities.[175] Yang's motivations in writing history, as in writing everything else, seem to have been complex. Over the years, his admirers have keenly regretted the loss of so many of his historical pronouncements and allusions. Clearly, Yang's writings in conscious emulation of the Ancients, like his serious philological work, cannot fairly be condemned as mere impulses toward nostalgia or antiquarianism.[176] Yang's construction of the Qin First Emperor as a double for Han Wudi drastically revised earlier assessments of these two larger-than-life rulers, setting the tone for later treatments of the centralizing program of the Qin and early Western Han rulers.[177] With equal force, Yang laid out the unmistakable outline of a program of institutional reforms that resonated with the Mencian prescriptions for benevolent government.[178] To bolster such reforms, Yang's "imitations of the Ancients" were designed to appeal to those with the requisite cultural sophistication and access to power or wealth, so that they might reinvent themselves and their society.[179] His contributions to classical learning lent color and substance to the Eastern Han and post-Han discussions of the Classics and paracanonical works. Due to Yang's influence, settled convention within a century of Yang's death virtually required a person to measure himself and others against the magnificent example set by the Ancients, especially when judging the quality of a set of writings. Chiefly through such comparisons between past and present, the wise person could both "see himself" and "make himself seen" (*zi xian* 自見).[180]

Yang as Philologist, Playing with Old Graphs

Yang Xiong entered court at a time when knowledge of the Classics had long been held to be one qualification for office-holding, even if birth and connections played far greater roles in selecting the candidates for prestigious posts.[181] All manner of polemical pieces, official pronouncements, and policy proposals alluded, if sometimes perfunctorily, to the Classics. Still, classical learning before Yang's era had been too often relegated to the prosecution of treason trials or to portent interpretation.[182] A few portent specialists were even manipulating hallowed interpretive traditions to predict the imminent collapse of the dynasty.[183] Yang Xiong despised as mere "clerks" and hacks the *wenli* 文吏, the scribes and functionaries who kept the penal system functioning, and he advised followers to look to the present court, not the stars, to ascertain the likely course of future events.[184] Fortunately for Yang, the word *xue* 學 ("study" or "learning") still retained its old sense of "emulation," in addition to "learning to govern" or "learning a special technique" (for example, swordplay, dice, or divining).[185] The concept *xue* extended to serious study of a master or model,[186] not just "studying" texts, even the texts of the Classics and masterworks. And by visual and aural puns, learning and emulation of the classical masters (*jiao* 教) was supremely efficacious (*xiao* 效), as well.[187]

Despite Yang's standing as one of the best, if not the best of the court classical masters, it is doubtful that he, with a few likeminded classicists, could have swayed his peers of the necessity to embark upon a coherent *haogu* program had not multiple chance discoveries of pre-Qin commemorative bronzes inscribed with archaic script generated so much excitement while posing serious challenges to experts claiming to know the distant past, in particular its rites and music (Figure 6.4).

One sparse account in the dynastic history of a find made near the capital a few decades before Yang's arrival at court registers the promise and predicament of the new finds: "At this time [ca. 51 BCE], at Meiyang [part of Fufeng County, in the capital district], they found a tripod, which was duly presented to the throne. The emperor ordered his officials to discuss it at court. The majority thought it ought to be presented to the ruling line's ancestral temple, since this had been done with a similar object during Wudi's reign." Zhang Chang 張敞, then governor of the Metropolitan District and one of the few early experts in the *Cang Jie pian*, a word list composed (at least partly) in pre-Qin script,[188] "was fond of reading archaic graphs," and so he tried to make out the inscription. He believed it had to do with the founding of the Zhou dynasty, an event worthy of such commemoration, which supposedly took place near the Han capital of Chang'an. He then summarized the main import of the inscription for the court: "The king orders the official in charge to make his office at Xun City [near Bin, the ancient Zhou capital]. The king confers the official banners depicting the *luan* bird; ceremonial robes with ax designs on them; and spears with engraved designs. The official in charge, in accepting the king's decree, said, 'I make bold to carry out the grand decrees of the Son of Heaven.'"

Figure 6.4. Western Zhou bronze wine vessel, excavated from a late Western Han tomb in Jiangxi province Guodun Shan 墎墩山 (2011–), first reported in 2016, height 360–390 cm, diameter 120–390 cm. For reference, see *Wu se xuan yao: Nanchang Handai Haihunhou guo kaogu cheng guo* 五色炫曜: 南昌漢代海昏侯國考古成集, ed. Jiangxi Sheng Wenwu kaogu yanjiusuo, Shoudu bowuguan (Nanchang: Jiangxi Renmin chubanshe, 2016), p. 63.

A Western Zhou bronze excavated in 2011 and published in 2016, from the tomb of a deposed Western Han emperor who reigned briefly in 74 BCE, for twenty-seven days, and was buried, far from the capital in 59 BCE, in his fief in today's Jiangxi Province. The inclusion of this bronze among the munificent Haihunhou grave goods suggests a fledgling interest in the antique, at least among some connoisseurs, already in Xuandi's reign (74–59 BCE), which would continue during Yuandi's reign (49–33 BCE) and become a full-fledged *haogu* movement during the reign of Chengdi (33–7 BCE).

Zhang hazarded the guess that this tripod in all likelihood recorded a Zhou ruler's decision to reward a high-ranking official and that the official's sons and grandsons had the tripod cast to honor their ancestor's merit, in which case, the tripod originally would have been stored in that family's ancestral temple.[189] Notwithstanding Zhang's cautious rhetoric underscoring his determination to obey the throne, Zhang's discomfiture is obvious. Evidently, he felt he had better apologize for his inadequate understanding of archaic script, since he was a recognized expert in pre-Qin writings.

Incidents such as these attest the enormous difficulty erudite men and women trained in clerical script experienced when trying to read the small and large seal transcriptions that had been in use less than two centuries before, especially when new readings might trigger politically sensitivities. Although moderns tend to overlook this fact, the imperial decision made shortly after unification in 221 BCE to replace the more elaborate pre-unification seal scripts with a simpler, unified clerical script in official administrative documents (and more widely, under the two Han dynasties) constituted the single greatest break in Chinese history in reading and writing practices before the massive conversion of "complex characters" to "simplified" in the mid-twentieth century.[190] Yet antique artifacts continued to surface during Yang's own lifetime, including a remarkable cache of sixteen old musical chime stones (*qing* 磬) in Chengdi's reign, widely hailed as a highly auspicious sign from above.[191] Since the pre-Qin bronze inscriptions were typically regarded as omens sent from Heaven to the throne, and Yang Xiong was a highly respected authority on omen interpretation,[192] Yang, in company with the other members of the court, must have been alive to the new possibilities of reading the hoary past via pre-Qin inscriptions.[193] Plainly, Yang could hardly *not* formulate a response to such discoveries, when he was among the scholars selected by the Han throne to devise the imperial holdings for the new palace libraries and he had moreover been commissioned to summarize the results of a state-sponsored conference sparked by the pre-Qin writings.[194] In addition, Yang had received expert training from several Shu masters in the reading of "strange scripts," that is, rare graphs in use prior to unification.[195]

The result: Yang, along with several of his most learned peers, a group that included his mentor, Liu Xiang, and his occasional rival, Liu Xin, set out to devise the "most reliable, upright, and good method" to decipher classical texts so as to determine the lessons handed down from the Ancients. They intended to put to good use their knowledge of the pre-Qin writings,[196] as well as their privileged access to old texts and archaic scripts,[197] in service to a new endeavor: textual criticism. (Note that their general preoccupation was with understanding old scripts. Only Liu Xin was singularly enamored of *some* parts of what is now anachronistically dubbed the Old Text or Archaic Script corpus.)[198] As an acknowledged classical master who was particularly steeped in philology, Yang Xiong himself spent a total of twenty-seven years compiling a major philological work that he deemed fundamental to the responsible decipherment of the Five Classics and the necessary excision of extraneous accreted readings attached to the Classics by careless scholars in previous generations (see below).[199]

The *haogu* reformers proposed at least four constructive solutions for what they deemed a regrettable inversion of the true values of competing textual traditions. Liu Xin asked the court to insist that its Academicians avail themselves of a handful of writings once transmitted in earlier script forms, even if newer recensions of the same texts employed modern script.[200] The response from Chengdi and his Academicians was predictably hostile, for to patronize a brand-new set of texts would be to abandon two centuries of court-sponsored learning. Obviously enough, the *haogu* movement, in offering itself as a viable alternative to the reigning culture of learning at court, also represented an implicit rebuff to it.[201] Liu Xiang spent the last years of his life implementing quite another strategy, that of activist editing, though we cannot know what hermeneutical principles (or even political calculations) he acted upon when he decided to excise passages and occasionally entire texts from the record.[202] Two additional strategies—a specific method for reading and an epigraphical turn—we now associate with Yang Xiong.

In one, Yang repeatedly asked his followers to commit to a set of reading practices capable of fostering a common set of values or orientations.[203] The key, he said, was to spend more time in reading the Five Classics and the neoclassics, partaking of their spirit until one became cognizant of the special pleasures that they offered. Yang boasted that there was hardly a book that he had not read,[204] and extraordinarily wide-ranging reading was a requirement for those who would cleanse the pre-Qin writings of the numerous flaws, interpolations, and lacunae that the texts had acquired as they "were copied over and over again and passed down through the generations."[205] Accordingly, Yang Xiong's ideal lover of antiquity would be well versed in the many "arts" (*yi* 藝), meaning the Classics, and conversant, too, with the *Erya* 爾雅, a lexicon organized by synonyms and written, at least partly, in archaic script.[206] Still, broad learning was always wasted unless it went hand in hand with a rigorous selectivity in choosing among the various early traditions in writing.

Selectivity for Yang meant, above all, elevating the Five Classics to premier status among the pre-Qin texts. For that reason, Yang's *Exemplary Figures* presumes to assess the real strengths and weaknesses of various works deemed "classics" by Yang's peers, including some of his disciples. These books included the *Laozi*, the *Zhuangzi*, the *Lüshi chunqiu*, the *Huainanzi*, the *Shiji*, the *Cycles of the Yellow Emperor* 黃帝終始, and the poetry attributed to Qu Yuan, as well as such military classics as the *Art of War* and two pieces of unknown length (now lost) entitled the *Swordsman's Treatise* 劍客論 and *Army Regulations for the Colonel* 司馬兵法. Some texts accorded the title of "classic" Yang rejected on the grounds that their topics were too trivial or their style too prolix or insufficiently dignified.[207] Some he discarded on the grounds of the author's personal failings (the writings serving as a stand-in for the author, in Yang's view), as with the poems ascribed to the suicide Qu Yuan. And still other candidates for the status of "classic" Yang disdained because they promoted antisocial or self-destructive behavior. But most of the books that Yang refused to accord canonical status lacked, in his opinion, either the Five Classics' delineation of social refinement or a comparably reliable and accurate account of things as they really are. As Yang concluded of his list of works to be downgraded (though he was speaking more narrowly of the *fu*), "I saw that the various masters all used their knowledge to gallop off in different directions.... Some engaged in strange convolutions, with hairsplitting arguments and paradoxical language that sowed confusion in the affairs of this world."[208] Thus to "restore" the preeminence of the Five Classics corpus—in fact, to elevate the Five Classics for the first time *far above* other rival classics and masterworks—became one of the avowed goals of Yang, as it was with the other *haogu* leaders.[209]

Echoing the complaints of Liu Xin, Yang insisted that the Academicians, acting like petty "clerks and scribes," wanted at all costs "to protect the flawed and deficient readings" of the Five Classics in order to advance their misbegotten interpretations. From the Qin dynasty onward, the Academicians had merely "maintained their own families' theories," he said, neglecting their other duties "to set out the ritual platters and vessels" in preparation for the rites of the Confucian Way.[210] He therefore reserved his sharpest attacks for the unthinking court officials who considered the Academicians "the most broadly learned scholars of the age."[211] In Yang's day, the Academicians tended to be appointed according to their allegiance to a particular interpretive line (*shuo* 說), rather than to a Classic or *jing* 經, and in the eyes of the *haogu* reformers, the correlation between career advancement and expertise in a *shuo* diluted the impact of the "core" texts on cultivation, interposing an intermediary between the ancient author and the receptive reader and so throwing up a potential barrier to their enlivening exchanges.[212] But a stricter mental separation between texts and their associated traditions, ideally to be reflected in imperial patronage, was not only necessary to prevent the gradual submersion of the core texts (*jing*) beneath the ever-accumulating piles of interpretive readings (*shuo*). Lamentably, many of the influential *shuo* explanations, in Yang's view, were ill founded to begin with — based on little but hearsay or misconstructions of ancient graphs.

Yang therefore queried the common wisdom that gathering all the pseudoclassical and paraclassical texts, along with their disparate *shuo*, "as if in a net," would reveal the overarching meaning of the Five Classics. After all, such indiscriminate gathering missions had not to date ushered in greater clarity of expression or intent. But Yang, not content with simply urging greater selectivity when choosing texts to memorize and copy, alleged that in blessed antiquity, the Classics were routinely mastered without the aid of such massive annotations.[213] A case in point: Kongzi himself had preserved the *texts* and *lessons* of the Classics, but he had done very little by way of explaining those texts.[214] Feigning an innocent air, Yang Xiong joked that were the greatest exemplars of antiquity such as Kongzi to return to life at the Han court, they would either flunk their "Class A" exams or join the miserable ranks of "candidates awaiting appointment," *if* they were lucky enough not to be dismissed outright as incompetents.[215]

Since the ultimate persuasiveness of any strategy depends in part upon the appeal of its rallying cries, Yang Xiong must have been a brilliant tactician, for it was he (if Fukui Shigemasa is correct) whose writings coined or propagated the term "Five Classics,"[216] the rubric that neatly substituted for the notion of competing interpretive traditions backed by rival factions the idea of a cohesive corpus that would promote unified empire. Yang pressed the point that Five Classics learning lent a writer a "perfection of eloquence" that was quadruply powerful, being a convergence of all the best that could be achieved in the literary, imaginative, political, and moral realms.[217] Yang compared the Five Classics to high mountain ranges that inspire awe, to high towers permitting wider vistas, to the profundity of seas, to the sun and moon in their dazzling incandescence, and, perhaps most movingly, to warm and sensuous furs that keep off the chill of winter. For Yang, the "really eloquent disputations" in the Five Classics were a virtual feast for the ears and eyes, even as they served the supremely pragmatic function of acquainting the attentive reader with all the significant prospects of the human and cosmic realms. ("Were we to discard these texts, all disputation in rhetoric would become paltry affairs," he wrote.)[218] With that notion established, Yang argued cogently, if quietly—never pressing the point beyond what was seemly—that his own philosophical masterworks merited consideration as a supplement to the Five Classics corpus.[219] He had presented a supremely cultivated form of homage to the Ancients that offered all the enticements of an intriguing game in hopes that his exegesis might finally dislodge the error-ridden oral traditions promoted by the Academicians.[220]

Yet another *haogu* strategy, which we might call "the epigraphical turn," in retrospect seems central not only to the larger *haogu* movement, but to Yang's individual projects, as well. Supposedly, Liu Xiang, before his demise in 7 BCE, had proposed to the court that he "correct" the Classics using the *Erya*;[221] after all, he had already "corrected" many masterworks in the palace repositories that could be traced to the pre-Qin period. But on apprising Chengdi of his eagerness to undertake such a project, Liu found the emperor understandably loath to sponsor a project that could potentially upend long-standing interpretive traditions.[222] (Recall Chengdi's similar reaction to Liu Xin's proposal, reported above.) Yet the early histories tell us that even without imperial patronage, interest in strange script and strange graphs was on the rise, thanks to the works of Yang Xiong and his close contemporary Du Lin 杜林 (d. 33 BCE).[223]

One special aspect of Yang Xiong's writing was reportedly its inclusion of rare graphs (*qizi* 奇字),²²⁴ and Yang was known for teaching those graphs to adults, including the adult son of the classical master Liu Xin.²²⁵ Evidently, Yang Xiong also agreed with Liu Xiang about the seminal importance of the *Erya*, believing that its ancient origin made it the right text "to correct the Six Arts [i.e., the Six Classics]." The *Diverse Records of the Western Capital* (*Xijing zaji*) tells of a learned official who asked Yang Xiong about the dating of the *Erya* and its intended use. Yang allegedly replied, "This was recorded by the likes of Kong's followers Ziyou and Zixia, in order to explain the Six Arts.... [As they say,] "'The origins of the *Erya* are remote indeed!'"²²⁶

Admittedly, the *Xijing zaji*, in which this passage appears, purports to date to the Western Han, but actually dates to the early sixth century CE, centuries after Yang Xiong, yet it manages to convey the excitement engendered by the *Erya*, which appeared at the Western Han court no earlier than Wudi's time, or even decades later.²²⁷ For the *Da Dai Liji*, a late Western Han compilation in Yang's own era, describes the same lexicon with equal fervor: "The *Erya* is the way to see what is ancient, such that it is surely sufficient to use to discern the proper phrasing. Its transmitted phrases employ images [whose graphic forms preserve older characters], which if reintroduced in one's speech can all be deemed [marvelously] 'simple'!"²²⁸

Lexicons had long been the mainstay of "elementary learning" and philological training (both *xiaoxue* 小學 in classical Chinese), and word lists the premier teaching tool for those wanting to turn fine phrases. Until about Yang's time, however, no one seems to have upheld the tradition that Confucius employed one specific lexicon — presumably an urtext of the current *Erya* allegedly written in pre-Qin script — "to explain the Six Arts" (that is, the Five Classics plus the lost Music Classic), though this belief became settled convention by the Eastern Han.²²⁹ Lexicons could help define the classical corpus narrowly, insofar as they drew upon a limited range of sources. Furthermore, the *haogu* reformers believed that *certain* types of lexicons, such as the *Erya*, painstakingly compiled, supplemented, and carefully digested, could elucidate age-old textual problems and restore integrity to the hermeneutical processes. Granted, some lexicons were useless, being mere word lists with examples of contemporary usage, but a single type of lexicon originally composed in archaic script and grouping words by their underlying categorical connections might reveal the deep structure of human society and the entire universe. Already, in 51 BCE, Xuandi had summoned experts who supposedly knew the correct pronunciations for the "ancient characters" included in the *Cang Jie* 倉頡 primer,²³⁰ causing a sensation in academic circles and creating an imperial precedent and sanction for such endeavors. Not surprisingly, then, in Yang Xiong's time, in 5 CE, under the boy emperor Pingdi and his regent Wang Mang, over a hundred experts were summoned to the main imperial residence, the Weiyang Palace in the capital, to explicate the graphs in the same *Cang Jie*, which by this time, thanks to Yang's additions, numbered roughly 5,340 graphs.²³¹

Meanwhile, Wang Mang and his circle of *haogu* supporters saw to it that renowned teachers of the *Erya* received a special summons to take up office in the capital and interpret omens via the *Erya*.²³² Pingdi—probably under instructions from Wang Mang—then ordered Yang to draw up the final report on the teachers' findings. Over time, Yang's philological projects were to yield at least three major lexicographical works: the compilation we now know under the title of *Fangyan* 方言 (usually translated as *Dialect Words*, but just as plausibly *Correct Words*);²³³ a glossary; and a one-*pian* summary and expansion of the *Cang Jie*. The last two works are now lost, almost certainly because later scholars, including Xu Shen, built their own word lists and philological studies on Yang's efforts and so obviated the need for anyone to preserve Yang's earlier and less complete versions.²³⁴ Notably, Yang Xiong devoted some twenty-seven years to compiling his lists of "obsolete expressions" and "regional sayings,"²³⁵ a figure that dwarfs the time he spent on each of the two philosophical neoclassical works for which he is chiefly known today. This should give us some rough sense of the relative importance to Yang of hard-won "truths" garnered from etymology and epigraphy.

Since modern lexicographers generally put little faith in tracing words back to their origins, it behooves us to consider what Yang and his peers thought they might accomplish by investing so much time and effort in that sort of project. Under the influence of what might be called an "etymological fallacy," the history of words—far from representing a neutral record of changes in meanings over time—took on the appearance of a full-blown narrative correlating different values with diachronic differences.²³⁶ Yang, as one of the *haogu* reformers, hoped his research would recover at least traces of the pristine state of language before misuse, dilutions, distortions, and inadvertent errors had corrupted it; through hard work, a prelapsarian language might yet be found to close the gap between sign and signifier. For in the view of the *haogu* reformers, old texts bearing ancient scripts promised more than a straightforward and efficient means of communication, insofar as they were much closer to the primordial ideograms devised by the first sages. Likewise, expressions from the peripheral places that were more immune to changing fashions than the polyglot metropolises could conceivably serve as repositories of older meanings, which might guide would-be reformers in restoring an urlanguage in its purest and most powerful form.[237] Evidently, Yang intended his philological works to function like an *Erya* in reverse: rather than listing the sorts of elevated expressions used in high literary Chinese (virtually a foreign language to most Han subjects) and gathered in the *Erya*, Yang compiled lists of obsolete expressions and regional patois, both of which he thought might preserve early forms of language anterior to the *Erya* itself.[238] Yang and his peers would finally be "rectifying names" in the manner adopted by Confucius — and on an unprecedented scale.[239]

Yang's philological works were part and parcel of a larger ethnographic impulse in the Han asking whether local folk traditions outside the capital region preserved earlier forms of learning.[240] Not to engage in such textual spadework was dangerous, for unless the criteria by which to correct the Classics were found outside the texts of the Classics themselves, contradictory readings of the Classics would inevitably multiply over time while diverging ever more sharply from the sage's ideas and intentions. At that point, the gross inaccuracies of the Five Classics readings would render them fit for little else save citation by the staffs of the hanging judges (*yi xian yu wenli zhi yi* 以陷於文吏之議).[241] From that unhappy fate, Yang and his fellow reformers were determined to rescue the Five Classics, so as to foster a better meeting of the minds with the antique sages whose spirit informed them.

That Yang Xiong, like his fellow *haogu* reformers, had a major impact on the course of Chinese history is undeniable. Historians note that the founding emperors of the Eastern Han, Guangwu (r. 25–57) and Mingdi (r. 57–75) were the first, so far as we know, to portray themselves in highly choreographed settings as patrons of classicism and donors of texts, to ornament the other imperial guises.[242]

(Some of this may have happened with Wang Mang, while he occupied the throne from 9 to 23 CE, but we do not see that in our extant texts.) Enthusiastic adherents of Yang Xiong's vision included most of the major thinkers of the Eastern Han. Xun Yue 荀悅 (148–209) spoke for many when he said of the Five Classics that one should "recite them, sing them, strum them, and dance to them!"[243] Yang had succeeded admirably in making a rarified form of pleasure-taking and pleasure-giving synonymous with high aspirations to embody erudition in one's writings while modeling the morality of the antique sages, honing some of the links first proposed by Mencius between human nature and the pleasures of sharing.

The Pleasures of Reading and Classical Learning

Analects 2/10 argues for a character analysis that closely examines a man's motivations, and, more specifically, what imparts to him a pleasurable sense of ease and security.[244] For Yang Xiong, the supreme pleasures were undoubtedly the allied pleasures of reading and classical learning. Yang's own writing suggests that much of the pleasure he derived from reading good writing lay in discovering the lexical and thematic links that animated it, also the degree to which authors could push experimentation with the formal structures they had inherited.[245] Yang's intention to "use sex and beauty (*se* 色) to illustrate ritual [as an instantiation of morality]" in the manner of one stream of the early *Odes*' commentaries, was itself an attempt, at once teasing and serious, to induce likeminded souls to his cause.[246] It was Yang Xiong's fond hope to cultivate in others a kindred taste for, even an addiction to, the Way (*dan dao* 耽道). Yang's ideal followers would then happily perform their lives in the prescribed dance of playful conventions, precedents, and social desiderata that he and his fellow reformers laid out.[247] Yang's mature neoclassical writings openly solicit reader responses as preludes to self-awareness, self-fashioning, and heightened social engagement.

Because Yang delighted in such play, he dedicated his later years to "warming up the old"[248] in entirely novel ways: composing neoclassical masterworks on the model of the *Analects* and the *Yijing* while putting together what seems to have been the earliest etymological dictionary, the precursor to Xu Shen's famous *Shuowen jiezi*. It was that quality of innovation and openness in his own person and in his writings that eventually made Yang's works the centerpiece of the curriculum at Jingzhou, the most important private academy of the late Eastern Han,²⁴⁹ as well as the chief inspiration for many Eastern Han writers delving into Mystery Learning. Sun Jing 孫敬, for example, proclaimed his own "inner taste for learning"²⁵⁰ in lines that recall Yang, and the historian Ban Gu 班固, writing a hundred years or so after Yang's death, similarly equated "playing with the texts of the Classics" with the admirable Way of antiquity.²⁵¹ One of his self-styled disciples, Zhang Heng 張衡 (78–139), a century after Yang, may have gotten even closer to Yang's mature spirit in a *fu* entitled "Returning to the Fields" (*Gui tian fu* 歸田賦), which describes the pleasures of "bidding farewell to worldly affairs" and "rising above the dust and dirt."

> I am so enthralled by the perfect pleasure to be had in rambling and roaming,
> Even as the sun sets, I am oblivious to fatigue.²⁵²
> Moved by the warning left by Laozi,
> I turn my carriage back to my thatched hut.
> And strum the sublime airs of the five-stringed zither,
> Reciting the writings of Zhougong and Kongzi....
> If I let my mind roam free beyond the material world,
> Why need I worry about honor or disgrace?²⁵³

Yang Xiong would have been gratified by the allusion to the Duke of Zhou and Confucius, as well as the poem's linkage of old writings and exemplary lives to the good life.²⁵⁴ Largely thanks to Yang, it seems, solemn texts once performed at ceremonial banquets, funerals, and sacrifices gradually came to be regarded as a font of welcome refreshment for the spirit.²⁵⁵ Yang's powerful writings had succeeded in ascribing to reading the Classics and neoclassics the ineffable power to transport readers into the pleasurable company of the Ancients. Acquaintance with Yang's writings equipped the reader for a rewarding journey backward in time. Visions of serried delights and the real possibility of setting aside mundane cares attended the thought of conversing with the sages of old. "Rapt in the Way," the heart would be free of the confines of time and space.²⁵⁶ Thus did ideas about the edifying influences of texts and morally exemplary persons become interwoven into the very fabric of early medieval thought.²⁵⁷

Four great biblioclasms occurred between Yang Xiong's era and the Wei-Jin period (220–420), during which Yang's influence continued unabated.²⁵⁸ Indeed, in the intervening centuries Yang's fame only grew, so much so that Yang's insistence on the pleasures of reading and classical learning came to define the person of genuine worth in the post-Han period, at which point reading became a virtue in itself. Li Yanzhi 李琰之 of the Wei, for instance, declared reading a pleasure incomparably greater than a fine reputation or more tangible benefits.[259] Shu Xi's 束皙 (d. ca. 300) "*Fu* on Reading,"[260] clearly took off from Yang's reading program. The polymaths Huangfu Mi 皇甫謐 (215–282) and Ge Hong 葛洪 (280–ca. 343) depicted the sage as someone marked by his love of reading and men of cultivation as preoccupied with textual learning.[261] Yang at the same time figured largely in the imagination of two celebrated poets of the Wei-Jin period: Xie Lingyun 謝靈運 (385–433) and Tao Yuanming, the subject of Chapter 7 in this book.[262] That Yang's own readers fully grasped the main thrust of his proposals is clear from Lu Zhaolin's 盧照鄰 (638?–684?) lyrical evocation, "Silent now and lonely is the place where Yang Xiong lived / Over the years, nothing left but a bed full of books."[263]

The modern disinclination to undertake the difficult task of restoring the rough contours of the distant past is a perennial problem, as Jean-Pierre Abel-Rémusat, writing in 1825, noted: "A bright light illuminated high antiquity, but hardly a few rays have come all the way to us. It seems to us that the Ancients have been in shadows, because we see them through the thick clouds from which we have just come out. Man is a child born at midnight; when he sees the sun rise, he believes that *yesterday* never existed."[264] Yang Xiong composed the stylish *haogu* writings to recover the bright light of day for the Ancients.

APPENDIX TO CHAPTER SIX

Brief Historical Background to the *Haogu* Movement

Before the late Western Han, the courts of the powerful had been unabashedly enamored of the new. In the dominant discourse, new institutions and policy proposals had to be created for the new realities of a newly centralized empire laying claim to vastly more territory than any one of the old pre-Qin states. Frequent wars, population transfers, treason trials, as well as the imposition of unified weights and measures and a single script for bureaucratic purposes[1] — all these threatened massive disorientation and dislocation. This the Han court initially proposed to ameliorate by reorienting society, under its supposedly beneficent lead, to the public good (*gong* 公).[2] By the late Western Han, however, many in the court were in sympathy with Yang's contention that aspects of highly centralized rule had actually undercut the Han throne's ties to its subjects, thereby alienating local populations and calling into question the very legitimacy of the dynasty.[3] The wars of expansion and the steady proliferation of imperial cults had proven ruinously expensive, and extra taxes and labor services had burdened the poorest in the land. If the Liu clan was to continue in power, it had to stanch the flow of imperial spending, even in the all-important state affairs of "war and sacrifice."[4] Austerity was firmly (if erroneously) associated in the *haogu* reformers' minds, if not in the standard rhetorical handbooks, with a return to earlier and supposedly simpler times.[5] Because the extravagant forms of public display inherited from earlier aristocratic eras were draining the resources of the relatively cash-strapped throne in the late Western Han, why not compensate for the privy purse's lack of cash by the creation of additional cultural capital? The construction of the emperor as scholar and evaluator of old texts, even the dynasty's collecting and sponsoring of Five Classics learning — these were relatively inexpensive ways to assert the court's preeminence. Hence the concerted efforts to portray good emperors as scholar-arbiters of the allied fields of classics and history.[6]

CHAPTER SEVEN

Semidetached Lodgings: The Pleasures of Returning Home in Tao Yuanming 陶淵明 and Su Shi 蘇軾

> The old homeland, the old city: just to gaze at it from afar makes the breast swell. Even when nine-tenths of your acquaintances lie underneath those hills and mounds with their tangles of weeds and brush, one still experiences a thrill when seeing it. —*Zhuangzi*, "Zeyang" 則陽 chapter

Home is the controlling metaphor by which we orient and anchor our senses of time and space, the lens through which we refract our recollections.[1] "Those places whose outlook matches and legitimates our own we tend to honour with the term 'home.'"[2] For ideally, "home" is the place that offers sweet release from pretenses and invites us to let out the characters we are or aspire to be, away from harsh scrutiny.

Some argue that "we need our rooms to align us to desirable versions of ourselves and to keep alive the important, evanescent sides of us."[3] Despite our lapses, deficiencies, and quirks, home is a welcome respite from the round of activities associated with work and duty. In addition, home is the primary place where we are schooled in the relative values of things, models, and sensations,[4] largely through ritualized occasions marking the inevitable passage of time that otherwise unsettles those getting on in years. In light of such cravings for the predictable and familiar, a love of home inadvertently highlights the degree to which our identities remain fluid. And since the graph for "coming home" (*gui* 歸) signifies as well giving ultimate allegiance to a person, an ideal, or a place, it evokes in all uses strong bonds forged from particular sensitivities and needs (Figure 7.1).

Figure 7.1. Jerry N. Uelsmann, untitled photograph, 1982. Reproduced by permission of the artist.

Our homes (mental and physical) grow roots whose profound depths tend to be exposed only under great stress. By unknown processes, the roots of one or more trees fused with the timbers of this wooden house and were eventually exposed as the valley eroded, presumably in a succession of storms.

Thus, the term *gui* signifies "lodging value in X," where X is often an beloved persona or place worth emulating. Not merely a safe haven, letting us "come in from the cold," home holds out the promise of "rest" and "approval" and "the auspicious," all *xiu* 休 in Chinese.[5] Homes, in the end, elicit both poetic and philosophical reflections on the infinite curves of human response;[6] these are all rich associations found in the writings of Tao Yuanming 陶淵明 (aka Tao Qian, 365?–427) and Su Shi 蘇軾 (1037–1101), the two subjects of this chapter.

For many readers of classical Chinese, Tao remains the most potent advocate of the pleasures of returning home. For centuries before Tao, poets — many now unknown — had bemoaned the pain of separation from family, lovers, and friends during transfers of posting or periods of exile; these poets had detailed the ordinary insecurities attending travel through strange lands that tended to prompt a heightened awareness of the transience of life. After all, "to be alive is to be vulnerable."[7] But Master Quiet (Jingjie xiansheng 靖節先生), as Tao styled himself, was the first poet, so far as we know, to speak at length about the ardent pleasures he invested in his return home,[8] with home the base and basis of all meaningful relations with family and friends (Figures 7.2 and 7.4). Through such writings, Tao and like-minded peers were able to establish the principle that the deepest concerns do not necessarily spring from the place where the body happens to reside, but rather from the community that most quickens the imagination and the emotions.[9] For most of his adult life, as Tao tells us, he was lucky to have those two locations coincide. And perhaps because so little is known of Tao's life aside from his celebrated return home at age forty,[10] Tao's ascription to his home of an inordinate range of pleasures looms all the larger to readers alive to the newness of his project and the freshness of his images. (The contrast with earlier poets "returning" to landed estates purchased while in office, *gui tian* 歸田, was unmistakable.)[11]

In conjunction with Tao, this chapter considers the case of Su Shi, since part of Tao's impressive fame was surely due to the focus on Tao's poetry almost seven centuries later by a second highly celebrated poet, the court official Su Shi. Su constructed a vivid persona that he ascribed to Tao and then used that persona as a psychic home in a set of "matching poems." There Su took refuge, in view of the Song court's repeated refusals to allow *him* to return to *his* true homes: the site of intimate family gatherings with his brother Su Che 蘇轍 and close kin,

Figure 7.2. "The Three Laughing Masters at Tiger Ravine," a silk scroll by the Zen master Soga Shohaku 曾我蕭白 (1730–1781) [aka Miura Sakonjirō], depicts Tao Qian 陶潛, aka Tao Yuanming (365–427), in conversation with the Buddhist monk Huiyuan 慧遠 (334–416) and Daoist Lu Xiujing 陸修靜 (d. 477).

Later traditions cast Tao, Huiyuan, and Lu Xiujing as the best of friends, despite their philosophical differences and the marked discrepancy in the ages of Lu and Tao, making it nearly impossible for the two to have known each other.

and yes, the court itself, where Su longed to hold office and engage in the suitably refined pursuits afforded by the capital.[12]

Together, the works of Tao and Su limn a world in which the person's past personal history and situatedness figure substantially in the assignment of ultimate value, just as much or more than in calculations of future prospects — the current preoccupation of most Western philosophy. The comparison of Tao and Su allows an exploration of a serious question raised by Tao (and Mencius and Yang Xiong before him): Can imaginative immersion in another figure known from history ever succeed in providing a "true home" and refuge for the weary soul?

Ironically, in view of the importance of home as a prime site of profound pleasures, educated men living from the fourth through the eleventh centuries tended to measure conventional success by the number of government offices held and hence the frequent, if usually temporary, transfers from home. To be stalled in one's career and thus not constantly on the move threatened the sustainability of the comforts of home. Prolonged stays at home (unless for mourning), like exile, whether self-imposed or court mandated, further complicated finances. Not coincidentally, then, in poem after poem, Tao Yuanming reveals his poverty, as did Su in exile. Tao also mentions the constant recriminations from his nagging wife, who was all too aware that authenticity did not put food on the table.[13] Meanwhile, the expectations brought to homecoming could become so high as to be easily dashed. A person, a prospect, or an object is not as remembered.[14] The desired accommodation for personal eccentricities is not forthcoming. Then, as now, such cruel disappointments did not usually discourage people with sufficient resources from trying to seek their true homes in one place or in one cultural icon. For to abandon that search would be to forgo something profoundly human, the longing for a physical or mental site in which to lodge those memories that constitute a person's specific geographies of desire: the paths most traveled, the corners deemed most comfortable, the best-known landscapes and vistas.[15]

To tease out such ideas, this chapter lingers first over the pleasures attending Tao's idea of "coming home." Just as Tao looked to his home to afford him a zone of comfort in which he could drink with friends, compose his poems, and peruse manuscripts (some of them illustrated), before venturing forth on outings in attempts to offset his sense of the surety of his own death and its erasure of personality, Tao's spare and unobtrusive language inscribes a poetic space within the receptive reader's soul, a safe space that encourages felicitous, if brief, encounters with aspects of the wider world. To highlight the distinctiveness of Tao's compositions, which both impart pleasure and lodge sustained propositions about pleasure's repercussions in daily life, this chapter takes as a foil Su's matching poems for Tao's oeuvre. In the intricate game of "matching poems," wherein Su's end rhymes for each couplet had to repeat Tao's exactly, there was the expectation that the poet performing the matching would give the characters a new twist, making for a lively back-and-forth conversation with the earlier persona; imitation per se would be too clumsy. There, Su laid claims to Tao as a kindred spirit and abiding source of inspiration over long years. And since the speaking subject in good poetry is the entire subject, with the Chinese poet believed to be abandoning himself unreservedly to his images, the resonance-reverberation between these alter egos seems at once subtle and pronounced.[16]

Any honest comparison between Tao and Su must foreground the greatest contrast between these two cultural icons at the outset. Tao, by his own decision, came to reside permanently in his beloved old home, in contented withdrawal from a minor provincial posting. Su, by contrast, was committed to serving the state with his brother, and he had received acclaim during his early years at the capital for his examination essays and memorials to the throne. Seldom at home during the last three decades of his adult life, Su "matched" Tao's poems in formal technique, particularly during the ten-odd years when he was condemned to three terms of exile in increasingly unfavorable circumstances that carried him ever farther away from his family and the center of his ambitions, the Song court. Arguably, Su "chose" exile, insofar as he had often been incautious in his writings and was adamant in his refusals to acknowledge any guilt during an era of vicious partisan politics.[17] Still, his status as an exile weighed on Su, and the brave face he put on in his exile poems, not always successfully, is at odds with his earlier literary self-presentations. The difficulty for Su was to reconcile enforced exile from all that he loved with the vision he sometimes projected of himself, that of a remarkable outsider in voluntary exile from a wayward court. Hence the ambivalence he expressed toward both continued service for the court and reclusion, the only two options usually available to men of comparable standing.

An added complication: what modern readers value most in Tao's work does not invariably tally with what Tao's contemporaries and later readers such as Su saw in it. Initially, Tao was pigeon-holed as a pastoral poet, though a "patriarch" of the lesser "fields and gardens" mode of lyric poetry.[18] Few credited him with having much to say regarding the most serious parts of the human predicament, such as how to maintain integrity while pursuing a career.[19] In the late Six Dynasties and Sui-Tang periods, Tao slowly gained adherents among aesthetes who deplored the dominant ornate style of poetry,[20] but only among those not put off by Tao's pervasive air of "haggard dryness" (kugao 枯槁). As time went on, especially from the Northern Song period, readers of Tao sought in him a model for a lofty detachment from sordid careerism, linking this to Tao's palpable love of his home,[21] although Tao's poetic persona seldom, if ever, commends disengagement from local society. Not only does Tao confess to being somewhat puzzled by his own lack of worldly ambitions, he also professes to be deeply gratified by the homely pleasures afforded by the ordinary, with no desire to "rise above" them. So to ascribe to Tao's poetry an ardent love of political engagement is quite bizarre.[22] Yet Su Shi, to take one example, at one point lauded Tao's "historical poems" (nine eight-line rhymed pieces entitled "Written after Reading History") as his finest works, casting them as coded assertions of Tao's unswerving political loyalty to the former Jin dynasty and his concomitant hatred for the usurper Liu Yu.[23] Su had good reasons to invest Tao with loyal motives, because this would justify Su's ongoing interest in a figure renowned for his hostility to public office. A lyric corpus as fine as Tao's invites a wide range of possible readings, and Su had made a similar hagiographic turn with the poet Du Fu.[24]

This chapter offers an occasionally unorthodox summary of Tao's ruminations on the pleasures of returning home, which frequently juxtapose manifest insignificance and true magnitude, in the hopes of imparting some of the characteristic savor of Tao's writings and Su's responses to them. As readers will soon discover, aspects of Tao's and Su's poetry take up themes that have been articulated in earlier chapters. Tao's insistence on remaining grounded in the miraculous present builds on points made in the chapter on the *Zhuangzi*, a text he greatly admired. Su's desire to find a home in Tao assumes the resonance theory explored earlier in Chapter 2 on music and friendship, also the allure of close friendships so many masters recognized, and Yang Xiong's connection with antique worthies, laid out in Chapter 6. As well-trained classicists, both Tao and Su knew the writings of Yang Xiong by heart, and the question of desire's relation to pleasure (central to Mencius and Xunzi) wove together the works of Tao and Su, as well.

Tao in His Place

No one familiar with the great display *fu* 賦 of the Han period—*fu* such as Sima Xiangru's and Yang Xiong's, expressly designed to show off the poet's vast erudition and total command of diction—can fail to note the striking contrast with Tao's poetry,[25] for its effect generally relies upon the simplest vocabulary composed in lines of regular length in short poems on a modest scale. Out of this seemingly naïve language and structure, Tao succeeded in conjuring all the subjects that passed vividly before his mind's eye, constructing a comfortable space for the reader inclined to enter and inhabit it.[26] I say "seemingly naïve," because the poems purporting to relate facts or events in Tao Yuanming's meager biography continually register the poet's fluency in the rarified realm of high cultural literacy.[27] Nothing in Tao's writing approaches modern notions of "self-expression," of course, because that would imply a measure of "autonomy" and a preexisting cult of "creativity" and innate "genius." For centuries, if not millennia, men of the governing elite such as Tao had presumed that human beings achieve their full potential only through forging strong connections with others, living and dead, as well as with the world at large.[28] And so the ostensible subject of Tao's writing—his experiences as he by turns seeks and despairs of finding security—is tied to a firm belief in the inherent goodness of people and the integral beauty of the cosmic order, even if the iconoclastic plainness of his poetry made unprecedented demands of the reader.

Two words borrowed from the English poet Gerard Manley Hopkins can elucidate aspects of Tao's writing. The first is "inscape," Hopkins's term for the rich and revealing "oneness" appropriate to an object, person, or pattern (what some call the "thinginess of the thing").[29] The second is "instress," Hopkins's name for the configurative energy or "moulding force" intrinsic to each living thing that implants in each its vital and distinctive form. For Hopkins, as for Tao, the cosmic whole is the sum of the mysterious resonating orders throughout the phenomenal world, including those within people, conceived as a fundamental unity, ample and fecund — "the dearest freshness deep down things," as one of Hopkins' most memorable lines puts it.[30] The flood of impulses and percepts issuing from the host of integral inscapes duly impress themselves upon the senses in ways that reactualize them in the hearts and minds of those prepared to celebrate this delectable connectedness, which surpasses any form of ratiocination.[31]

"Inscape" and "instress" bear mentioning because Tao, like Hopkins, sought to capture both the deep structures of things, which "nourish their appearances and maintain their essences,"[32] and the outward visible signs of a profound harmony coursing through people and the extrahuman world. In classical Chinese, the concept of *qi* 氣, the configurative energy that shapes and propels the basic stuff of life (also somehow *qi*), approaches Hopkins's notions of "instress" and "inscape," for *qi* supposedly ensures that each aspect of the cosmic order, itself utterly unique, participates in that fluid and resonant macrocosmic order in its own way, in its own time, and on its own terms, that is, in *ziran* 自然, literally "self-so-ness," a concept badly translated either as Nature or as "spontaneity." "Apposite intrinsicality" comes closer. For Tao, human beings, despite their marked inability to understand themselves fully or the things around them, do operate from an integral core capacity rooted in the divine, as the poet's references to *qi* and *ziran* attest. Moreover, the responsive potentials imbedded in the very stuff of *qi* ultimately conduce to the "rich fusion of emotion and scene" (*qingjing jiaorong* 情景交融) that facilitates the integration of the poet's domestic and social lives with his surroundings.[33] Certainly, Tao's poetic persona is enthralled by the shifting visual forms different phenomena assume, reading in them an ineffable beauty and mystery. There is pleasure to be had in following each glance of Tao's, since in his sights, the world is a wondrous "traveler's inn" to which the weary can return for lodging for no cost at any time. This metaphor seems especially apt when locating (physically or imaginatively) a semblance of true feelings and decent conduct in a home, in kindred souls, or in any other sort of safe place. And so in the poems, Tao the narrator looks up and down, inside and outside, like a sage-king of old,[34] beckoning readers to follow him to one scene after another, scenes designed to elicit a wealth of aesthetic responses predicated on a strong trust in the marvels of connectedness.

Tao's most famous poem, entitled "Returning Home," dated to December 405, opens with an autobiographical account that successfully merges Tao the historical figure with Tao the narrator.[35] (If the best lyric poetry blurs author and persona skillfully, Tao is an acknowledged master.) According to Tao himself, Tao lacked the skills to make a good living from the fields he possessed, so in middle age, relatives and friends urged him to accept a minor official post in a town some thirty miles distant from his home in Chaisang 柴桑, in today's Jiangsu Province. Probably Tao's well-connected uncle had brokered the appointment for him. However, it took but a few days on the job for Tao to realize how ill-suited he was to the bowing and scraping of bureaucratic life and how homesick he was: "My natural inclinations could not be coerced, even when hunger and cold cut me to the bone. To go so very much against the grain would merely court various afflictions,"[36] he reasoned. By his own telling, Tao was depressed and "thoroughly ashamed to be dishonoring the aspirations he had earlier harbored in life."[37] His original plan, he tells us, had been to stick it out for at least a year, earn some money, and avoid the worst social censure for quitting a post granted as a favor. But when Tao's beloved sister died, Tao could no longer stand it. "Using the event of her death as an excuse to follow his heart's desire," he fled at once. In all, he had spent some eighty days away from Chaisang. Tao's account then turns to the question of whether the body's urgent calls do not often hobble a person's spirit, which bows to the necessity to do whatever it must to scrape together a living. Tao recounts his internal dialogue:

> ... Why not return?
> Once the heart is enslaved to bodily form,
> Why continue so downcast and, solitary, grieve?
> I know "what's done is too late for censure."
> Still, "the future can be an object of striving."
> Yet I feel I haven't missed the road by all that much.
> I sense that today's right and yesterday was wrong.[38]

The poem then relates Tao's immense joy at first catching sight of his old home, where servants and children are waiting by the gate: "So excited am I that I break into a run."[39]

Taking the children's hands, I enter the house,
Where there's ale to fill a jug.
Lifting the jar, I pour myself a cup.
A glance at the courtyard trees brightens my brow.
Leaning on the south window fills me with pride:
I see how a space barely broad enough for the knees is
Easy to feel secure in.[40]

Mindful of Zhuangzi's capacious narrow space,[41] Tao then goes outside to observe the movements of the clouds and the birds, who, "weary of flight, know enough to turn back."[42] "Stroking a solitary pine" with affection, he ponders the implications of his return:

Let there be an end to ties, with no more wandering.
Let the world and me take our leave now.
Were I to harness my carriage again, what *would* I seek?
My family's affectionate talk gratifies me.
Pleasure in the *qin* and books dissolves my cares.
The farmers will tell me of spring's arrival.
That means work tending the western fields.
I celebrate how well the many things follow the times,
Alert to and grateful for my life's travel on to rest.
Enough, then!
Lodging in form within the cosmos, turning to the little time I have left:
Why not entrust the heart to it, and let things go or stay as they will?
Why be so vexed about where I go?[43]

For Tao, then, the best sort of life that a person with his temperament could hope to achieve was release from irksome official tasks, with a return home to the fine scenery and warm relations he knew from his beloved hometown, an assessment all the more convincing in view of the political hazards of officialdom. In one poem, he sketches his own character: "And since I never reveled in favors, / ... I drank my fill and made my poems."[44] Fervently proclaiming that the court's bestowal of "riches and honor" no longer holds any allure for him, despite the continual references in his poems to wretched poverty, Tao lists the pleasures now at his disposal: he "wants a fine morning to go out alone" (like Confucius) and the chance to "plant his staff and weed and hoe" (like a second legendary hero of his).[45] He asks for time "to climb the eastern hill and let loose long whistles," and last, but not least, he wants to walk to a clear stream where he can compose verses. The stunning goodness of daily life has been reaffirmed, with Tao's dread of mortality mitigated by his bold decision to trust to and appreciate the ever-mysterious processes of change. "Then I can go to my final home, riding on Change / Taking pleasure in my lot, with no recurring doubts."[46] This attitude allowed Tao, equipped with a robust sense of his own limitations, to set his own priorities (in his metaphor, "to travel his own road"), despite the ordinary ills that beset him ("Man's life is truly hard").[47] To the erudite, such lines recall Confucius's definition of sagely behavior: to "recognize destiny and accede to one's lot."[48]

By frequent resort to the single character *xin* 欣 ("to be heartened," even thrilled) Tao, more than any other Han or Six Dynasties writer, signals appreciation for the beauty and rightness inherent in the cosmic order, the immanent model for the social order. Tao uses this word in nearly a quarter of his poems, in some poems more than once. Tao's signature poem "Returning Home" ("Gui qu lai ci"), for example, includes the word no fewer than four times, as does a poem about the sublime pleasures of friendship.[49]

Usually (mis)translated as the anodyne "cheerful" or "happy," *xin* does more than designate just any pleasant affective state. Instead, it refers to two overlapping and allied *relational* pleasures (*jiao xin* 交欣, literally "doubled or compounded pleasures"): a profound trust lodged in a higher power or in an inclusive order that fosters flourishing and a correspondingly heightened sensitivity toward the fine, often accompanied by an appreciative laugh, smile, or glance. Allegiance to this superior power or order (yet another form of "return" or *gui* 歸, in Chinese) means willing service in the furtherance of its goals and a gratifying sense of community engendered by gratitude for the good life that it enables and enhances.

Describing the heart eager to serve a larger cause, a number of passages by Tao invoke a world where "those of one heart" (that is, likeminded folks) reside, travel through life, and may even die together in conscious fellowship with willing hearts.[50] Frequently, *xin* appears in descriptions of ideal ruler-subject relations, where the ruler has earned his subjects' undying commitment, because he has labored on their behalf.[51] (Shades of Mencius, who uses *xin* but twice, in a single speech in Book 1, devoted to pleasure.) *Xin* also connotes the cordial relations among loving family members and fast friends bound together by congenial ties.⁵² (Needless to say, no coercive group merits such devotion.) In light of family bonds, *xin* may even be used of reverent duty to the dead, reaffirming the inviolability of the bonds across the generations. Relatedly, mention of *xin* often crops up in discussions of longevity, insofar as communal endeavors promote greater serenity and better health,⁵³ with appropriate attachments buttressing strength and vigor.⁵⁴ In short, *xin* bubbles up wherever people find added zest in working toward the common good.⁵⁵ Such feelings of trust can attach to a place, however small, and by definition that place becomes a home, as Tao notes in the lines, "Flocks of birds appreciate their perches / And I, for my part, love my cottage."⁵⁶ Hence the perception that the *xin*'s interactions are capable of moving "the very finest particle" of a person's being⁵⁷ and so improving the well-being of parties "heartened by" each other's company.⁵⁸

All the antonyms for *xin* bespeak a sad lack of mutual support. They are "weariness," "worries," "troubles," "shame," "terror," "resentment," "suspicion," "dejection," "lassitude," "grudging service," and so on. For an inability to situate oneself comfortably within a larger scheme of things that is recognizably beautiful and good (both are *mei* 美 in Chinese) is apt to leave the person feeling downhearted, dispirited, or worse.⁵⁹ By contrast, to have found one's place in the larger scheme of things in a supportive environment — to belong — is the precondition, one suspects, for feeling "at home" in one's own skin.⁶⁰ In that happy situation, ordinary tasks become fulfilling endeavors,⁶¹ their simple pleasures recast as blessings, regardless of religious persuasion or specific lot in life.⁶² Tao the narrator in part dwells upon a disarmingly sly pun conflating "pleasing oneself" and "being Tao himself" (*zi tao* 自陶),⁶³ while conceding that such linguistic plays did not invariably clarify much about real life. Affecting "no great concern about what might come," he reasoned that a less than ideal turn of events had at least the merit of being simply the way things were.⁶⁴ As he wrote in a metaphysical poem entitled "Soul," "Do not return to lonely overbrooding."⁶⁵

In early writings in classical Chinese, Tao is admired for knowing who he was and what he needed. By consistently addressing the prospect of death when living, Tao worked his way toward the all-important question of what possibilities exist for leading an authentic life, one suited equally to personal inclinations and to the way things are, variously dubbed as the cosmic order, timing, Heaven, or fate.

> Suddenly, I'm over forty,
> My body bears the traces of where I'm going now with Change,
> Still, the soul's seat for long has been singularly at rest.
> True and steady, gaining ample substance on its own.
> Neither jade nor stone can claim to be as strong.

At the same time, Tao never appears to think his world "the best of all possible worlds."[66] He considers his house burning down not with sunny indifference, but rather with an ineluctable sense of how dramatic loss strangely can make the world come alive: because his beloved shelter is a ruin, [67] he has no alternative but to entrust all he cares for to a single frail boat, and yet he cannot resist noting the night splendor all about him: "Far and clear, the evening of new autumn / High, so high, a moon nearly round."[68]

Tao leaves his readers in no doubt about the viability of competing views.[69] On the one hand, he maintains that physical immortality is impossible ("Whatever flourishes will decline" is a constant refrain).[70] Yet in more optimistic moods, he hints — as Su would centuries later — that a decent man relying on his "true feelings" can transmit his heart whole and intact via his writings. ("My bodily frame has long since been transformed, / But if my heart is constant, what more need I say?")[71] In the end, Tao could hazard no more straightforward answers to the questions he so insistently posed than his one statement of fact and one profession of faith. (As we will see, at some points, Tao credits posthumous fame with giving meaning to life and at other times denounces hope of fame as the last illusion plaguing estimable men.)[72] Depending on his mood, context, chosen genre, and maturity, Tao the narrator tried out a range of contradictory, yet plausible formulations to the following questions, without ever settling on one. Is it wiser to remember or better to forget? Is Heaven's much-vaunted impartiality ultimately an advantage or disadvantage to humankind?[73] For if Heaven really has no biases or predispositions, then its boons will fall randomly on evildoers and the good alike. How can one hope to express the inexpressible? Why does a yearning to find one's true home not inspire a similar longing for the final return to nonexistence?

All these preoccupations of Tao circle back to ultimate questions about pleasure, for any pleasure calculus requires a person to assess when a given pleasure might reach its height, what possible role one's situation or timing might play in providing occasions for pleasure, and the degree to which both the object of pleasure and the seeker's personality determine the level of enjoyment that can reasonably be attained.

Tao's poems, again and again, speak of the mature joys he derived from the most ordinary sorts of exchanges with neighbors and friends. In his verse, he seldom regretted his decision to forgo the pursuit of riches and high rank, though he readily complains about eking out a meager existence "in rags beneath thatched eaves."[74] That hardly counts as a high perch. But, as one modern critic has noted, Tao "was by no means without second thoughts about the more engaged life he had given up, and he fretted about his [present and future] reputation."[75] Still, "pulling in the reins" was infinitely preferable to "losing the Way," in Tao's estimation. And *if*, moreover, he was deeply perturbed about the coups d'état at court — and that is hardly certain, despite Tao's likening himself to a "bird, frightened" by the "wide snare" of politics — that perturbation would only have confirmed for him the wisdom of his choice.[76] And *if* Tao's poems convey the historical Tao's reality, a large caveat, these friends were mostly local officials, even if the conventions of the pastoral genre translated them into fishermen and woodcutters. Few if any people of lower status and incomes would have had the leisure or the ability to read books and solve literary puzzles with Tao.[77]

Tao loved wine to the very last, as the twenty-odd "Drinking Poems" cycle attests.[78] Drink — as much a metaphor for an immersive attitude toward life as a habitual condition for Tao[79] — promised to transport the poet to that state of dreamy exuberance in which youth could be recovered and cares suspended in the company of intimates bound by music and laughter,[80] or, at the very least, an obliviousness to the monumental difficulties of life demanding total surrender to the imperatives of the here and now. Wine, we are told, loosened Tao's tongue, which enhanced edifying conversations about the Classics ("What we discuss are the writings of the sages. / Sometimes with several gallons of wine") and composing poems.[81] Wine lent Tao an enviable measure of "steadfastness in adversity" (*gu qiong* 固窮),[82] he said, a stiffened resolve in the face of troubles and the easing of his frustrations. Drunkenness could even provide ready-made quips directed to those who thought Tao should adopt a more ambitious course, as one selection from the drinking poems shows:[83]

Clear dawn, I heard a knock on the door.
With clothes askew, I went myself to open up.
I asked, "And who, sir, are you?"
An old farmer[84] with a kindly heart
And a jar of wine had come a long way to visit.
Suspecting that I and the times were at odds,
He said, "In rags beneath thatched eaves —
That hardly counts as a high perch!
Since all the ages count the same things good,[85]
I wish, milord, that you'd take thee to the mire."
"Deeply grateful for your words, old man,
My spirit's out of tune with what you say."
Pulling in the reins, in truth, is something we can learn.
But thwarting ourselves, we surely lose the way!
For now, let's enjoy this drink together.
Having come this far, my carriage can't turn back.
(part 9)

The sages were gone long before my time.
In this world, few return to what's true....
If I'm not then happy in drinking,
I will have worn *in vain* a commoner's cap.[86]
I regret my many mixings-up:
You, sir, must forgive a drunken man.
(part 20)

Best of all, indulgence in wine quieted Tao's nagging doubts about human justice, as we read in parts 5 and 6 of the "Drinking Poems":

I built my hut beside a traveled road
Yet hear no carts and horses passing.
You want to know how such a thing is done?
With the mind remote, the place slides far away.
Picking chrysanthemums beside the eastern fence,
I catch sight of far South Mountain:
The mountain air by day and dusk is fine.
In their numbers flights of birds return.
In this there is a deep and simple truth.
I'd expound it, but I've lost the words already.[87]
(part 5)

> There are a million pretexts to go on or stop,
> Who knows which is wrong or right?
> For right and wrong give each other form.
> "Alike as thunderclaps," men's praise and blame.
> (part 6)

Any reader of Han and Six Dynasties stelae knows that "South Mountain" was the principal euphemism for the tomb.[88] Geomantic considerations put cemeteries on the south-facing slopes, whenever feasible, in hopes of introducing *yang* light into the pitch blackness of the grave. Such locations consigned the dead to a space far above the fields and orchards needing to be worked.[89] For Tao, more than for most other poets, rambling among the hills and mountains brought the grave to mind, and yet such climbs helped him somehow to put his own anxieties into proper perspective, easing the transition from the autumn of life (the chrysanthemums) to the wintry grave, since life in all its wondrous complexity was keeping to its rounds.

Wine was hardly the chief comfort associated with home. There was much to revel in, especially the amiable companionship and back-and-forth betokened by those flocks of birds at their return. Wine was but a prelude and pretext for that, as Tao reports in two of his most famous poems, written in reply to Adjutant Pang and Magistrate Ding of Chaisang, the first of which reads in part,[90]

> Beneath a rough cross-beam door
> There is a lute and also books.
> By turns I play and I recite,
> And thus find my amusements.
> Surely you do not think I lack for other fine things?[91]
> I revel even more in this secluded life that has me
> Mornings watering fruit trees and
> Evenings at my ease in a thatched cottage.
> With all that others treasure, they
> Overlook something there I prize.
> Had we two not shared the same likes,
> How could I have drawn so close to you?[92]
> In my search for a good friend,
> I have come upon a man worth cherishing.
> Happy hearts in great accord,
> With houses in the neighborhood.

> That man I hold dear:
> Without fail, virtues delight.
> "I have some good wine,
> a pleasure to drink with you."
> Only then we can set forth fine talk,
> Only then compose new poems.
> One day of not seeing you —
> "How could I not be full of longing?"

The two stanzas of the "return" poem sent to Magistrate Ding in Tao's hometown of Chaisang likewise describe friends sharing their basic values and a taste for good talk.[93] They remind us that Tao lived close to Lu Mountain, a cultural center, so he did not lack for refined companions. Eased by free-flowing wine, the heartfelt exchanges culminated not in further indulgence, but in the mutual decision to return home to rest:

> "I have a guest, a guest!" say the *Odes*.
> Sometimes he turns up and stops.
> A blessing to all in his province,
> He takes each in from the cold, as if they had come home.
> He hears each good deed, as if for the first time.
> It is not only the harmony between us,
> There are the many fine ties that bind.[94]
> Now talking, now just looking around,
> "And so we dismiss our cares."[95]
> Brimming with fellow-feeling as soon as we meet,
> Once drunk, we make our ways back to rest.
> Truly heartened by our close accord,
> Loose and easy is our roaming.[96]

Particularly fine is the way that Tao manages simultaneously to hint at the qualities he looks for in those in his circle (superior men who can make others feel "as though they had come home") and those they discern in him (his receptivity, "as if for the first time"). By Tao's account, it is friendships such as these that sustain him in his troubles, both emotional and practical, since they allow him to be himself, whether drinking or resting, together or alone. It comes as no surprise that a capacity to strike up an intimacy immediately with a good man is the capacity that Tao finds most impressive in his acquaintances."⁹⁷ And surely one of the chief pleasures of reading Tao is the gratified sense that his work, while sweetly refusing court carriages, generously invites us, as new acquaintances, into his narrow lane, to join his pleasurable gatherings with intimates.⁹⁸

Another poem, "Living in Retirement, Composed on the Double Ninth Holiday,"⁹⁹ encapsulates in a mere six lines that mingled sense of wonder, satisfaction, and relief that followed in the wake of Tao's growing certainty that retirement (hence "reclusion" from court) was the right choice made at the right juncture:

> Life is brief, with too many goals always,
> But men like us are pleased simply to be living long.
> Adjusting my robes, I sing alone at leisure.
> Brooding on the past awakens deep feelings.
> Resting on my perch means, in fact, much amusement.
> Just to linger awhile — surely *that* counts as success!

Tao's last couplet expresses the poignant sense of surprised relief at receiving the gift of yet another day to live. For Tao, since life is inherently good, merely to witness another day counts for more than a brilliant career; bare survival sometimes represents a formidable achievement in and of itself. Best, then, "to linger awhile," fully alive to the deep feelings aroused by the day's ceaseless rounds of activities. In a similar vein, Tao gently mocks the career-minded when saying of himself that lacking the court's sponsorship, "I'll cultivate what's genuine in my humble home / And so earn myself a name for goodness!"¹⁰⁰ And in yet another poem, Tao disposes of "man after man" who so "constrain their true inclinations /... Since all *they* care for is their names."¹⁰¹ Because of this appreciation of the way things are, and hence the enjoyments that life affords outside the court, we believe Tao when he writes, "I exerted all my strength for a full belly, / And a little seemed, in fact, a surplus to me!"¹⁰²

At the same time that Tao, on a good day, is fully cognizant of the joys to be wrested from "this single life," he holds out no hope whatsoever for a life beyond the grave: "Man's life is but a conjuror's illusion or a dream / That reverts to nothing in the end."¹⁰³ That being the case, the wise man will concede that

> What gives worth to this bodily frame
> Surely doesn't lie outside this one life!
> But a single life is not long, after all.

It's brief as a flash of lightning.
The staid and stolid within their spans
Cleave to this — but can desires *ever* be fulfilled?[104]

How are we to live our lives to the full, undistracted and undaunted by transformation? Should we rely on the thought that something will live on after us, something worth striving for, as our hope of continuance? Most days, Tao thinks not. Posthumous reputation is the final delusion of otherwise clear-thinking people,[105] and he scorns those foolish enough to barter away this life's ordinary pleasures in the expectation of lasting fame. As Tao's "Poem of Resentment in the Chu Mode" complains, "The Way of Heaven is dark and remote, / Ghosts and spirits inhabit a shadowy realm /... / Alas! Fame after death / That's no more to me than drifting smoke."[106]

Tao's most sustained rumination on the pleasures to be wrought from life — the single piece that has earned him the honorific of "poet-philosopher" in many a secondary study — is his long work entitled "Form, Shadow, Spirit," translated nearly in its entirety here.[107] The preface to the poem reads:

> All men, high or low, wise or foolish, strive and strive to husband every bit of vitality they have. How utterly deluded this is! Therefore I have set out in full the bitter complaints of the bodily form and its shadow. Then, to resolve and relieve the matter, I have given voice to the spirit's own discerning sense of the cosmos.[108] Honorable men who are busy at court should partake of the heartfelt ideas in this and take heart from it.[109]

After the preface, three set speeches ascribed to corporeal Form, Shadow, and Spirit proceed by turn. The first, by Form, is a sober reminder of death; the second, reiterates, this time with a laugh, the impermanence of life as seen from Shadow's own volatile experiences ("Resting in the shade, we seem to separate. / Stopping in the sun, we are never parted. /... This union of ours is hard to make permanent!"), before tentatively broaching an equally painful subject, the impermanence of fame; and the third, by Spirit, after denouncing Shadow for his waffling, forthrightly regards death as the end, but refuses to have death overshadow life.

Here is corporeal Form, giving Shadow this advice:

Earth and heaven endure forever,
For streams and mountains, no changing time.

> Plants observe a constant pattern:
> Frost withers, dew restores it.
> But man, allegedly most sentient,
> Is by no means their equal in this.
> Present here in the world today,
> He'll leave abruptly, with no date set for his return.[110]
> No one remarks that there's one man less.
> Friends and family: will they *really* think of him with longing?
> The things he once used are all that's left
> To catch their eye and move them to tears.
> I have no technique for transcending Change.
> That death is inescapable is beyond further[111] doubt.
> You, sir, I hope will take my advice:
> When wine is offered, don't refuse it with polite phrases.[112]

Form's deep disquiet about its impending demise leads to carpe diem: enjoy the pleasures of this life, since we know of no other. That was hardly a new trope in Chinese poetry. One of the anonymous Nineteen Old Poems once dated to Eastern Han[113] enjoins readers to "Enjoy life while there is still time / How can one wait for what is yet to come?" Form's anxiety was bound to evoke strong responses in readers, for little perturbs people as much as their own mortality. Shadow then responds with a somewhat ambiguous restatement of the arguments in support of the conventional equation between postmortem fame and eternal life:

> To maintain life — no use to talk of that,
> When merely to protect the life one's got is hard!
> In truth, I would love to roam in paradises,
> But they're far away, the roads to them cut off.
> Ever since the time I joined up with you [Shadow says to Form],
> I've known no other's sorrow or delight.
> Resting in the shade, we seem to separate.
> Stopping in the sun, we are never parted. Still
> This union of ours is hard to make permanent![114]
> In the dark time, both you and I do perish.
> And once the body goes, fame likewise ends.
> The mere thought makes my emotions burn.
> "Do good, and you will leave some love behind."
> Follow *this* saying, and you'll wear yourself out![115]

Wine, they say, can drown all sorrows,
Compared with this search for renown, how can it be any worse?!¹¹⁶

Those struggling to ascertain Tao's message often suspect that he means to say, "We die of having lived. We die whatever we do. So we've nothing much to lose if we think of doing a bit of good." Most of Shadow's response refutes the credible, if conservative argument in favor of posthumous fame, said to render a person "incorruptible" (*buxiu* 不朽) to memory,¹¹⁷ regardless of the decay to which the mortal body is prey. For Shadow is more acutely aware, perhaps, that fame shadows a person and is no less fickle a presence, here today and gone tomorrow; when a person dies, whatever fame he or she enjoys in life will soon come to an end. And the slim hope that "doing good" may inspire some lingering affection after death is killed in the caustic line "and you'll wear yourself out!" (more literally rendered as, "How could you *not* use yourself up [in this sort of futile effort]?").¹¹⁸ In the end, Shadow, evanescent and obscure, prevaricates, for if fame dies soon after the body, what sense does it make to weary the body and exhaust its limited resources for such a transient theoretical good? Note Shadow's deep multivalence in Tao's last line, the line that should, by convention, clinch the poet's meaning, for the two-character phrase that begins that line (*fang ci* 方此) could equally well mean "Being focused on this [wine drinking]," "Playing fair and square in this,"¹¹⁹ "To find medicine in this," "A stratagem in this," "To neglect this." "To possess this [insight]," and so on. Astute readers will doubtless notice, too, that Tao places the defense of fame in the mouth of an insubstantial Shadow. Certainly, Shadow's stance has no more heft than its spokesman, and that Spirit is bidden to answer both Form and Shadow underscores an equal dissatisfaction with the proponents of fame and virtue.

Like any good persuader, Spirit begins by asserting unparalleled authority: if indeed Shadow depends upon bodily Form, which is mortal, Spirit must be the one thing exalting men and women far above the rest of the animals, making them one of the three cosmic powers in the universe, on a par with Heaven and Earth and joined to them via the mysterious interlocking cyclical operations of *qi*. The Great Wheel of life, Spirit says, "is an impartial force," vastly superior to Form or Shadow, if irrevocably bound to them. Yet Spirit's advice is plain enough: from time immemorial, even the longest-lived and most renowned of all people have never managed to evade mortality: "Young and old alike face one death / Wise or not, there's no recalibration [of a lifespan]."[120] Wine is no permanent solution. At best, "In daily drunkenness one may be able to forget." At worst, wine in excess may itself "hurry the end of one's allotted years." Spirit then turns to demolish Shadow's justifications for establishing a reputation for morality:

> They say "to heap up good" may hearten a man
> But who will be there to give you the praise?
> Too much thinking harms our lives.
> We'd best trust ourselves to the cycle's going.
> Floating free on waves of great Change,[121]
> Neither delighted nor stricken with fear.
> "When it's time to go, go quickly."[122]
> Do not return to lonely overbrooding.[123]

For a "poet-philosopher" to advise readers to think less about life's most serious questions — the very sort of questions typically addressed by the philosophers — that in itself compels readers to accept the sheer impossibility of breaching life's mysteries through any process of ratiocination. A poetic image eludes causality; it keeps watch.[124] To expend one's limited resources of energy and intellect on an endless series of problems of infinite complexity seems less worthwhile than resolving to entrust oneself to the waves of inevitable change so as to use one's resources in more rewarding fashion during the short time on earth. Tao's formula, "Floating free on waves of great Change /... Do not return to lonely overbrooding,"[125] supplies as good an answer to the intertwined puzzles of morality and mortality as we are likely to get. No activities forestall death, and, beyond the grave, of what use is trifling reputation or any other ordinary good? Better to confront death squarely, in small homeopathic doses, persuading yourself to welcome it — but not before its time. Clear-eyed acceptance of the world may indeed temper a few exquisite joys, but it leaves us more open to others. Only after "recognizing destiny and acceding to one's lot" can a person truly *live* the days in a life.[126] In any case, life's small, sustaining pleasures are not inconsiderable gifts. Thus, Tao urges a series of quiet "returns," each of which seems "like coming back home." Fortunately for the poet, a trace of the vital spirit somehow lingers in his writings, "preserving his own person" (*zishou* 自守) without causing anyone harm.[127]

Tao did not hesitate to intersperse his most lyrical flights with lists of the frustrations and disappointments met in retirement: twenty-two years of eking out a frugal existence, his children growing up undisciplined and incurious, his friends dying or leaving the area, his own house burning down (twice!), his wife's angry disapproval of his feckless decisions. At times, neither family nor friends kept him from being "bitter...lonely, and disconsolate" (*ku...wang wang* 苦惘惘).[128] Tao could only do his best to prevent such difficulties from destroying his appetite for life. For that reason, the motif of forgetting of various kinds threads through much of Tao's poetry, figuring as a precondition for both insight and pleasure, as it had done for Zhuangzi. As Tao says in the fifth song in the "Drinking" cycle, quoted above, "In this [a scene that includes a sunset, a flock of birds, and also the probable site of his future grave] there is a deep and simple truth. / I'd expound it, but I've lost the words already."[129] Evidently the beauty and pathos in the scene have left Tao dumbstruck. But other alternatives crowd in: that Tao saw no reason to communicate a truth in so powerfully affecting a scene; that he thought the scene would prove more moving if uncluttered by facile rhetoric; or that he laments the sheer inadequacy of words to awaken or transport overwhelming emotion, words being but provisional tools employed for temporal and temporizing ends. Better to watch or read than talk, as Tao suggests in his "For My Sons"[130] in lines recalling the sage Confucius:

> In my youth, I studied *qin* and scrolls
> And chanced upon my love for calm leisure.
> When I opened a scroll and found something to my liking,
> I'd be so taken with it that "I forgot to eat."
> And when I saw the trees meshing in shade
> Or heard the birds changing their songs by season,
> This likewise restored in me a sense of sheer delight.
> Whenever late summer rolled around,
> I dozed beneath a window on the north,
> And when a cool breeze came suddenly,
> I'd scorn to change my state with kings.[131]

Rapt attention to the familiar suffices to pique curiosity and hearten the soul. When Tao the narrator, on the point of death, declares, "I had the good luck to become a man...among the myriad things,"[132] we are convinced that he has known this to his core.

A different kind of forgetting obliterates conventions while quieting the ego and feelings of self-importance. For example, in the biography (hagiography?) he wrote for his maternal grandfather, the former chief of staff, Meng Jia, Tao praises Meng for his ability to forget the court, because Meng sought to avoid an imperial audience and refused a post.[133] Elsewhere, Tao states his belief succinctly: "No coach-and-four will ransom you from care, / But poverty and low estate are double joys to share."[134]

Thus, in memorable line after line lauding the humble, Tao equates an ability to forget vainglory with a disentangling from burdensome tasks and outworn ideas and a consequent opening of a path for making contact with one's own past and living in concert with others. Speeding through life in a frantic hunt for glory is no way to enjoy life. As Tao writes, even though some men like him are incapable of *this* striving for wealth, status, and prestige, he would emulate those who "devoted their whole hearts to *that* [this sense of belonging]!"[135] No fewer than forty-eight times does Tao's corpus mention "cherishing" an experience (literally, "holding [it] in one's heart" [*huai* 懷]). Memory's reconstructions of pleasurable events, people, and things are vital to visualizing future pleasures, and with luck, similar pleasures may come again.

A favorite word of Tao's, *mian* 緬 — a word seldom used before him, if the extant sources are any guide — conjures "thinking fondly" of an encounter and "recalling something with affection," with that something quite possibly remote in time or space. Always the resources exist for a good life, even in old age and sickness, when "recumbent journeys" (that is, perusing books on a couch) can supply as many delights as the spring excursions of old to scenic spots.[136] Indeed, "by reading... he [Tao] is transported to a different place — a place no more and no less real" to him than the phenomenal world.[137] Evidently, trust in past associations as presage of what is to come has led Tao to discern the true value of his relations with old friends and close kin. Tao's short poem "Stopping Clouds" clearly shows the migratory birds' return to Tao's courtyard and the budding of flowering trees igniting in Tao an intense longing to be reunited with an old intimate, so that the cycle of return might mirror among men the extrasocial world replete with extraordinary beauty. Stanza 3 reads like this:[138]

Trees in the eastern orchard
Their branches laden with buds again.

> These renewed attractions vie
> To call my feelings out.
> People have a saying for it:
> "With time on the march, in sun and moon,
> Where will I find a close companion
> To talk over these days of ours?"

For all his seeming artlessness, then, not to mention his harping on the benefits of forgetting, Tao did not hesitate to alert readers to the crucial role that memories in the form of classical allusions play in the pleasures of reading his poems. His "Seasonal Round" poem, for example, hinges upon the reader's knowledge of the *Analects* 11/24 passage wherein Kongzi and Zeng Xi, one of his disciples, agree that the keenest pleasure of all lies in setting out in spring in new outfits for a day spent singing, river bathing, and paying respects to the local rain gods at their altar.[139] The third stanza of that poem reads:

> I strain my eyes to the middle reaches,
> Yearning for the clear Yi River [of Kongzi's day],
> Where the boys and newly capped youths, fasts over,
> Took their time to go singing before going home.
> It's the quiet in the scene I love.
> "Awake or asleep," such contacts stir me.
> But those were other days, I fear,
> Too long ago and far away to recapture now![140]

The scene opens with Tao musing on Kongzi's time and place, glimpsed through the beloved passage where Kongzi confesses that he prefers a stroll on a fine day in congenial company to the finest official posting or opportunity for doing good. The pleasures of contemplating exemplars of the past is hardly unalloyed, however, since Kongzi's antique era cannot be "captured now," and some grief attends the apprehension of the gap that separates true greatness from one's meager self: "There's no reaching the ancient sage-rulers, / Sorrow and loneliness are mine."[141] Tao is inexorably drawn to the heroic, even as he is mindful of his own inadequacies.

The question of what and when to forget or remember links to another theme on Tao's mind: whether Heaven is "impartial" or not, since "impartiality" itself had two contrary senses: that Heaven ignores human conduct and leaves it to society to reward or punish and that Heaven is "impartial" in meting out good or ill fortune according to people's merits. Tao ponders this conundrum of Heaven's distance in his poem "Endless Rain, Drinking Alone"[142]:

> The old fellow who sent this gift of wine
> Said, 'To quaff this will make a man immortal!"
> One draught and all my cares are gone.
> Another cup and whoosh! Heaven is forgot!
> But is Heaven, after all, so very far from this?

The proverb "Heaven's Way has no favorites"[143] furthers the suspicion that disaster afflicts the upright as often as the malefactor. For if an anthropomorphic Heaven "constantly supported goodness and aided virtue," such exemplars as Bo Yi 伯夷 and Shu Qi 叔齊 of Shang, Yan Hui 顏回 of Zhou, and Generals Li Guang 李廣 and Wang Shang 王商 of Han would never have suffered so during their brief lives.[144] Clearly, "Azure Heaven is remote," and "human affairs just keep going on."[145] That said, Tao usually overcame the crippling resentment and self-pity associated with Bo Yi and Shu Qi.[146]

Tao's contemporary Dai Kui 戴逵 (d. 395) had written in a *fu* that the wisest of the wise cannot but be perplexed that virtue and intelligence are so seldom recompensed.[147] Tao readily acknowledges his bafflement. But Tao's *fu* "Moved by Good Men's Failures to Meet Good Fortune"[148] shows how very far he had moved beyond the conventional *fu* of frustration, those stylish laments for a topsy-turvy world composed more often than not by those with meteoric careers, rather than by those who'd been thwarted.[149] Tao's implicit response celebrates those who, in or out of office, cherish their principles and labor on behalf of others, from early adulthood until the last breath. In defiance of genre conventions, Tao begins his poem on such gross inequities by recalling once again the singular good fortune of living as a human being equipped with unique sensitivities and sensibilities (*du ling* 獨靈). For only human beings can choose between "handing down their names" and "hiding their brilliance" in reclusion, between idly enjoying themselves and spending their days "aiding all humankind." No fate or lot determines the directions the individual human life may take; people can "act in accordance with their feelings" and take pride in doing so.[150] In consequence, "men of insight" (*da ren* 達人) have been known to "flee salaries and return to ploughing," just as Tao did. "Finding poverty and low status sweet" (that is, to their taste), such men "renounce glory" with relative ease.¹⁵¹ "Seeking a source of value in all the various forms of conduct," men of sufficient insight may even discover that "doing good is pleasurable."¹⁵²

Who cares that one's cultivation counts for naught in the world's eye, unless one is lucky enough to meet with good fortune? Of course, Tao does not pretend ignorance of the ways of the world: "One's bosom may hold jade; one's hands, orchids, / Clean and fragrant but alone: who then will manage to shine?"¹⁵³ But at least, when "Carriage and cap are no glory / Poor hempen robes can never be a source of shame!"¹⁵⁴ Tao's writings astutely answer the conventional *fu* of frustration, in that frustration need plague only self-involved idiots who "proffer a true heart in fond hopes of reaping renown."¹⁵⁵ Besides, doing good for instrumental reasons, rather than in a disinterested way, invites misfortune, as well as frustration.¹⁵⁶ Having proven that Heaven guarantees no reward, Tao concludes with some resolve: "Let trust and appreciation [*xinran* 欣然] in the return home bring a stop to striving" for fame and fortune. He has done well to have "turned down a good price" for his services "from market and court."¹⁵⁷

Having so deliberated, Tao proceeds to illustrate the pleasures of family life and amiable companionship with neighbors and friends. Hence Tao's lovely poem commemorating the naming of his son,¹⁵⁸ also the poem dedicated to Magistrate Ding, which casts the meetings of true friends as "the fulfillment of our wishes" and the basis for serene pleasure.¹⁵⁹ The equally lyrical "Reply to Adjutant Pang" (quoted above) bespeaks Tao's longing for a good friend with whom he can live in perfect harmony.¹⁶⁰ Longing, in fact, in Tao's writings bridges sorrows and blessings, since life's riches can be enjoyed to the full, so long as one learns how to long properly for established connections; on the other hand, with ordinary riches, one can only anticipate loss while scurrying to pile up more stuff. "My anguish grows from my world's changes,"¹⁶¹ Tao writes at one point. Give up the anguish, and there are marvels to behold.

Three more poems—the first two half historical and half autobiographical, with the third a lyric of communion—impart the pleasurable sensations induced by Tao's sense of belonging in his world. Juxtaposing past and present, the poems "speak to us" and, in giving language its full value, call up contact with other things and moods:

Yuan An, stuck at home, in piles of snow,
Let his thoughts wander off, lest he cause trouble.
Master Ruan, seeing the money pour in,
That very day quit his post, knowing
In straw there's always warmth, and
Picked taro's fine for breakfast.
I don't say they felt no distress, but
They never feared their own hunger and cold.
Poverty and riches are ever at war.
Yet when the Way prevails, there are no furrowed brows.
Supreme virtue crowns a kingdom or a lane.
Clean hands bring light to the most distant lands.
(Poem 50, "Lauding Impoverished Gentlemen," part 5)

Master Shang, in former days, held office cheap,
Content to share mornings and evenings with family.
In poverty and low rank, riches and honors,
He recited the *Changes*, mindful of waxing and waning.
Master Qin liked to wander about,
Wandering, each day farther afield.
Going, surely, in search of the "mountain of fame."
The mountains climbed, did he know to return?
(Poem 64, "Appraising Shang Zhang and Qin Qing")

With heights and far-off places, amusing myself.
Screened is the rough beam door.
The brook flows on and on.
I call on my lute, I call on my books,
Looking around, I see a few buddies, and as for
"Drinking from the Yellow River," there's water enough.[162]
Beyond this, all cravings are put to rest.
I recall fondly those of a thousand years ago.
Trusting to my bonds with them, I can wander alone.
(Poem 63, "Appraisal on Paintings," from line 40)

On "Returning Home" (Gui) in All Its Senses

Two main sources now exist for students of Tao Yuanming's ideas of home and pleasure: his corpus of poems and a group of later paintings (from the Song and after) depicting Tao in his home setting, paintings that underscore the later Romantic readings of Tao the culture hero, the detached transcendent, if you will, more than any particular line in Tao's poems.[163] (One recalls in this connection several of Su Shi's letters, especially one sent to his brother, and the thirteenth letter to Zhu Kangshu 朱康叔.)[164] As Tian Xiaofei has observed in her secondary study of Tao in relation to manuscript culture, Tao's style of talking, as well as his themes, "grow out of conventional literary themes ... closely intertwined with concurrent cultural practices," despite the "unique and unusual" feel of his images. Neither Tao nor anyone of his time would have belabored his solitary genius; instead, Tao constructed connections with an extraordinarily dense network of people and things, all of which signified "home" to him.[165]

Judging from Tao's poems, certain stock scenes are critical to Tao's setting a mood: clouds scudding across the sky, flocks of birds in fluent murmurations, a smallish cottage engulfed by pine woods, five willows, a stand of bamboo. Somewhere an oriole sings. At the edge of the southern wild, near the thatched "cottage" of ample proportions, a garden has been cleared and planted with mallows or with sunny clusters of summer sunflowers and autumn chrysanthemums (these last for long-life tea) contending with rows of beans, and, in the farther fields grow mulberries, wheat, and hemp. Neighbors, mostly gentleman farmers themselves, wander over for desultory chats by Tao's rustic brushwood gate, approached by a narrow lane with no deep ruts for carriages. (This gate, needless to say, marked the line of defense between Tao's inner sanctum, center of all pleasures, and the social world figuring as terra incognita.)[166] The vast silence of the sky is pierced by someone whistling. Perhaps a boat moored nearby portends a welcome visitor. Occasional sightings of geese mark the change of seasons or the imminent arrival of heavy frosts and dews.

Most likely, Tao loved cloud imagery not because clouds seemed disinterested,[167] but because clouds are shape shifters and mobile; they epitomize the evanescent, like wisps of smoke or the short-lived mayfly. Clouds also, no less than birds, signal the seasons, though this observation is often lost on city folk. (Tao's poetry often brings to mind the Ecclesiastes line, "to every thing there is a season.") Birds, for Tao, are either "caged" or "uncaged," and while he himself has chosen to go through life "uncaged," flying free did not necessarily guarantee for Tao the autonomy that most Euro-Americans and East Asians in concert seek so avidly in his poetry. Always, for Tao, soaring birds in the wild "belong to a flock," as Tao makes plain in his poem "Returning Birds,"[168] their flight paths synchronizing in marvelous formations until the lines resolve themselves into a single astonishing image.[169] Waterways, by the standard tropes, represent the enduring propensity of the extrahuman world to give of itself, for water slakes, refreshes, and cleanses all things. In giving life, water is thus the great equalizer, giving priority to no one and no thing in its path, adapting to the shapes of the things it meets as it flows, expressing no determination to have its own way. Boats on water foster the expectation that one can and will be ferried onward across vast distances, though boats, especially for Tao, tend to be "returning boats" that seem to know how to follow the currents to make their way home.[170]

Crossing the threshold and entering the house, the saccades multiply: Tao mentions eight or nine rooms, when most farming houses of the time had but one or two. Probably Tao did not often do the hard work of farming, even if he mentions shouldering a hoe for tilling. One of his fields lay at quite a distance from his home, so perhaps he used an overseer or tenant farmers. Servants magically appear as needed. And they are needed, for five boys and at least one daughter inhabit the house with Tao and his wife. Inside the house, we find a jar of ale and cups for guests, a lute, manuscript copies of the *Odes* and *Documents* classics, the *Zhuangzi* and the *Analects*, an illustrated *Mountains and Seas Classic*, and perhaps a sutra or two, since Tao, admittedly by rather late reports, befriended the Buddhist master Huiyuan 慧遠[171] (see Figure 7.2). Possession of this number of manuscripts alone marks Tao as anything but a country rustic, since fair manuscript copies in his time entailed huge expenditures of time and labor or wealth, especially when they contained illustrations. The domestic scene is set in readiness for "laughter and talk." (Shades of Confucius and other learned gentlemen.)[172] "I play, I chant, / And so I find my delight."[173]

Because the imagination fires better when asked to fill in the blanks, scanty props are all that's needed to imprint on the mind's eye[174] a sense of what drives Tao and keeps him alive:

    climbing heights to gaze at the "view into the distance,"[175]
    sharing thoughts with likeminded men about good writing,[176]
    composing new poems by and for himself,[177]
    pouring wine for friends, especially on feast days,
    alternatively, drinking alone while thinking up sly jokes,

perusing picture books and fantasies (that *Shanhai jing*),[178]
reveling in the freshness of the day,[179]
enjoying the notion that he is still alive, seeing it through,[180]
yielding to his ingrained inclinations to "return to authenticity" (or what's true),[181]
resolving to let things pass without holding onto the emotions,[182]
letting himself embrace self-acceptance.[183]

It would falsify Tao to assign to any homely pleasure in this constellation of satisfactions the label of "higher" or "supreme" pleasure, because all merged seamlessly into one another, but one notable omission from Tao's list is the conflation of home with a wife and children. True, Tao's famous "Returning Home" mentions his delight in seeing his family members, his servants, and his old friends, but several other poems plainly refer to disharmony in his home life so severe that it drives him to drink or to seek sanctuary among male friends. Tao's "Reproving My Sons," to take one example, communicates genuine dismay, despite the surface jocularity:

> White hair covers my brow.
> My flesh has thinned out.
> Although I have five sons,
> Not a single one likes paper and brush.
> Ashu is already twice eight years.
> For laziness, he has no peer.
> Axuan's near the age of "setting the will to study,"[184]
> Yet he loves neither books nor learning.
> The twins are thirteen,
> But they do not know how to sum six and seven.
> Tongzi is just on nine.
> He only hunts after pears and chestnuts.
> If my destiny be such, then
> Bring me what the cup holds![185]

Tao's exasperated tone may signify little more than that it was not yet fashionable to trumpet happy domesticity in writing (which probably explains why the Song hagiographies, crafted in an era that valorized domestic bliss, worked so hard to depict him as exemplary in this regard).[186] But if we continue to probe such dislocations in his poems, it may eventually occur to us to try reinsert Tao's seemingly artless collection of images attesting his simple life into their precise literary context: for a century before Tao, at least, a whole poetic genre dedicated itself to detailing the ostentatious pleasure gardens of the rich and famous who were parading themselves as self-declared hermits.[187] That one contrast leads to another, for Tao's treatment of home—that conventional symbol for *lack* of movement (that is, stability) and sufficiency—resonates so powerfully in part because quite a few of his poems suggest that Tao is always on the move and lacking what he craves: in his youth, basic food, clothing, and shelter, but as time goes on, any sense of lasting permanence.[188] (The one instance where the graph for "home" [*jia* 家] is not colored by this air of restlessness and unease proudly lays claim to his descent from General Tao in formalistic language.)[189] A small sampling of Tao's finest lines conveys this unease with impermanence.

> Except the Way, what can we rely upon?
> Except for goodness, what should we strive for?
> (Poem 3, "A Tree in Bloom" [a poem on morality])

> Although the basket employed is small,
> Going on to the end, one makes a mound.
> (Poem 4, "To My Grandfather's Cousin")

> Aware that "Man is one of the Triune Powers...."
> I have had joy in our union.
> (Poem 10, "Form, Shadow, Spirit")

> Just to linger a while—surely that counts as success!
> (Poem 11, "Living in Retirement, Composed on the Double Ninth Holiday")

> In Nature's changes, difficulties may come or not,
> But to follow one's ideals is a level road.[190]
> (Poem 19: "Composed on the first day, fifth month, in answer to Registrar Dai's poem")

> If one trusts in the True, one does not strive for first place.
> (Poem 20, "Drinking Alone in the Rainy Seasons")

> In the flowing illusion of life,
> Cold and heat always jostle and *push*.
> (Poem 37, "Returning to My Old Home")

> In this there is a deep and simple truth.
> I'd expound it, but I've lost the words already....
> "Men must understand what is beyond their ken."
> (Poem 42, "On Drinking")

Always, Tao wrote movingly of the perils of warm attachments to adults, especially friends departed (in both senses of the word), aware that such leave-takings raised the specter of his own end. Tao's "Tree in Bloom," a poem in archaic four-character lines, reiterates Tao's sense that "Man's life is like a traveler's stay, / With ample time for suffering," because "For good luck and ill, there are no separate gates."[191] Even family and dearest friends could disappoint.[192] Traces of disquiet bleed into even Tao's most lyrical celebrations, such as his "Excursion to Xie Brook":[193] ostensibly a carefree description of a fine outing to a scenic spot, the same "Excursion" poem ends on a gay note that rings hollow, after including the somber lines:

> I offer my companions the flask,
> Filled to the brim, each cup is offered round and pledged,
> There is no knowing whether after today
> There'll be the return of such a fine sight or not....
> In our cups, we give free rein to our emotions,
> Forgetting death's "eternal cares."
> Making the most of this morning's pleasures.
> Let us not surety in the morrow seek.[194]

One imagines Tao, with death approaching, turning ever more often to contemplate the full range of meanings assigned the multivalent graph *gui* 歸, which designates in Chinese not only a return home, but also a return to one's final home in the grave.[195] For Tao, "brooding awakened deep feelings."[196] (Similar ambiguities surround the word for "stopping" in Tao's work, as well as travel/travail.)[197] Tao acknowledged, with an oddly warm and comic resignation, that his relatives and friends would bury him, gorge themselves on baked meats and drink at the funeral, and then quickly forget him.[198] Hence the sense of suppressed feeling in Tao's philosophical poem "Form, Shadow, Substance" (as above): "No one marks there's one man less / Friends and family: Will they *really* think of him with longing?!"[199] Singularity of temperament may have reinforced his sense of loneliness. Tao's "Resentful Poem in the Chu Mode" complains in the penultimate line, "Moved, on my own,

I sing a sad song."²⁰⁰ When he was young and vigorous, Tao's sense of singularity had undoubtedly lent him confidence: "In my youth, I was strong and tough; / Grasping my sword, I wandered alone."²⁰¹ But as old age approached, Tao experienced the reverse: "going out alone" no longer necessarily meant to him "going on an outing or a pleasure excursion." Pleasure still meant intensity, but now it was coupled with the thought, "Will there be a tomorrow?" Tao foresaw the final "going out" and return that is death, and yet he could sometimes defiantly speak of death as but a new adventure: "I relish a fine morning to go out alone. / Riding on Change, I'll go to my end."²⁰²

Still, to provide suitable comforts for the home and the tomb as "eternal home"²⁰³ required considerable planning and forethought. Though the pleasures of family and friends are inevitably flawed, the human urge to connect with others, living and dead, like the human urge to essay the inexpressible, is surpassing strong. It is such contrary drives that some of Tao's best poems evoke. One, "Reply to Senior Officer Yang,"²⁰⁴ recalls the thrill of "meeting the Ancients"—the legendary Bo Yi and Shu Qi, along with a host of other sages.²⁰⁵ Such literary encounters occur as Tao contemplates the traces left behind; supposedly they swell Tao's breast and furnish his mind, thereby enlarging his capacities, but Tao nonetheless by poem's end has to arrive full stop at the impossibility of pursuit or exploring the inexpressible: "Such men have departed, their age seems far off. / Yet they swell my breast,²⁰⁶ many ages after. / I've said what I can. My thoughts I can't unfurl."²⁰⁷

For Tao, to point to the inexpressible mystery was to gesture beyond the aesthetic qualities of the classical Chinese language to life's vital energies,²⁰⁸ also to close friendship, with each gesture worth a thousand words and words hardly ever adequate. All the same, to "step back" to *try* to express the inexpressible was only natural (as natural as for birds to cheep to one another, according to *Zhuangzi*'s "Seeing All as Equal"), so Tao wrote poetry as a way to connect his life with those of others. Tao's solution (there is no finer) is framed for likeminded souls: "I'll let my message ride outside the words—/ Who but *you* can interpret this pledge?"²⁰⁹

Three famous poems by Tao—elegies for his beloved dead—maintain that the grave forecloses any possibility of the deceased's return home to the living. This observation would be wholly grim, were it not for the ineffable sweetness wafted by lingering memories of warm relations. "Grief for My Late Cousin Zhongde," "dwelling below at the Nine Springs," declares,

> My mourning robes mark me just another cousin,
> But my love for you once made us brothers.
> When last we clasped hands before the gate
> I never thought that you'd fall ahead of me.
> Through the many realms, there's no evading this.
> Into the dark you've gone, borne on wings of Change,
> To the end of time, a shape never to return.²¹⁰

The intimate love as virtual brothers betokened by the clasped hands—that, no death can ever tarnish or erase. The second verse of the elegy "Offering for My Late Cousin Jingyuan" (Poem 67) similarly expresses Tao's chagrin and puzzlement when Jingyuan, "a decent and humane man," by Tao's reckoning, dies unexpectedly before his time ("Then abruptly you bade the world farewell").²¹¹ The same complaint—you were good, so why was it *you* who had to go, never to return—appears in Tao's "Offering for My Late Sister, Madame Cheng" (Poem 66), which asks plaintively, "I had heard that doing good brings / Good fortune in its wake. / Why, then, was azure Heaven so unfair / That this one went so unrequited?"²¹² That Tao refuses to deem the deaths of good people just and fair makes his poems all the more believable. This is the context in which to analyze Tao's poem "Sacrifice for Myself," apparently Tao's way of composing himself before dying²¹³ and an unprecedented attempt by a poet writing in classical Chinese to have the last word on his own death and obsequies.²¹⁴

Through a cognate, *gui* 歸 ("return" or "death") means *kui* 饋, "to feed by means of sacrifice,"²¹⁵ so let us recall what every middle-period Chinese (like medieval Europeans) knew but many moderns would now deny: achieving "a good death" is one of the Five Blessings available to the noble of spirit, according to canonical works, with the wake bringing welcome relief in a wild revel.²¹⁶ Tao in this final poem ("Sacrifice to Myself") oversees and superintends his own transition from living person to a spirit-ancestor, eliding the usual sharp break between the living and the dead while asserting the enduring character of his own distinctive bundle of inclinations. The relatively brief eulogy opens with a short preface dated to the fall of 427 CE, the season and year of Tao's death, though quite possibly the piece was prepared long before and only revised or published when Tao knew he was on the point of dying. It begins, "Master Tao is about to take leave of this 'traveler's inn' [i.e., his bodily frame] to return forever to his base and proper home," the grave. As the faces and voices of friends gathered for his funeral feast grow ever fainter to the ghostly Tao, Tao orchestrates the following complex response to his imminent death:[217]

> Vast, how vast is the Stuff of the Great Cosmos!
> Far, how far is High Heaven!
> These give birth to the myriad things.
> I was permitted to become a person.
> From the time I became a man . . .
> I have met poverty as my portion.
> "The basket and gourd" were "often empty,"
> And "thin spring clothes laid out in winter."
> . . . .
> Taking pleasure in Heaven, I trusted to my lot,
> And so lived out my days to the end.
> These paltry "hundred years" humans cling to so.
> They live wide-eyed with fear, lest they not have them all.
> Coveting the days, begrudging the seasons.
> Simply holding on is prized by the world.
> And absent, they would further be remembered.
> Ah! I have gone my one and single way,
> Never agreeing with what they deemed right.
> And since I never reveled in favors,
> How could mudslinging sully my person?[218]
> In my poor hut, I remained somewhat aloof,
> Drinking to the full and making poems.
> Knowing the cycles and comprehending fate,
> Who can be utterly free of cares?
> But now that I go to be so transformed
> May I go unregretting, knowing that
> Long life has reached its bourne.
> Alive, I craved and strove for a fat and easy life.[219]
> From old age I have wrested, at least, a good end.[220]
> Wherefore should I long to return? . . . [221]
> Dark is my travel, and
> Desolate the door to the tomb. . . .
> Into nothingness, already sunk,

> From great passions, removed...
> Before I put no prize on fame,
> So why give weight to obsequies?
> Man's life is truly hard, with
> Nothing to do about death,
> More's the pity!

On the one hand, Tao acknowledges that his life has been comparatively "fat and easy," and in death, he will have come to a good end, his integrity intact and his reputation unsullied. But he is not so dishonest as to pretend that he can relish the isolation of the grave and the cessation of normal life. So while Tao's most famous poem, "Returning Home," may speak lyrically of death ("Lodging in form within the cosmos, / Turning to the time I have left") — Tao was much younger — more poems show resignation, as in "Excursion to Xie Brook," seen above.

The *Analects*' injunction to be "steadfast in adversity" — the phrase repeated no fewer than seven times in Tao's poems[222] — urges men of goodwill to stay the course they believe in. This, for Tao, represents the highest ideal that one can achieve in life: to preserve one's own dignity. Tao's archly autobiographical "Tale of Master Five Willows"[223] disavows any hope for lasting fame ("Nor do we know his family or courtesy name."). It sketches an admirable person who comes to the end of his days:

> In his ease and quiet, he speaks few words,
> Envious of neither fame nor fortune.
> He likes reciting books, without
> Searching for involved explanations.
> But each time he gains an insight,
> "In sheer delight, he forgets to eat."
> By nature, he is prone to wine,
> Though his family's too poor to procure it often....
> He has forgotten how to cling to gain,
> And so comes to the end of his days.

In "drinking wine and versifying, in order to please his own ambitions," Tao describes in the piece's final appraisal (*zan* 贊) a man who had learned not to "fret over poverty and low rank, nor pursue empty wealth and rank."[224] That Tao makes such remarks so often highlights an inward struggle on his part. Erudite readers have always noticed that when Tao read history, the figures he most admired were those with the fewest options,[225] who, like Tao, were always "making merry 'neath the cypresses" at their tombs.[226] Tao's "Returning to My Old Home"[227] confesses as much. Coming back to his former haunts, the protagonist sees that while the raised paths between the fields are where they always were, some old homes in the village (and thus some old acquaintances) are gone. The visit back leaves him with mixed emotions, happy and sad, as fond memories of the past jostle with fears for his own future:

> Few of the old neighbors still remain.
> Step by step, I search for traces of times past.
> Some spots especially tug at my heart.
> In the flowing illusion of life,
> Cold and heat always jostle and *push*.
> I'm always afraid of the end, the Great Change,
> Though my energies have not begun to wane.[228]

Tao's preface to his famous "Record of Peach Blossom Spring"[229] reiterates the fragile hope of life restored, coupled with the hard truth of time's ineluctable passing. Peaches, as every reader of Chinese knows, symbolize the conferral of immortality, for the Queen Mother of the West feasted her mortal lovers on peaches to fit them for eternal life in her abode. Written in gently mocking imitation of historical tales, the "Record" tells of an old fisherman who "forgot how far he had traveled" on his journey. Suddenly, he found himself amid a forest of blossoming peach trees that lined the banks of a stream for several hundred paces. Continuing, the fisherman came to a small opening in a hill from which light seemed to emanate. Beaching his boat, he squeezed through the narrow opening, which fanned out onto cultivated fields and cozy houses, with men, women, and children who were somehow startlingly different: "All were at their ease, enjoying themselves, each in his or her own way." Cut off for nearly six centuries[230] from the outside world, to which they had never tried to return, they asked the fisherman what dynasty held sway. He was everywhere treated with perfect courtesy and invited to stay, but soon in his new home he missed the old. His hosts begged him never to reveal their whereabouts, but following the route by which he had come, he carefully noted the way, not so *he* could return, but rather because he intended to divulge the secret to the governor, in hopes of securing a rich reward. But the route the fisherman took could never be found again, not at any rate by the sort of people willing to repay kindness with treachery. (See the cover for one version of the idyllic Peach Blossom Spring.)

The people of Peach Blossom Spring, who had evidently escaped time's depredations, looked not so very different from those the old fisherman knew. True, they used "older" types of ritual vessels and clothes, but they tilled and ploughed and harvested like men of recent eras. But the residents of that curious place had avoided not only the usual terms of mortality itself, but also the customary anxiety occasioned by time's passing, content as they were to let the years roll on; thus, the inhabitants traveled as if wafted on a light breeze.[231] Tao relates the people's sense that they have "more than enough pleasure,"[232] because they feel no need to accomplish everything within a short life span in a world of "dust and noise." And yet even in the midst of paradise, Tao's fisherman yearns to return to his former home, fondly believing — contrary to all his experiences — in the greater pleasures yet to be wrung from life's brief sojourn on earth.

If challenged to name one couplet that best summed up Tao's wariness, for me it would not be the oft-quoted couplet from the fifth poem in the "Drinking" song cycle ("Picking chrysanthemums beside the eastern fence, / I catch sight of far South Mountain" [his grave site] in the distance.").[233] Agreement with Tao's general observations — that one climbs mountains to gain a broader perspective and views one's own life best from its final vantage point — does not entirely redeem such commonplaces. It might rather be the wry couplet conveying Tao's tangle of emotions in his "Living in Retirement" poem: "Resting on my perch means, in fact, much amusement. / Just to linger awhile — surely *that* counts as success!"[234] The same longing to accept change, however much it runs against the grain, comes in a poem describing the beauty of hollyhocks or red hibiscus: " So very lovable up to now, / But nothing to be done when they fade away."[235]

Ideally, people know when it is best to move on, just like the birds and the beasts. (Recall *Zhuangzi*'s story of the piglets, as recounted in Chapter 5.) In one poem, Tao writes, "During farm work, each returns on his own. / Once rested, we suddenly think to invite one another for a drink."[236] He has a deep-seated trust in the nature of things: "Clothing and food must be provided, / And hard ploughing won't let us down."²³⁷ Such lines set Tao apart from the aristocratic Seven Sages of the Bamboo Grove, the Wei-Jin aristocrats famed for their elegant indolence and stylish unconventionality. Literary histories often link them to Tao, but the accounts and tone of the Sages' writings could not contrast more with Tao's. Not for them were impromptu gatherings with friends and neighbors, pleasure in a fine day or in a jar of good wine, or such mixed feelings about home and death. Importantly, Tao never expected unalloyed happiness at all times; by his lights, he, like all other mortals, took pleasure and felt sorrow by turns, even as he sought to realize the full magnitude of life's simple gifts. One aid in this, no doubt, was Tao's refusal to hide the volatility of his moods; expressing them, he thought, might help rid him of the harmful propensity to cling to outworn emotions, as Zhuangzi had advised. The final couplet of eight in one of Tao's least-known poems, "Reply to a Poem by Clerk Hu," reads, in two versions here: "I can't control this keyed-up mood, / And hence my unabashed complaint." "This persistent thought I cannot control, / And wildly indulge a long-lived grief."²³⁸

Parting is often painful, and death, while preventing further partings, worse. At the same time, taken together, Tao's lines on the final return are redolent with "the beauty of things":²³⁹ "For goodness and evil are two things and still variant, / but the *quality of life*, as of death, and of light, / As of darkness, is one, one beauty, the rhythm of that Wheel."²⁴⁰

Perhaps this is the single mood that Song paintings of Tao would extract from all Tao's contradictions. A form of exegesis in visual form,²⁴¹ paintings of Tao from the Song — none earlier survive — lead us to imagine a less complex Tao than we discern in his poems, a Tao closer to Su's later construction of Tao, as we will soon see, in that the Song paintings neatly occlude all the grimmer aspects of Tao's life and writings. To take but one famous example (Figure 7.3), that by Liang Kai 梁楷 (active early to mid-thirteenth century),²⁴² we see that Tao is more feminized, less haggard, more casually elegant, as Susan Nelson has noted.²⁴³ His attention is directed to a point far away, even as he holds a flower, usually the attribute of a bodhisattva. Aside from his lifted chin and distant gaze, his attributes send mixed messages through chrysanthemums, trailing scarves and robes, and a staff in hand. In keeping with the Buddhist overtones, Tao figures in the scroll by Li as a "diviner mover," at the still center while the air

Figure 7.3. Detail of a scroll entitled "Dongli Gaoshi" 東籬高士圖, "Gentleman of the Eastern Fence," attributed to Liang Kai 梁楷 (ca. 1140–ca. 1210), dated to Song or post-Song, silk, length 71.5 cm, width 36.7 cm, National Palace Museum in Shilin, Taipei City. Reproduced by permission of the museum.

Here we see a transformed Tao, transcendent and elegant, dressed in the finest of clothes, looking to the far distance; nowhere in sight is the pervasive air of "haggard dryness" (kugao 枯槁) that we find in his poems. This transcendent figure is similar to how Su would have portrayed himself, in all likelihood.

swirls around his figure (but not those of secondary figures, such as his servants), in the manner of Ma Yuan's *Riding a Dragon*. By convention, the passions are often constructed as "female," in China as in the West, and it's females who typically appear with swirling robes and scarves. (Famous poets are usually treated more soberly, as in the portrait of Bo Juyi 白居易 [772–846] in the Muto collection in Japan.) But the biggest departure in some Song paintings like that ascribed to Li Gonglin is that they depict a Tao "all alone with his shadow," and not enjoying others' company, as Tao did enjoy, greatly. Reportedly, in a second painting illustrating Tao's "Returning Home" (now lost), Li Gonglin posed Tao by a passing stream, sign of life's evanescence and possibly of weeping,[244] rather than in his rustic garden, where all earlier painters had positioned the poet.[245] But it is a third painting of Tao by Li, a "simple" outline drawing in monochrome ink (*baimiao*), whose loss seems most regrettable at this remove, for good reasons: Li was especially famous for those outline sketches, which he thought of as "artifacts of friendships.[246] One wonders, then, whether Li in the third rendering did not finally return Tao to Tao, with the *baimiao* style especially suited both to Tao's lack of showiness and to his steady focus on friendship.

In any case, Tao's sense of fellowship receives far more emphasis in one section of a Freer Gallery handscroll, also ascribed to Li Gonglin (Figure 7.4), which shows Tao entertaining intimate friends. He leans forward, as if to hearken to his guests' conversation, with a large wine jar at the ready, as other members of the household look on.

Su Shi on Tao:
The Consolations of Immersing the Self in Another

Students of Chinese literature know of Su Shi's astounding feat in providing matching poems for nearly all of Tao Yuanming's oeuvre; also of Su's continual expression of his "authentic feelings" for Tao as a man and a writer (Figure 7.5).[247] Su Shi apparently began his "matching poems" in casual fashion while holding an important administrative post at Yangzhou in 1092. He then continued to write them in earnest when forced into the second and third of three unpaid and humiliating exiles from court, in Huangzhou (1080 to 1084), Huizhou (1094 to 1097) and Danzhou (that is, the western side of Hainan Island, 1097 to 1100).[248] The eight years when the vast majority of the matching poems were written, from the autumn of 1092

Figure 7.4. Section 4 of the handscroll entitled "Tao Yuanming Returning to Seclusion," attributed to Li Gonglin (ca. 1041–1106), ink and color on silk, height 37 cm, length 518.5 cm. Freer Gallery of Art and Arthur M. Sackler Gallery, Smithsonian Institution, Washington, D.C., Gift of Charles Lang Freer, F1919.119.

This section of the handscroll shows Tao at home, entertaining his intimate friends, who are pleased to see that he has come home. Horses and bored servants wait outside the gate, and an important-looking guest (Wang Hong?) is about to be welcomed in.

Figure 7.5. Shitao (Chinese, 1642–1707), *Reminiscences of the Qinhuai River*, album leaf, ink and color on Song paper, image 25.5 x 20.2 cm, overall: 33 x 24.3 cm, The Cleveland Museum of Art, John L. Severance Fund, 1966.31.8.

The colophon on the painting reads: "Along the river with its forty-nine bends, I search for the remains of the Six Dynasties along the Qinhuai. Who walks in wooden clogs after the snow has cleared on the East Mountain and composes poems while the wind roars through the west chasm? One must sympathize with the plum tree's lonely state, forever without companions. Its branches proudly stand, with only a few remaining. Blanketing the ground with fallen flowers while spring is not yet over, the plum's sour pit, now still small as a pea, causes men to ponder."

to 1100, less than a year before Su's death,[249] saw dramatic reversals in Su's position vis-à-vis the court, and the dreariest of the exiles, those spent at Huizhou and Danzhou, could not but prove particularly irksome, because the harsh sentences came after more than thirty years of distinguished service in posts at the court and in the provinces. When Su, at the age of fifty-seven *sui*, first set out to imitate Tao's drinking poems, rather casually and with no particular reverence for Tao,[250] he enjoyed some favor,[251] but by the time he had finished, Su had endured long years in exiles so awful that he celebrated the imperial pardon that released him with a final burst of song, in a *ci* 詞, rather than in regular meter.[252] During banishment to Huizhou and Danzhou, matching Tao eventually came to absorb nearly all of the limited energies Su had left after attending to the pressing problems of finding a place to live and a way to feed himself and his family at a time when the court forbade local officials from offering him any assistance. While Su's collected poems suggest that he was sent and responded to hundreds of poems over those nine years, Su still had very few other civilized outlets for his time and energies.

This section compares matching poems by Tao and Su in order to explore a larger question of what kind of writing results when a later author consciously treats earlier writings as a secure lodging place for his own thoughts, lacking nearer alternatives? After all, Su claimed to find in Tao's poetry a personality that functioned as psychic refuge, but given the inevitable differences in situation and temperament, in what sense and to what degree did Su, capable of matching Tao in technical matters, find it easy and natural to "merely entrust himself to them" (Tao's poems)?[253] Can reliance on such remote relations via the well-stocked memory or library — making "friends in history" in Mencius' fine phrase — ever function as sufficient hedge against current feelings of dread and loss?[254] Since Su Shi claimed to model himself on Tao as a person and as an author, Su's writing about Tao offers a perfect case study through which to frame provisional answers to these important questions about remote influences.[255] Judging from Su's poems, the consolations of reading and writing could never equal the pleasures of being with Su Che and friends. We find the "matching" poems, despite all of Su's elaborate artistry, mirroring the disparities between Tao and Su. The intent here is not to take Su Shi to task for failing to match the tone and content of Tao's writings in his "matching" poems.²⁵⁶ The intent is rather to consider why Su's pleasurable "meetings" with through poetic exchanges did not—indeed, could not be expected to—entirely allay Su's anxieties, longstanding literary claims for the transformative powers and consolations of writing to the contrary.²⁵⁷

Su's Possible Motivations in Matching Tao

Since Su's own day, people have speculated about his motives in matching Tao Yuanming's poems. In some theories, Su Shi chose Tao Yuanming for his alter ego because he hoped to style his forced departure from the court as the voluntary reclusion by a noble man above the fray.²⁵⁸ Such a stance would allow Su not only to appear thoroughly at home in places far from home, but also to register his utter disdain for the careerists and sycophants who remained safely at court. Deeming himself high above the common run of men, Su was inclined to take Tao's plaintive cry that he was "frequently at odds with things and circumstances"²⁵⁹ as confirmation of Su's belief that most men displayed their glories on the outside, but were dry as dust within.²⁶⁰ Surely, on some level, Tao's more acerbic remarks resonated with Su's own sense of the illusory nature of life (more Buddhist-inflected), as with Su's contempt for the stupor in which most people stumble through life: "The times we're drunk are the very times we talk most soberly."²⁶¹ Certainly, Su sought in Tao's poems an air of uncompromising authenticity (*zhen* 真),²⁶² with Su ready to credit Tao with the same lofty and transcendent attitude toward life that Su's era made de rigueur for the so-called "great man."

Because Tao was a man who had found a compelling voice while living apart from the court, any concerted effort to "match" such a man constituted a quiet, but defiant assertion of writing's power to transcend time and space, leaving "a lingering flavor that goes beyond the matter at hand,"²⁶³ with or without the court's backing. Su's profession of faith came in response to the court's harsh review of Su's writings, we should recall.²⁶⁴ Su, during his three exiles, was forbidden to write about government affairs, and the court dispatched clerks to review copies of his poems, combing them for satirical lines with seditious content. Luckily for Su, "meeting the Ancients" in poems was one socially approved and tried-and-true method for carving out a moral space where one's own self-image was inviolate. Sharing reminiscences or observations with revered heroes, one could hope to leave a lasting impression of impressive authority.[265]

Others have emphasized how important it was for Su to put a brave face on his own situation, to cheer up himself and his loved ones (especially those living apart) for the sake of their collective mental and physical health.[266] Composing, reciting, and reading good poetry could restore balance to the emotions and even cure physical ailments through a welcome release of tension and nervous energy (*qi*). Thus, the flagging writer was to seek out poetic subjects among exceptional objects in the natural world. Plants that flourished when all else decayed, for example, supposedly had a restorative effect for the vital energies that aided creativity, even if infusions of such might imbue the drinker with too-strong energies.[267] (Let us not forget that Su Shi was one of the two most highly regarded medical authors in the Northern Song, who may have believed in following his own prescriptions.)[268] Then, too, poetry transcribed on paper was wonderfully efficacious at papering over problems; it could minimize disasters and reassure friends and family members that all was in order or soon would be. The air of sunny bravado that so many of Su's matching poems in exile feign — Ronald Egan talks of "determined contentment" — was crafted in the wake of bouts of anger, frustration, and even despair, as we know. And when distances were very great between those aching to reunite, poems provided a relatively inexpensive bridge by which to meet up in spirit with intimates.

Imitations of old poems were a well-established genre by the mid-third century CE, in company with the "reminiscence" (*huaigu* 懷古) pieces expressing nostalgia for a lost and presumably better past.[269] By Su's day, the late eleventh century, it was fairly common for poets to "match poems" by adopting the same theme and the same end rhymes for each couplet. But Su's decision to "match" nearly every one of Tao's poems — not to mention composing multiple counterparts for Tao's "Returning Home" poems — was highly unusual for at least three reasons. First, in Su's day, most matching poems were composed during semipublic occasions, official banquets, for example, in response to extemporaneous poems by the others who were present. Second, Northern Song poets, for public consumption, seldom "matched" more than a few poems by their contemporaries,[270] and certainly none besides Su ever aspired to match the entire corpus of another poet. Third, when the Song poets *did* match more than a few poems by another poet, they typically chose to match the poems of a patron or of one of a handful of canonical poets — Cao Zhi, Li Bo, or Du Fu being the usual choices. True, for centuries before Su's final burnishing of Tao's image, Tao's ranking within the poetic canon had been steadily on the rise, especially during the Northern Song, partly because Tao's taste for quiet understatement suited the prevailing aesthetic in all the arts.[271] Ouyang Xiu 歐陽修 (1007–1072) had singled Tao out for particular praise,[272] as had Zeng Gong 曾鞏 (1019–83), and Sima Guang 司馬光 (1019–1086).[273] Still, when Su undertook his poetic experiment, Tao was not yet routinely ranked among the premier poets working in any genre, just among the best minor pastoral poets.[274] So, while Su's choice of Tao as poetic model was hardly wildly unconventional (contrary to Su's boast), Su's determination to so thoroughly emulate Tao's poetry and character was surprising. But then, creating an eccentric public persona was all the rage by Su Shi's day.[275]

Su made others keenly aware of his claim to have embarked on a novel course when he decided, *after* first composing his matching "Drinking" poems, to "pursue [perfect] attunement to and harmonizing with" (*zhui he* 追和) a figure from the hoary past, matching *all* the poems ascribed to Tao.[276] Tao was a fine poet and a human being sufficiently self-aware for Su to regard him as kindred spirit, potent doppelgänger, and source of inspiration,[277] as Su acknowledged in a letter to his brother Su Che written in 1096, the very year when all the members of the Yuanyou "faction" close to Su were exiled:

> Although the poets of the past have written works that imitate the Ancients, none so far has actually sought to match the poems of an Ancient. The effort to match [all] the poems of a man of antiquity thus begins with me. I hold no poems of a particular poet in particular esteem save those of Tao Yuanming, which are not too numerous [to accomplish the task]. Apparently plain, Tao's poems are actually quite refined, and while they appear to be spare, they have a rich, full substantiality. No one is Tao's equal.... But how could I be fond only of Tao's *poetry*? I have great feeling for him *as a man*.[278]

Based on this passage, we should not discount the excitement Su may have felt at the prospect of undertaking such a venture, despite Su's professions of modesty, as when he writes once that he "is ashamed" before Tao, unable to duplicate either Tao's artistry or his equable temperament. Yet elsewhere, Su proclaims, he "need not be *very* ashamed before Tao," since he is "able to capture [in his own writings] this life of Tao."²⁷⁹ In an optimistic mood, Su claims that he was up to not only the "peerless" Tao Yuanming (an oxymoron that), but other celebrated models, as well. As one line by Su reads, "Paint him [Ge Hong] with Tao Yuanming and me. / We can make a "Three Scholars Portrait." / I am sadly late to study the Way. / But in writing poems, am I not as good as they?"²⁸⁰

To supply some needed background: long centuries before Su, literary theorists had named three levels of difficulty in matching poems, depending on how many lines were "matched" and in what specific way. (Additional rules governed tones and the placement of stresses, but they are too complicated to explain here.)²⁸¹ Naturally Su, who prided himself on his versatility as a poet, often took up the most demanding sort of matching, which required the later poet not merely to employ the same set of end rhymes for his couplets, but also to place the same end-rhyme *graphs* in precisely the same order — a level of artistry beyond all but the very most technically proficient and inventive of poets.²⁸² Unquestionably, Su enjoyed showing off in this way. His boast that he was the first to engage in such a sustained literary exchange must be taken with a grain of salt; strictly speaking, he is pointing to his innovative way of treating an ancient as a contemporary social intimate in an extended series of verse "exchanges." Of the 2,337 poems written by Su, some 785 (roughly a third) "match" all the rhymes of another poet. In addition, it gratified Su to think that his set of matching poems would "enlarge" the reputation of a fellow poet, all the more so, given the restraint and pathos that Su read into Tao's life story. For no one doubted that it took a man of superior insight to identify greatness in others and bestow the proper commemoration.²⁸³ Mei Yaochen 梅堯臣 (1002–1060), Su's senior contemporary and a fine poet himself, once warned of the difficulties of "consigning oneself to poetry," especially poems distinguished by the "understated subtlety" (*ping dan* 平淡) promoted in Song scholar-official circles above the florid and dramatic High Tang style.²⁸⁴ But this advisory Su Shi chose to ignore, for by his own ideals, his poems had to be every bit as visual as his paintings, "light and subtle ink, as if in the mist or a dream."²⁸⁵ Only in his last years did some of Su's poetry become more understated, partly because he had spent so much time with Tao.²⁸⁶

And truth be told, Su may have sensed that the challenge he set himself would be all the greater when it came to Tao, for several reasons. Tao is one of the most unpainterly of lyric poets, whereas Su had once wanted every poem of his to be visually compelling; Su would have to train himself in a new poetic style, and a gradual shift from the verse form using seven words per line that he preferred in his youth to simpler poems with five words per line "in the archaic mode" followed in due course.[287] Tao had waxed eloquent about what he "could not find the right word for,"[288] making it very hard to recapture that indefinable quality in Tao's corpus. Nonetheless, Su believed that making nearly every single couplet rhyme with Tao's poems in some hundred cases (with some "poems" consisting of long poem cycles with up to twenty different parts) would testify not only to Su's extraordinary virtuosity as a poet and his unparalleled command of technical skills, but also, most importantly, to the rarefied aspect of his moral imagination.[289] It might even lead to a new way of writing in forging a new mode of authenticity.[290] For to excel in total matching, Su would have to go well beyond matching rhymes to consider the distinctive aspects of Tao's vocabulary and images with the utmost care. In other words, to produce sufficiently arresting variations, Su would have to delve into the core of Tao's inner being, the source of his inspiration, with all the effort and discipline that entailed. With Tao's writings especially, Su could show others that the highest level of expertise was mere child's play to him. Not only would the same graphs appear for end rhymes in the same order. Often Su created delicious *jeux de mots*, subverting or even reversing the original meaning of the graphs deployed in Tao.[291] Doubtless, the desire to cap his reputation in this way figured among Su's motives to match Tao, along with the host of other motives already cited, for Su was no less complex a personality than Tao. What is relatively clear at this remove is this: so completely did Su intend to adopt parts of Tao's identity that he probably derived his sobriquet "East Slope" from Tao's poems, according to the critics.[292]

Su, as he tells us, was forcibly struck by certain similarities in the two poets' predicaments, especially when both men were out of office and living in out-of-the-way (and to Su) increasingly uncongenial places.[293] Nonetheless, modern readers tend to be more struck by the quite different lives led by Tao Yuanming and Su Shi, even if it was Su's harsh dismissal from court that shifted his attention so decidedly to Tao.[294] Impatient with the pettifogging required in a minor bureaucratic post and possibly worried about the likelihood of coups d'état at court,²⁹⁵ Tao had abruptly abandoned a local post (his first?) after less than three months on the job.²⁹⁶ Tao's momentous decision may well have posed dangers at the time, and some in his own day, as well as later, judged Tao's decision a serious blot on his character.²⁹⁷ Tao's poems, virtually our only source of information about his state of mind, do not clearly state his motivations for leaving, although they supply the pretext in his sister's death, and certainly Tao felt compelled to justify his impulsive move during the nearly three decades that followed. But Tao was resolute. In the aftermath of his decision, he resumed service only briefly in one or another post, each time under duress, either because as a nonofficial he owed the dynasty a set period of corvée service or because dire poverty temporarily compelled him to accept a regular official salary.

By contrast, Su had never wanted to leave office. He was a good administrator, and he knew it, so during most of the time that Su was writing poems to "match Tao," he was more or less openly angling for a return to the court that would lift the dark cloud under which he and the members of his tight circle had been condemned to operate.²⁹⁸ Quite a few of his matching poems aimed to flatter the emperor into recalling Su to active service,²⁹⁹ and Su himself realized that he was continually vacillating between his desires to serve and to abjure all worldly ambitions. As an ex-official out of favor, but hoping against hope to return to the court he viewed as "home," Su could hardly afford to parade the sort of insouciance in which Tao exulted.³⁰⁰ Only in Su's very last matching poems, toward the end of his third exile and shortly before his death in 1101, did he declare that he had finally abandoned all hope of another official position and even of returning to his native Sichuan, which was then plunged in turmoil.³⁰¹ The best conceivable outcome at that point, he felt, was to return to his first place of exile, in Huangzhou (in present Hubei Province), where he owned a mountain retreat in a scenic spot. When the imperial pardon finally arrived in Danzhou, Su quickly reversed that decision; he was determined to make the journey to court, despite age, infirmity, and the prospect of long distances.³⁰² Thus, Su never did manage to get free of the world of the court, even at his most disillusioned.³⁰³ Chronic illness and dysentery from dirty river water finally killed him during his final attempt to secure a new official posting. It proved small comfort to him that the slogan "Return Home" had become a rallying cry for the members of Su's defeated Yuanyou faction,[304] who were no less determined than he to "return" to high posts after their opponents and rivals were undone.

This difference in temperaments and situations surely compounded the difficulties when Su determined to give himself entirely over to Tao's sentiments and writings. For with Su, if not with Tao, "home" signified "family" primarily, especially his beloved brother Che and his own sons. He had written of his sons and grandsons in happier times with pride:

> Noisily the six male children
> Musically intone each a single classic.
> They will, in their turn, grow to adulthood
> In harmony, each a real grown-up by age twenty-one.[305]
> Returned to the fields, they can compose an entire household.
> And for the state, they can fill the service ranks.[306]
> Pu'er has just been learning how to talk,
> Brainy, his jade bones opened to heaven's court.[307]
> Huai Lao, a crane chick broken from its shell, lets out a long cry.
> I raise my wine and survey the scene, far and wide,[308]
> A rush of happiness, I'm ashamed of such ordinary feelings.[309]

In exile, Su would have taken to heart Tao's poems devoted to the theme of returning home because it was Su's fondest hope to go back to his birthplace in Sichuan or live together with his brother Su Che; this sentiment alone can account for Su's lines "In dreams perceiving, in stupors aware, / Surely I am the reincarnation of Tao Yuanming."[310] But Su's terms of exile explicitly forbade contact with family members (during his third and last, he was permitted to have only one son accompany him) and dimmed the career prospects of the males in his inner circle, putting his brother's career on indefinite hold, when Su had wanted his family members to "fill the service ranks," as the poem states.[311] Su could never really bring himself, when matching Tao, to make light of domestic woes.[312]

Meanwhile, let us not downplay class and status differences or the changed outlook that the civil service examination had wrought in the elite consciousness by the Song. Seven centuries earlier than Su, even a low-ranking aristocrat such as Tao may have felt little embarrassment at consorting with inferiors, since heredity secured his superior place in society.[313] Living in the highly commodified and competitive elite culture of the late Northern Song, the bourgeois Su would have found it hard to transgress class lines with equal lightheartedness.³¹⁴ Tao styles himself as a good friend to anyone who did not positively play him false; his basic nature, he said, was disinclined to discriminate.³¹⁵ In contrast, Su, while enjoying a wide circle of allies, clients, and admirers, prided himself on his fastidiousness in his choice of company. In a poem matching Tao's "Moving House," Su writes, "Going out the door, I have no place to pay a visit," content to advertise a mood that might have perplexed Tao.³¹⁶ In a second "match," Su recounts, ³¹⁷

> A guest knocks at my door,
> Tying his horse to the willow in front of the door.
> In the empty courtyard; the magpies have scattered.
> For a long time the guest stands, the door closed.
> The host, using writings for his pillow, is lying down,
> Dreaming of friends in his former life.
> Suddenly I hear the tap at the door.³¹⁸
> Startled from my stupor, I spill a cup of wine.
> With clothes in utter disarray, I rise to apologize to the guest.
> In my sleepy state, I feel the burden of two hard clods.
> We sit and talk variously of things old and new.
> Sometimes I don't reply, my face blushing less and less.³¹⁹
> He asks me from what place have I come?
> I say I've come from a land called Nowhere.

Whereas Tao's original poem had spoken warmly of the keen pleasures to be garnered from friendship, the visitors to his house being all too few ("Before we spoke, my heart was already drunk. / Not from taking repeated cups of wine," but from "loyal and generous" friends), Su encounters friends in his dreams, but neglects a polite caller at the door (those "two hard clods"). Relatedly, Su Shi's constructed Tao also prefers the hills and valleys that were "ownerless"—possessed by no one—since they afford temporary release from the burdens of human entailments.³²⁰ Had Su really been a second Tao, he would have been more at ease with talking with the locals, one suspects, so long as they were "plain and honest people" (*su xin ren* 素心人).³²¹ But even during his harrowing exiles, Su found it impossible to forget his celebrity standing among elite office holders, whose members were, by most of the latest political theories, nearly equal to the emperor in authority.³²² And surely it is relevant also that members of the Northern Song dynasty scholar-official elite regularly looked for inspiration to such elegant Wei-Jin aristocrats as Wang Xizhi 王羲之 (303–361), oblivious to disparities in the two groups' values and stations. Ironically, the company, real or imaginary, of men of status such as Wang Xizhi was precisely what Tao in an earlier age had *not* bothered to seek out. So *if* these two authors' limpid writings mirror their innermost feelings (a huge caveat), it would have been surprising if Tao and Su, living in such different eras and situations, were very much alike, Su's protestations to the contrary.

Ergo Su's dramatic retrofitting of Tao's temperament in many of his writings, among them the matching poems, already evident in his "matching" of Tao's "historical poems" during the second year in his second exile, in Huizhou (1095).[323] Tao had characterized the decision by two ministers to abandon court life as a wise act to be emulated by the wise men in later generations: "Whistling high, they returned to their old homes.... / Who would say that these men were lost? / Until the very end, the Way prospered whole and intact" in their persons.[324] Su, out of favor because of incautious remarks (many made in his poems), was understandably anxious to avoid any appearance of subversion. Yet perhaps unwittingly, Su chose to dwell upon the gross injustices inflicted through dismissals of good men, writing with a bitterness about interruptions to meteoric careers akin to his own that is wholly alien to Tao's cast of mind.[325]

The Matching Poems' Mismatches
With matching poems, Song readers expected the new poetic images to reverberate with, not echo, the major tropes from the earlier poet.[326] As a supremely aware connoisseur of the arts, Su knew full well that his fellow men of letters[327] and later generations would continually weigh his matching poems against Tao's own productions. That, too, was undeniably part of the stated reason Su decided to match Tao so thoroughly, to demonstrate a consistent meeting of two souls and minds. Arguably, if somewhat paradoxically, when Su was still in relatively good odor with the court,[328] Su's "matches" did a far better job of carrying the thrust of Tao's message and spirit, despite Su's contentment in office and comparative sobriety, whereas later attempts made during Su's longer second and third exiles seem more stilted. Su's early matching poems (the twenty-part "Drinking" cycle, for example) are a case in point.

Let us first examine, then, a sampling of poems that match very well, paying particular attention to Poem 42.

"Matching Poems"
by Tao and Su

POEM 12, STANZA 1

Returning to the Orchards and Fields to Dwell 歸園田居

Tao writes

From my youth, I've been out of step with the common rhyme.
My nature from the first[1] loved hills and *mountains*.
By mistake I fell into the dusty net,
Away from home for thirty *years*.[2]
The tied bird longs for its old forest.
The fish in the pond yearns for the old deeps [*yuan*].[3]
I have cleared the land at the edge of the southern wild.
Keeping my clumsiness,[4] I return to the *fields*.
I've got a plain square house and ten-odd plots,
A grass roof over eight or nine *mats*.
Willow and elm give shade in the back orchard,
Peaches and plums, their branches plaited *before*.
In the heat of the sun, far from the villages of men,
Hamlets full of dust and *smoke*.
A dog barks from the deep lanes;
A cock crows from a mulberry *overturned*.
In my courtyard, nothing kicks up dust.
An empty house makes for more than enough *ease*.
For too long I was prisoner in a cage,[5]
But now I'm back again to what is *so*.[6]

---

1 The word means "root," so it can mean "from the first" or "at the root."
2 Variants on the line make it "more than ten years," "thirteen years," or "thirty" that Tao was away from home.
3 *Yuan* refers to "deep pools," but it is Yan Hui's personal name as well.
4 Either "kept to my old rustic ways" or "continued in my ineptitude," with the former considerably more positive than the latter.
5 Compare *Zhuangzi*, chap. 4, as analyzed in Chapter Five.
6 Meaning, he is at one with the cosmos and with the particularities of his own nature.

Su responds

Surrounding this place, much clear water,
At ocean's edge, it's all dark *mountains*.
To this limitless view
I entrust my numbered *years*.
The family to the east has a Confucius,
And that to the west, a Yan *Yuan*.
In the city there's no double pricing,
Nor do the farmers quarrel over the *fields*.
Zhougong and the brothers Guan and Cai must
Regret not thatching mere hovels of three *mats*.
is enough to sate me.
Ferns and brackens, I filled up on *before*.
Disciples bring me kindling and rice.
(Otherwise the kitchen would have no *smoke*.)
With a dipper of wine, a single fowl,
Sweet songs for a head now *turned grey*.
How would the beasts and fishes know the Way?
I just fit myself to things and feel at *ease*.
In the long run things may not always be as I like,
But for now there's pleasure in my life being *so*.

POEM 13

Excursion to Xie Brook 遊斜川

Tao writes

The preface includes: "Joy at the scene was not enough. Spontaneously we composed poems, lamenting the passing of days and months, and the lack of fixity that is our years."

With this new one, fifty years have slipped away,[1]
My life proceeds to its final return and *retiring*.
This thought stirs my feelings deep within.
It should not be missed — today's *outing*.
The air is mild, and the sky all clear.
As we sit together by the stream's far-*flowing*,
In the gentle current race striped bream.
In the secluded valley soar crying *gulls*.
Over distant marshes, we cast roaming eyes.
And far off, we see the tier upon tier of *hills*.
It may not compare with the Ninefold Peak's perfection,
Still nothing else in sight for it is a true *match*.
I offer my companions the flask,
Filled to the brim, each cup is offered round and *pledged*,
There is no knowing whether after today
There will be a return of such a fine sight or *not*:
In our cups, we give free rein to our emotions,
And forget death's "eternal *cares*."
Making the most of this morning's pleasures
Let us not surety in the morrow *seek*.[2]

---

1 A. R. Davis takes this quite differently, noting the variants (fifth day) and (fifty years of age). In addition, we do not know if this is the *xinchou* year (401) or the *xinyou* (421). For more on the history of reading this poem, see A. R. Davis, who believes that the most influential post-facto account of Tao's visit (by the eighteenth-century commentator Tao Shu) is probably wrong.
2 The influence of the preface to the Orchid Pavilion Poems is obvious.

Su responds

Su's preface includes the lines: "It is quite the erstwhile scene of Xie Brook. Now I am old, I'll entrust to it my remaining years.³ [This is neither a *ci* or a *fu*.]⁴

Quiet living in a place demoted, with much leisure,
Is really not different from old-age *retiring*.
Although I am past the age of Yuanming,
I have not lost the desire for a Xie Brook *outing*.
The spring river is calm, with nary a ripple.
As we recline, the boat on its own follows the *flow*.
I have never had any plan about where I should go;
I just float, following the call of the *gulls*.
In mid-current, we suddenly encounter an eddy.
Leaving the boat, we walk up the terraced *hills*.
I'll drink with anyone with a mouth and inclination.
Why should I need to seek out my true *match*?
My son Guo's poetry resembles his old man's:
I sing and unceremoniously as he pours a *pledge*.
I wonder if old Tao of Pengzi
Rather knew this kind of exquisite joy or *not*?
I ask my own Dian what he now thinks.
He shares not the sage's *cares*.
When he asks me why I laugh [in reply],
I say it's surely not because of You and *Qiu*.⁵

---

3 *Dongpo yuefu jian* 2/2a.
4 In Yuan Xingpei (2003), 2145; Lin Yixun (2008), 1299. The excursion was made in 1099, in Danzhou, two years before Su's death. He went in the first month out with his son, Ziguo.
5 See *Analects* 11/24 for this allusion in the last four lines. Dian is the disciple who wished to go out bathing in a spring excursion; "Qiu," the name of another disciple of Kongzi, is the same graph for "to seek."

POEM 21, STANZA 1

Moving House 移居

Tao writes

For long I yearned to live in Southtown —[1]
Not that a diviner told me the proper *site*.[2]
I had heard of many plain-hearted men,
With whom I'd gladly count the days and *nights*.
This I've had in mind for quite some years,
And today finally have begun the *task*.
A makeshift cottage: who needs it to be spacious?
Just enough cover[3] for beds and *mats*.
Neighbors' songs may come from time to time,
And good straightforward talk about the distant *past*.
Rare scripts and writings we might praise,
Puzzling passages we'd open a *crack*.[4]

---

1 Possibly another reference to South Mountain, where the graves are put. Nearly every line in the first stanza seems to contain a double entendre.
2 Or, "What I do *not* want is [to have someone] divine a proper site [there]."
3 The character could mean also "ragged." The line would then mean, "Just enough for ragged beds and mats."
4 To "crack" them open; figure them out, thrash them over. Compare "split" in Su's response.

Su responds

Long ago, when I first came,
East of the river there was a secluded *site*.
Dawns, I'd be up with the crows and magpies.
There with the oxen and sheep for *nights*.
Who ordered me to move to this nearby town?⁵
For days I'll need help for this my *task*.⁶
Songs and calls in the alleyways.
Drumbeats, singing at pillow and *mat*.
Going out the door, no place to visit, and
For pleasure, no friends from the *past*,
Sick and thin, alone in recent years.
Bundles of kindling, but none to help me *split*.⁷

---

5 Of course, Su's in exile, so it's the emperor who has ordered him.
6 To survive days filled with busywork or makework; menial tasks.
7 Meaning, who really is my true companion?

POEM 21, STANZA 2

Tao writes

Spring and autumn bring many fine days
To climb the heights and compose new *poems*.
Passing by, we greet each another,
And if there's wine, we pour *it*.
After farmwork, each goes home on his own.
At our ease, it's company we *want*.
When it's company we want, we put on coats.
Of talk and jokes — we won't tire in *time*.
This kind of life is matchless.
No reason to hurry to quit *this*.
Clothes and food must be provided,
But hard plowing won't *let me down*.

Su responds

The eddy swirls along the crooked banks.
I'm writing my "Yangzi suburbs" *poems*.
And now I would make this piece of land fit to rent.
Only this one field was worth *it*.[8]
For good measure, I'll start a strict fast [9]
I want the me with nothing to *want*.
The old lookout tower has long been gone.
The White Cranes' return — who knows the *time?*[10]
How could I be a barbarian[11] brave,
Who after a thousand years returns to *this?*
Coming back again to this blessed land of rivers and hills,
The old friends[12] will not *let me down*.

---

8   Meaning, only with this field will I get the chance.
9   Literally, "with nothing awry." He makes a solemn fast, careful to do everything right.
10 Of course, the White Crane Lookout is where Su built his retreat. Lin Yixun (2008), 1229, thinks that a mention of "white cranes" means that the poem anticipates a move there and should be dated shortly before that time.
11 As mentioned in the *Shanhai jing*.
12 I.e., men he meets in books (?) or friends in the old places; not the locals.

POEM 34

Early Spring, in the *Guimao* Year, Yearning for the Ancients

癸卯歲始春懷古田舍

Tao writes (in 403)

PART 1

"Southern fields,"¹ places I've read of often in the past,
In my prime are still places not yet *walked*.
"Often empty"² — with Yan Hui as my high example,
And with work to do this Spring, why be *excused*?
Early morning, I ready the carriage to set out,
But starting on the way, my thoughts remain *distant* [*mian*].
Chirruping birds rejoice in the new season;
A cool breeze sees off the last of the *chill* [*han*].
Winter's bamboo has overgrown the track.
This land, lacking people, feels *remote*.
And I — a second "old man who plants his staff,"³
Who travels far, never again to *return*.
Ashamed as I feel just now, before perfect knowing,
Still, what I've held to I know cannot be *shallow*.

---

1 An allusion to the Mao Ode no. 102, which urges the person "not to cultivate fields too large," nor to try to win over people living far away; rather one should think of one's child.
2 Kongzi's best disciple, Yan Hui, who was desperately poor. Tao says he is too.
3 Sign of a recluse, with an allusion to *Analects* 18/7.

Su responds (in 1097), in Huizhou[4]

PART 1

For life in retirement, there's an apt saying:
"When old, we find the useful path is still to be *walked*."
My path, unlike Tao's, has not led to homecoming,
Instead, office to office, I find I'm *excused*.[5]
Though our calm freedom from rank be akin,
In reckless thought, our distance leaves me *shame-faced* [mian].[6]
What clear self-knowledge I have, let me use.
Let my family slowly build up the *good*.[7]
East of the city live two native young sons:
Their rooms nearby it, their minds *remote*.
They've called me to come fish in their pond,
And men and fish alike forget to *return*.
The local prefect has visited, too, by carriage,
And disdains the wood's pond as too *shallow*.

---

4 Su's poem speaks of two brothers, members of the Danzhou tribesmen, who helped him build his house.
5 Feng Yan 馮衍 was ousted from his office as local prefect, on account of his writings and factional discord, so Su identifies with him.
6 Tao has used *mian* 緬 (distant); Su uses the homonym *mian* 靦 (shame-faced).
7 This is not a match; it is a rhyme, Tao's *han* 寒 becomes Su's *shan* 善. It echoes lines in the "Xici zhuan" section of the *Changes* divination classic.

POEM 34 (CONT'D)

PART 2

An earlier Master left behind this precept:[8]
"Give thought to the Way, not to being *poor*!"
I honor this — though it be remote and unreachable.
Instead, I'd like to set my mind to ongoing *labor*.
Grasp the plow, glad of a season's tasks;
With a smile, encourage my farmer *folk*.
Level fields are wed to the winds from far-off
And tender shoots of grain embrace the *new*.
I can't yet measure this year's harvest.
Still, in the work at hand there's much to *savor*.
Plowing and sowing have their times of rest.
No traveller comes to "ask after the *Ford*."[9]
Sunset, and a return home all together;
A jug of wine to hearten my near *neighbors*.
Humming to myself, I shut the brushwood gate.
For now, I belong to dykes and fields, a *commoner*.

---

8 The earlier master was Kongzi (aka Confucius).
9 This means, there is no Confucius or even moral people searching for the Way in Tao's day.

PART 2

Reeds and thatches broken, unrepaired:
I sigh for my sons — who will be *poor*.
With only greens to fatten them, men grow leaner —
Stove unused; at the well, constant *labor*.
My own desires brought me to these straits.
I doff my robes, to encourage seated *folk*.
Near the pond, I've built a meditation retreat.
When rainfall looms, its clay-tile rattle sounds *new*.
Guests show up, laden with rare dishes
And ripened fruit we will deeply *savor*.
Red lichees opened show a jade-white pulp,
Tangerine juice, a virtual aromatic *ford*.
— A loan to us of three plots of land!
With this hut of tied grasses done, we become *neighbors*.
If the tongue of butcherbirds is something I can study,
I'll truly make myself a Danzhou *commoner*.

POEM 37

Returning to My Old Home 還舊居

Tao writes

Some time ago, I made my home at Shangjing
But after six years, I left to return *home*.
Today for the first time I come back.
Hard to see so much to make me *sorrow*.
Paths in the fields are unchanged,
But some houses in town remain *not*.
I pace around my former home,
But few older neighbors are *left*.
With every step, I search for traces of the past.
For some places, I still feel a *need*.
In the flowing illusion of life,
cold and heat always jostle and *push*.
I always fear the Great Change's end,
Before my spirit and strength are in *decline*.[1]
Cast that thought aside, for now not to think.
A cup of wine will make this feeling *flag*.

---

1 I wonder if this is not a mistake for *tui*, which also means "decline."

Su responds

An impotent man dreams of getting it up.[2]
How could I forget about going *home*?
I dare not dream of the old mountain.
I fear to waken grave-mound *sorrow*.
Life in this world is but a short stay.
This body yearns, though it should *not*.
Wild Goose City,[3] what does it have?
I chance upon a bit of down a crane has *left*.
A stranded fish would keep to the old pond,
It gasps for a bit of froth,[4] still in *need*.[5]
My eldest son guarding the gate[6]
Has given the generations a *push*.
In my dreams, I talk with an old neighbor.
He sympathizes in silence, he pities my *decline*.
Our back and forth comes from the Shaper of Things,
No use for us to raise another signal *flag*.[7]

2 An impotent man can't get it up, by Sichuan dialect. This line is frequently read in a tamer fashion to mean, "A cripple always dreams of standing up."
3 I.e., Huizhou, north of Mt. Luofu (at the base).
4 *Mo4*. This is a pun for "death" (?).
5 Ref. to a Zhuangzi anecdote (陸處之魚), to rely upon the pond for comfort.
6 Ref. to Du Fu's poem. The eldest son supports the family in Su's absence. Su left him behind at White Crane Lookout, in Huizhou.
7 Meaning: it depends on fate whether we will have more meetings. The phrasing refers to a story in the *Shiji*, where a person asks about Ji An's character. The assessment: he keeps to himself; he doesn't have anything to do with others.

POEM 42

On Drinking 飲酒

Tao writes

PART 1
Failure and success have no fixed abodes,
One man then the next, by turns, shares *them*.
Long and luxuriant the melons in the field.
Peaceful, on the eastern peak, our *times*.
Winter and summer in the ongoing cycle,
For each man's way is like *this*.
The man who's aware sees what's fated,
In past and future, no return to *doubt*.
Suddenly I share a cup of wine.
Night and day, together these *sustain me*.

PART 2
"Pile up good deeds, they say, and your reward will come.
But Bo Yi and Shu Qi starved on West *Mountain*.
Since good and bad go without just recompense,
What use is it all the empty cant they *talk*?
Rong Qiqi at ninety still wore a belt of hemp,
More cold and hungry than in his prime *years*.
If one does not rely on "steadfastness in adversity,"
Through the ages, whose names should we *pass on*?

Su responds

PART 1

I am no match for Mr. Tao.
The world's affairs — I'm ensnared in *them*.
They say, "How is one to find a suitable match?"
And how go back to Tao's own *times*?
A spot of land, no brambles or thorns,
A goodly place is found in *this*.
"To unburden one's heart"[1] and go toward the past,
So that in what I meet, there'll be no more *doubt*.
If I learn something from the wine,
Then an empty cup can always *sustain me*.

PART 2

Two rough fellows censure the drunkard,
Their *qi* swelling in their breasts, hard as *mountains*.
But, once dented, collapses like ice as it thaws.[2]
Only to return at a single word of *talk*.
The energetic *qi* I've conserved fills up my belly,[3]
They say it ought to prolong my *years*.
Drinking less[4] means a short cut to drunkenness.
This secret I would not have you *pass on*.

---

1 *Analects* 2/4, said of Kongzi at seventy, that he could "let go his heart."
2 *Laozi*, section 15.
3 Probably a reference to Daoist breathing exercises.
4 Compare the slogan "Lessen desires."

POEM 42 (CONT'D)

PART 3

The Way has been lost for nearly a thousand years,
And men all now so parse their *true inclinations*
That the wine they have they refuse to drink.
All they care for is their *reputations*.
What gives worth to this corporeal self
Surely lies within this single *life*!
But a single lifetime is, after all, how much?
It's brief as a bolt of *lightning*.
The staid and stolid through their allotted years:
Clinging to that, can their desires *complete themselves*?

PART 5

I built my hut beside a traveled road
Yet hear no passing carts and horses' *noises*.
You would like to know how it's done?
With the mind detached, the very place slips *aside*.
Picking chrysanthemums by the eastern hedge
I catch sight of the distant South *Mountain*:
The mountain air as the sun sets is fine.
Flights of birds in their numbers *return*.
In this there is a fundamental truth
To discern, but already I've forgotten what to *say*.

PART 3
Having lost the Way, the literati have lost themselves,
They spout words that do not reflect their *true inclinations*.
The people of Jiangnan are great poseurs.
Even in drunkenness they are seeking *reputations*.
Tao Yuanming alone was pure and authentic.
His talk and laughter, [he said] "got at this present *life*"
His person like bamboo rustling in the wind, bending
Back and forth, light and dark, like leaves in *lightning*.
Looking up or down, for each the proper bearing.
"With a little wine in me, my poems *complete themselves*."

PART 5
A small boat is really just a leaf.
Below are dark water's lapping *noises*.
At night, with a scull, I set off drunk.
Unaware when my cushion and seat slip to the *side*.
When dawn comes, I ask what's up ahead,
Only to find I've crossed a thousand *mountains*.
A sigh for me, why do I do this?
This winding road always goes and *returns*.
In future, I'd do better to plan a little earlier,
Once a thing's past, it's too late to know what to *say*.

POEM 42 (CONT'D)

PART 6
For going and stopping, there are a million reasons,
Who knows which is wrong or *right*,
For right and wrong give one another form,
"Alike as thunderclaps," men's praise and *blame*. . . .

PART 18
Yang Xiong had a natural taste for wine,
But being poor he found it hard to *get*.
He had to wait for sympathetic friends
To bring him wine; he would drive away their *doubts*.
He would drink up as the cup was passed,
And let his counsel go on to the *close*.
The only questions he'd refuse
Were those about attacking other *states*.[5]
The good man uses his own heart.
How did he err in adopting "candor or *silence*"?

---

5 Cf. Yang's memorial against attacking the Xiongnu.

PART 6
In a hundred years, sixty transformations.
The constant worry: is this, in the end, wrong or *right*?
This physical body is an empty shell.
Who is fit to receive praise or *blame*? . . .

PART 18
What man built East Terrace, from which
In one glance a whole kingdom is *got*.
Alone, alone, an old Buddha image,
Like a gnomon set above the masses' *doubts*.
This weedy city[6] has seen dynasties rise and fall,
Leitang has had openings and *closings*.
Next year, I will raise a flowery canopy there,
I will set out wine, and mourn for the lost *states*.
Forbidding thoughts of old Bamboo West Lane where
Songs and wind instruments have long been sunk in *silence*.

6 Yangzhou.

POEM 42 (CONT'D)

PART 20

The sages flourished long before my time,
In the world today few return to what's *true*.
Tirelessly he worked, that old man of Lu,[7]
To stitch and patch, until the age was *pure*.
Though while he lived no phoenix came to nest,
Yet briefly rites and music were *renewed*.
By the rivers in Lu, his subtle tones ceased.
Things drifted on as far as reckless *Qin*.
The *Odes* and *Documents* — now what fault was theirs
That one morning they'd be turned to ash and *dust*?
Detail after detail preoccupied all the old scholars.
They attended to court affairs with *care*.[8]
How is it, then, in later cut-off ages
The Classics have still had no sound *friends*?
All day speeding carriages go by,
But I see no one "asking for the *ford*."[9]
If I'm not then happy in drinking,
I will have worn *in vain* a commoner's *cap*.[10]
I regret my many mixings-up:[11]
You, sir, must forgive a drunken *man*.

7 Kongzi/Confucius.
8 Conceivably, "In their reconstructions, truly they took *care*."
9 Ref. to *Analects* 18/6.
10 What Tao wore instead of an official cap; a white cloth.
11 Or, My only regret: that conversation's so muddle-headed.

PART 20

Lord Gai happened to talk with Cao Shen,
Qi's Chancellor alone saw in him what's *true*.
Downcast, he did not attend to business.
But a guest came, and had him drink wine so *pure*.
Just then the civil wars had ended,
And the blighted lands were *renewed*.
With three cups the Warring States are washed clean.[12]
It takes a whole dipper to excise strong *Qin*.
Silent and alone, a thousand years later,
Lord Yang inherited in Tang this worldly *dust*.[13]
In a drunken sleep, resting on his client's breast,
Talking and laughing; quite in vain all that *care*.
At times I scan the old histories,
Alone with three men who seem intimate *friends*.
There's no time to sup on gleanings of rice.
With bitter heart, I study Marquis Easy-*Ford*.[14]
Drafting documents[15] — what use is it?
Drunken ink soaks my clothes and cloth *cap*.
At one go, I finish thirty silk squares,
Carry them off, to listen to the seated *men*.

12 It takes three cups worth of time to talk about the events of that era.
13 I.e., worldliness.
14 Gongsun Hong, Marquis of Pingjin (i.e., "Easy-Ford").
15 Or writing in draft style calligraphy (?).

POEM 43

Stopping Wine 止酒

Tao writes

For making a home, I'd stop in the city,
When roaming the country, I'd *stop*.
For sitting, I'd stop beneath a tall shade tree.
For walking, not past my own gate, but *inside*.
For favorite tastes I stop with my sunflowers.
My chief joys stop with my *children*.
In the past, I did not stop wine:
If I had stopped, I'd have felt no *delight*.
Had I stopped in the evening, sleep would not be easy;
Had I stopped in the morning, I could not *get up*.
Day after day, I wished to stop it,
But my system, if I stopped, would be *disordered*.
I only knew that stopping was not pleasant;
I didn't think that stopping would *benefit me*.
Now I have begun to see that stopping is good;
From this morning on I've truly stopped —*for sure!*
From now on, I've stopped and abandoned it.
I shall leave and stop on the Fusang *shores*.
This newly pure face will stop and stay.
Why should it ever stop in a million *years*?

Su responds

When the time comes, things depart and die.
When the road dead-ends, it is not I who made it *stop*.
Each of us has followed our own ideas, in
The same village, with the Southern barbarians *inside*.
With grizzled heads, the two of us part on our mounts,
Each leading a young *child*.
Your house has Mengguang, the Middle Light, while
Mine has only Faxi, Who In the Law *Delights*.[1]
Each of us met in a mountain valley.
For a full month, together we slept and *got up*.
How vast is the area north of Hainan; here
Rude conditions suffice to give our lives some *order*.
You encourage me to take Tao Yuanming as master.[2]
In health or weakness, surely he'll *benefit me*.
Slightly ill, I sit and pour a cup.
Were I were to stop wine, I'd be cured *for sure*!
I look hard at the Way, for I've not yet crossed,[3]
I make a secret vow: to see the ford's *shores*.
From now on, at my own East Slope house
I'll not make Dukang[4] offerings each *year*.

---

1 Probably a reference to a Buddhist believer, who is Su's female companion, his wife having died (?).
2 The idea is, Tao stopped drinking wine; Su and his brother Che should too.
3 This could also mean "not yet at the point of no return." Hexagram 64 ("Not Yet Crossed") talks of "drinking of wine in genuine confidence." The hexagram, last in the sequence, represents a transition to a new order; as the last in the sequence, it also may signify death.
4 A pre-Qin notable famous for brewing good wine "that relieved sorrows."

POEM 48

After an Old Poem 擬古

Tao writes

PART 5

To the East, there lives a scholar.
What garb he owns is never *intact*.
"Once every third day, he'll take a meal."[1]
"For ten years on end, he wears but one *hat*."[2]
For sheer hardship, none can equal him,
Yet he always shows the world a pleasant *face* [*yan*]
Now since I wished to meet the man himself,
I set out early, crossing rivers and *passes*.
Green pines grow alongside the road,
White clouds sleep at the eaves' *edge*.
You've seen my aim in coming here!
For me you set your lute to *thrumming*.
Your first song is the 'Lone Crane.'
Later, strings play the 'Lone *Phoenix*.'
I wish I could stay, lingering with you
From this day on, till the year turns *cold*.

1 Literally, "in nine times thirty days."
2 Phrases used of Yan Yuan, aka Yan Hui, the favorite disciple of Kongzi.

Su responds

PART 5

There lives a secluded man on Li Mountain,
Wizened in body, while his spirit stays *intact*.
Bringing down firewood to the market in town,
He laughs at my scholar's robe and *hat*,
Having never heard of the *Odes* or *Documents*,
Let alone Kongzi and his disciple *Yan*.
In complete freedom he comes and goes alone.
Over glory, disgrace — his thought scarcely *passes*.
Come sunset, birds and beasts return to hiding,
Heading home along a lonely cloud's *edge*.
Slow to understand his questions or replies,
I give up with a sigh, my fingers still *drumming*.
Does he say, "You, sir, of truly noble pitch,
Bed down in the wilds, like dragons or *phoenix*"?[3]
He's given me a bolt of cotton-tree cloth.
The sea wind at this season is *cold*.

[3] A variation on the question asked by a recluse of Kongzi.

POEM 49, STANZAS 1-2

Miscellaneous Poems 雜詩

Tao writes

Man's life lacks a firm root,
Is whirled along the road like *dust*,
Scattered as the wind wills.
Surely there is no abiding *person*
If all who fall to ground are brothers,
Why need it my longing be for *kin*?
When we find joy, we must make merry.
A measure of wine calls my *neighbors*.
The years of our prime do not return again.
No day can have a second *dawn*.
To meet the time, we must take pains.
The years and months await no *man*.[1]

I set out but had not yet gone far,
I turn my head, the grievous wind is *chill*.
The springtime swallows answering the season, rise
And fly up high, fanning the dusty *rafters*.[2]
The frontier goose, grieved that it has no place,
Yields and returns to its northern *home*.
The lone *kun* bird sings beside the clear pool
Through summer's heat into fall's *frost*.
For a grieving man, it's hard to put into words:
Interminable, the spring night is *long*.

---

1 Or this is advice: "Do your best." Cf. stanza 5: "To Ancients begrudged even an inch of time."
2 This is the same word as "bridge" in Su's version; the graph refers to a strong piece of wood that spans an open space.

Su responds

Slanted sun lights up my poor corner.[3]
I begin to see the air has *dust*.
A light wind moves in the myriad hollows.
Who thinks I've lost my physical *person*?[4]
With a laugh, I ask my little sons
"Who among you are really my *kin*?"
"Who will follow me to this Hainan?
In remote and cut off places, no close *neighbors*.[5]
Bright light,[6] like a sickle moon;
Alone with the Morning Star's *dawn*.
This Way of life will suit you.
Better not to blame any *man*.

Long ago, I climbed Mt. Qu. When
the sun came out, I contemplated the *icy* expanse.
I want to go to Donghai county, but
I regret to it there's no Stone *Bridge*.[7]
Now, about this Jimu kingdom:
How does it differ from Yugong's *home*?
The oysters stuck together in a mountainous pile.
The summer path, likewise, has fleeting *frost*.
What makes me glad, is that I do not lie to myself.
Nor do I resent that the road is *long*.

3 The setting sun is mentioned in Tao's own Poem 2 in this series.
4 Cf. Bo Juyi's line, "Not only forgot the world, but also forgot my person."
5 Tao's poem 1 says he doesn't need only relatives (kin) to feel good. Su disagrees; he specifically mentions his sons.
6 But *genggeng* also carries the sense of "disquiet"; "firm."
7 Ref. to the Nineteen Old Poems.

POEM 50

Lauding Impoverished Gentlemen 詠貧士

Tao writes

PART 1
All creatures can turn to their own places,
Only a lone cloud wisp has no *support*.
Dimly it dissolves into emptiness.
When did we notice its last remaining *light*?
The red of morning shears through late mist.
Birds flow together in *flight*.
One laggard on the wing leaves the wood.
Before dusk, it will find its *homecoming*.
Weighing our strength, we keep to old tracks:
How not to be cold and *hungry*?
If "knowing the sound" should fail me,
I give up too. And why *lament*?

PART 2
Chill and harsh, the year now comes to a close.
Clutching my thin coat, I sun myself on the *porch* [xuan].
The southern garden patch has no remaining green.
Only withered branches to the north, in the *orchard* [yuan].
I tilt the jar, to finish the last drops there,
Inspecting the stove, I see no sign of *smoke*.
With the Classics piled high about my seat, the sun
Goes down, leaving no time for further *study* [yan].
Retirement cannot rival "distress in Chen,"[1]
Yet to me "they came, all indignant *speech*."[2]
What means shall I use to soothe my breast?
I rely on thoughts of these antique men of *worth*.

---

1 Such as Kongzi experienced in his wanderings.
2 *Analects* 15/1–2.

Su responds

PART 1
Evening Star above the slender moon,
Each, bright, turns to the other for *support*.
Pondering at heart our brief span here,
I relish this one moment of clean *light*,
Though for blue Heaven there's no old or new.
Who knows the weaverbird Sun's *flight*?
I'm soon to die and fare to the Nine Fields,
And be alone with Tao Yuanming at this *homecoming*.
An ordinary man will not grieve for himself,
But look at the more careworn, needier, *hungry*.
"However grand we be, can we have longer lives?"
Yet, rushed like so many racing teams, we've cause to *lament*.

PART 2
Ashamed to eat the grain of Zhou, Bo Yi and Shu Qi
Lifted their voices, to sing of the sages Yu and *Xuan* 軒.
And which of the Empress' two nephews[3] could have
Enticed to court the Greybeards Ji and *Yuan* 園?
From of old, recluses have died; and only a few, once
Turned to ashes, left behind their names, like *smoke*.
At the end of this, my own road, I am ever more chagrined.
So the finest inks red and black, by hand I'll grind [*yan*].[4]
At first, Tao also served the court, but his
Strings and songs were based in honest *speech*.
And when service no longer pleased, he made for home.
Surveying the world, embarrassed to alone exhibit *worth*.

3 Nephews of Empress Lü, wife of the Han founder, who did nothing to support the rightful heir. The two men's personal names are puns for "property and rank."
4 Meaning, "I will honor them with fine calligraphy." The graph for "grinding" is the same as the graph for "studying," a type of grinding away.

POEM 50 (CONT'D)

PART 3
In old age, Master Rong wore rope for a belt.
Still, with appreciation, he played his *lute*.
Though Master Yuan's[5] shoes fell to pieces,
He gave his clear refrain to an old Shang *tune*.
Twin-Pupil Shun is far from us in years, yet
For its own poor gentlemen, no age need *search*.[6]
Those ragged robes out at the elbows!
That goosefoot broth, scarce enough to *pour*!
While we don't forget we once wore light furs,
To come by such dishonestly can't be our *hope*.
— As for you, Si, merely clever at talk,
You'll not take in all I have at *heart*.[7]

PART 5
Yuan An didn't like snow piling up round his gate.
He lived aloof, to no other man a *barrier*.
The day Master Ruan found money flowing in,
He sent his resignation off to the *palace*.[8]
In hay there's always warmth. Wild rice
And gathered taro will make a morning's *meal*.
Surely there have been those in real hardship
Who didn't fear hunger the most, nor *frost*.
Wealth and want will ever be at war,
Yet where the Way prevails — no furrowed *brows*.
High virtue may crown both farmland and village;
On the farthest pass, our clear ideals shine *together*.

5 Yuan Xian; i.e., Zisi, a disciple of Kongzi (not to be confused with Yan Hui). To Zigong, he remarked that poverty was no ailment, though ostentation and a neglect of virtue was.
6 More literally, And poor gentleman in every age continue (*xun*).
7 "All Zigong could do was quibble/ Into my mind, he had no insight," as Yuan Xian says.
8 Literally, Ruan sent his resignation to the higher official, and the cakes and wine in Su's matching poem are sent by the official in charge of the Palace Stores. To keep the matching rhyme, I have resorted to use of the synonym, "the palace."

PART 3
Would you tell me Tao Yuanming was poor?
He still had his one plain unstrung *lute*.
With mind at ease, so too his hand:
He lived on in a limitless *tune*.
Loved the Double Ninth festival, its
Fresh chrysanthemums: he'd make his own *search*.
He'd don a plain cloth cap and sigh,
Yet laugh at a dusty goblet, dipper with nought to *pour*.
Suddenly given a gift of twenty thousand
From Yan Yanzhi,[9] a help beyond *hope*,
He straightway sent off to town for some wine,
Not to fail or offend his old friend's *heart*.

PART 5
A lotus grows among tawny chrysanthemums;
Branches and leaf-sprays grow through the *barrier*.
From a great distance, I pity gentlemen at court,
For whom cakes and wine arrive from the *Palace*.
This far from court, how could one guess
Such fallen petals suffice to make a *meal*?
A poet pawned his robes for the Double Yang;[10]
By year's end, he'll fear oncoming *frost*.
Without warm clothes, we simply have to shiver;
Without wine, we simply knit our *brows*.
Poverty truly is cause for lament,
The two having housed so long *together*.

---

9 Another poet.
10 A festival in China.

POEM 54

Reading the Classic of the Mountains and Seas (*Shanhai jing*) 讀山海經

Tao writes

PART 1

In early summer, when the grasses grow,
My house is surrounded by greenery spreading (shu 疏).¹
Flocks of birds trust to find a refuge there.
And I likewise love the hut that is *my humble home*.
With fields plowed; the new seed planted also,
And now it's time once more to read my *books*.
This dead-end lane a bar to deep-worn ruts,
It rather tends to turn away old friends' *carts*.²
With pleasure, I pour the spring wine,
And pick in the garden some *greens*.
A light rain, from the east comes,
A gentle wind, it *accompanies*.
I skim through 汎覽 the *Tale of King Mu*,
Peruse 流觀 the *Seas and Mountains Classic pictures*.³
A glance up and down goes to the very end of time and space,
Were I *not* to take pleasure in this, what will the future be *like*?

---

1 This entire poem rhymes: *shu* 疏; *lu* 盧; *shu* 書; *ju* 車; *shu* 蔬; *ju* 俱; *tu* 圖; *ru* 如. Note also the internal rhymes *fanlan* and *liuguan*, said of reading.
2 This is sometimes read instead: "will still attract my old friends' carts."
3 Internal rhymes. The last image is startling, because *liu* usually refers to "flow," while *guan* usually refers to "staring fixedly" or in "rapt contemplation" (as with religious statues).

Su responds

PART 1

This very day, the first frost in the sky,
Many trees hunkered down and getting sparse (shu 疏).[4]
The recluse shuts the door, and lies abed.
The bright sunlight brings back to life *my humble home*.
Open is my heart,[5] but lacking a good friend here.
I cast my eye, and take up a valuable rare *book*.
In Jiande, there are people left behind the times.
But the road is long; and I have no *cart*.
Even without grain, there's still enough to eat.
Do you mean to say that I talk of grains and *greens*?[6]
I'm so grateful to old Ge Hong,
Who across a thousand years me *accompanies*?
Paint him with Tao Yuanming and me.
We can become a "Three Scholars" *picture*."[7]
I am sadly late to study the Way.
But in writing poems, am I not with Tao *alike*?

---

4 The same Chinese characters means "spreading" and "sparse" in different contexts.
5 Ref. to *Hou Hanshu* biog. of Ma Yuan, where he is described as "open-hearted" and appearing to have integrity.
6 Meaning, by now he's feasting on *qi*, as Ge Hong could be done.
7 For another "Three Scholars" portrait, see Figure 7.2. By Su's time, such imagined portraits of friendships out of time were clearly becoming commonplace.

POEM 54 (CONT'D)

PART 3
Far off lies the Huaijiang Range;
Near at hand, what they call Xuanpu *Mound*.
Southwest, facing paradise, on Kunlun Peak[8]
Shines its vital qi, grand and *peerless*.
How splendid the bright *gan* gem's luster
And, down-rushing, the clear Yao's *flow*!
I grieve that I could not go with Mu of Zhou,
Climb in my carriage and, beside immortals, *wander*.

PART 13
"Greater talent means greater chaos, if it goes unused."
Noble appears the minister in court and market,
For the true Emperor makes careful use of *talent*.
Did sage Yao not dismiss Gonggong and Gun
So that, to Yao, Twin-Pupil Shun might *come*?
When uncle Guan proffered sincere fair advice,
Old Duke Huan of Qi expressed *doubt*.
Dying, the Duke told fellows of his hunger and thirst.
Too late to avail him much, *I'd say*.

---

8 Kunlun Peak is the home of the immortals, where Duke Mu of Zhou was entertained by the Queen Mother of the West.

PART 3
While Yuanming had a modest lifespan,
His fine aspirations were fit for Cinnabar *Mound*.
Far-ranging indeed was Tao Yuanming the man:
Excellent, free of limit, *peerless*.
So rare the writing he spun before death,
Hasn't he gone beyond life's mortal *flow*?
After death, I'll stay among the Nine Fields.
We three,[9] alive in three eras, together will *wander*.

PART 13
[*no prefatory saying*]
Old Su[10] indeed is a man at odds with the world.
Get involved with it, and you scatter your *talent*,
Near Enemy Pond, I'm told, runs a road back home.
Tell me, Mount Luofu, not for nothing have I *come*!
Here I tread on snakes, must live on grubs.
My heart, emptied, no longer harbors *doubt*.
Hands joined with Ge Hong and Tao,
I'll go back! Let's go back, *I say*.

9 Tao Yuanming, Ge Hong (alchemist and reputedly an immortal), plus Su.
10 Su here names himself, by the sobriquet Dongpo.

POEM 56, LINE 81 ON

Moved by Good Men's Failures to Meet Good Fortune 感士不遇賦

Tao writes

Why are the good days so easily overturned?
Why does harm win out, always so quickly?
Azure Heaven is too remote to hear us.
Human affairs have been ever thus.
Now sentient, now unaware.
By principles unfathomed and unfathomable.
I prefer to be "steadfast in adversity," keeping clean my aims.
I will not trust in what's awry and entangle myself.
Once I saw that cap and carriage spelt no glory,
I never thought my thin robes a source of shame.
With the time's so out of joint, I took on a bit of awkwardness.
Happy to be alive, I then came home for good.
Keeping my solitude close, to live out my years,
Turning down a good price from market and court.[1]

---

1 It may well be that this poem was too bitter in its tone for Su, then in trouble, to imitate, especially when Tao writes, "Human affairs have been ever thus."

POEM 63, LINE 40 ON

Appraisals for Paintings on Fans 扇上畫贊

Tao writes

"With heights and far-off places, amusing myself."
Screened is the rough beam door.
The brook flows on and on.
I call on my lute, I call on my books,
Turning to the distant past to find a few friends.
When "drinking from the Yellow River,"[1] there's water enough.
Beyond me, all is at peace.
I recall fondly those nine men of a thousand years ago.[2]
Trusting to my bonds with them, I can wander alone.

1 Here almost certainly Tao means "from the huge store of legendary figures I've got to draw from," with "deep drinking" a metaphor for deep, imaginative immersion into their stories.
2 Eight tableaus would have been painted on the fan(s?), and Tao gave a short colophon for each scene. After the ellipsis, in Lines 41ff., Tao describes himself. Line 40 shows how Tao talks of an unidentified Master Zhou Yangkui, who was a recluse, like the others portrayed in painting and in Tao's verse, and like Tao himself.

POEM 63 (CONT'D)

There is no response by Su to Tao's poem, but one anecdote shows that Su knew fan-painting well:

When Su Dongpo was an official in Qiantang, there was a man who wanted to report another to whom he had lent 20,000 cash for non-repayment. Su questioned the defendant in the complaint, who said, "My family makes fans for a living. As it happens my father just died, and this spring, it has done nothing but rain since spring started. With the endless rain and cold weather, I can't sell the fans that we've made. It's not the case that I intentionally aim to not pay the money back."

Su Dongpo looked intently at the man for a very long time, and then he said, "Bring the fans you've made to me, and I will help you to get them sold." A little while later, the fan-seller brought the fans and Su Dongpo selected twenty of them that were completely white. He then used the same official brush he used to render judgments to inscribe them in running and cursive script; he also painted withered trees and bamboo and rocks. In a very short while Su had finished and he handed them back to the fan-seller and said, "Quickly! Go outside and sell the fans and then return the money." The fan-seller took hold of all of the fans in his arms, and burst into tears of gratitude. After he left the official compound, there were lots of people willing to buy a fan for 1,000 cash, and before long, the seller had managed to sell them all.

In such poems, there are lines where Su matches Tao so beautifully that it is hard to tell their voices apart.[329] Similarly, the third verse in the two poem cycles entitled "Lauding Impoverished Gentlemen" (Poem 50 in Tao), a relatively early matching poem, carries much the same atmosphere as Tao's original. Let us take a look. Both poems by Tao and Su feature as subjects noble men of old who represented radiant examples to their peers and those who followed, although the world largely ignored their merits as they endured dire poverty. In both poems, the cultural heroes are content with their fates and happy to be away from court, where dishonesty and double-dealing prevails. Both poems highlight the momentary radiance that infuses the most ordinary occurrences in the daily round, and both bear witness to death's approach in the midst of life.[330]

But even matching poems in similar moods can, upon closer reading, sometimes reveal divergence. Let us return to Tao's sublime "Excursion to Xie Brook" and Su's antiphonal response registering his passion for home. Su Shi mused while reading Tao's "Excursion to Xie Brook" that one did well to "draw back" from present concerns in order to register the inexpressible beauty underlying all creation.[331] True enough, but excursions and pleasure outings tend to afford the greatest pleasure to those most secure in their homes to which they will return — homes defended against unpleasant change. Tao's preface to the excursion poem is significant, then: "Our joy at the scene was not enough. We composed our poems in call and response, lamenting the passage of days and months, bemoaning our untarrying years." For Tao, the miraculous beauty of the day offers but a temporary respite from sorrow at time's passing, for all that he finds a true home within his circle of friends:

> With this new one, fifty years have slipped away,[332]
> My life proceeds to its final return and retiring.
> This thought stirs feelings deep within.
> It should not be missed — today's outing.
> The air is mild, and the sky all clear.
> As we sit together by the stream's far flowing,
> In the gentle current race striped bream.
> In the secluded valley soar crying gulls.
> Over distant marshes, we cast roaming eyes,
> And far off, see tier upon tier of hills.

It may not compare with the Ninefold Peak's perfection,
Still nothing else in sight could be closer to a match.
I offer my companions the jar,
Filled to the brim, each cup is offered round and pledged,
There is no knowing whether after today
There'll be a return of such a fine sight or not:
In our cups, we give free rein to our feelings,
And forget death's "eternal cares."
Making the most of this morning's pleasures
Let us not surety in the morrow seek (*qiu* 求).³³³

By contrast, claiming kinship with Tao's spirit, Su speaks of demotion, but not the prospect of death. Accordingly, his poem ends on a playful note distinctly at odds with Tao's last six lines, merrily dismissing career concerns: "I wonder if old Tao Qian of Pengzi / Rather knew this kind of exquisite joy or not?... When he asks me why, I laugh in reply, / Saying it's surely *not* because of You and [Ran] Qiu 求."³³⁴ This evasion of inauspicious reflections on death and dying, howsoever graceful, may tempt a reader to employ a couplet of Tao's own making to chide Su: "You, sir, are but a clever talker, / So you do not understand what is in my heart."³³⁵

Su, in part 9 of Tao's "Drinking Poems," provides another good illustration of emotional mismatches. Tao's meeting with an old neighbor ends with Tao amiably rejecting the man's homespun advice to learn to get along better in the world, as we have seen ("Deeply grateful for your words, old man, / My spirit's out of tune with what you say"). Tao the speaker in the poem is content, he says, to carry on as he is ("Having come this far, my carriage cannot turn back"). By Tao's account, he knows himself, and thus he is willing to bear the consequences of his refusal to reform his ways in pursuit of fame or rank. He never boasts of his own purity; it's just that people differ, and it is wise to accept the effects that those differences make for each in life. For that reason, Tao's notion of authenticity, riffing on Zhuangzi's insights, meant living within the constraints of his personality and situation. Su's matching poem, in contrast, advertises his purity, symbolized by the lotus in the excerpt below and earlier in the poem by his "orchid-pure heart." Su's consequent elevation from the "floating world" of illusion in which he travels, its final concession an allusion to struggle, replaces Tao's more robust notes of acceptance:

> One root, plucked and broken off,
> Forever cut off from the Yangzi or the lake.[336]
> Once cut, silk threads [its roots] are not to be reattached.
> A dipper of water, how could *that* suffice for "a safe perch"
>   for the likes of Tao?
> This ordinary lotus is not as good as the Jade Well lotus
> Rooted in a lake in paradise.
> Feeling this strongly, each time I comfort myself.[337]

Not infrequently Su indulged in talk of being "newly cleansed / feeling the body lightened"[338] by way of announcing that he had transcended all worldly care. (Recall Figure 7.3, where Tao's swirling robes indicate this.). Like Tao, Su endeavors to portray himself as conforming to the larger scheme of things, with the world he knows a microcosm of the greater whole. But unlike Tao, Su claims to need no convivial wine to experience or recreate heady pleasure.[339] So exquisite are his tastes that he can get all the satisfaction he needs from merely holding the cup in his hand and surveying a scene with the eyes of a true aesthete, superior to the common run of men.[340] Su sometimes boasts that he can "fill up his belly" with the *qi* he has wisely conserved, and Su's matching poem for Tao's "Reading the Classic of the Mountains and Seas" implies that Su has attained an enviable state of perfect inner equilibrium in the here and now: "The heart, once emptied [that is, detached from worldly concerns], can be without guesses or anticipations." Equally implausibly, Su occasionally converts poverty to a pleasing idyll or an object of curiosity, as in the following poem.

> Who says that Tao Yuanming was poor?
> He still had his one plain zither without strings.[341]
> His mind relaxed, his hand, too, was then at ease.
> So he could dwell in countless melodies....
> In a plain cap, he sighed, thinking there was no point in straining the wine.[342]

For these reasons, Su's version of Tao's "steadfastness in adversity" (that is, equanimity in the face of troubles) seems to have little to do with Tao's dogged determination to keep going until his last breath, in order to secure the pleasures of home.[343] As Ronald Egan astutely observes, Su "keeps outdoing Tao Qian in the virtue Su associates [most] with him," Tao's "steadfastness in adversity."[344] It is not that

Su does not occasionally torment himself with thoughts about what he might be missing at court or at home. ("Alas! What am I doing here?")[345] By and large, however, only hints of misery ruffle the surface of Su's matching poems,[346] though his letters to his brother Che and to clients and disciples betray dark mood swings.

Yet another comparison suggests, if inadvertently, the strenuous efforts made by Su to appear to transcend all present cares. While in the second poem of his "Returning to the Orchards and Fields to Dwell" Tao had expressed worries for the coming harvest,[347] Su responds with a dreamy surcease of all mundane cares that is foreign to Tao:

> The desperate gibbon has hidden in the forest.
> From the weary horse, the halter has been loosed.
> My mind empty, I am sated with newly got things:[348]
> The land familiar, I give my thoughts over to dreams.
> River gulls flock tamely beside me.
> Dan[349] boatmen sail back and forth.
> Green coins [that is, lotus] growing from the southern pond,
> Purple bamboo shoots tall on the north peak.
> How could a raise-a-gourd really know how to drink?[350]
> Yet his [Tao's?] welcome beckoning often eases my heart.
> The river in spring inspires fine poetic lines.
> As I drink, I slip into the vast void.

The sources of the two men's pleasure could hardly vary more. Tao, free of constraints of official service, roots his pleasures in this material world whose particulars cause him elation or concern by turns. Su, by contrast, would shed the mundane, "slipping into the vast void." It comes as no surprise, then, that in the third poem of the same poem cycle, Tao registers no complaint as he describes his sweat-drenched clothes after plying the hoe or that Su substitutes for this earthy image a delicious bath near a waterfall that leaves him feeling as light and pure as an immortal.[351] Likewise, in the fourth poem in the same cycle, Tao — who has set off on a sightseeing trip — is preoccupied with the men he's known and loved who have gone to their graves, "reverting to nothing at the end," whereas Su avers that he has everything he needs at hand.[352] By the fifth poem, Tao's energies have waned ("Depressed, I come back alone, gooseneck staff in hand"), while Su names himself a "dissipated romantic"[353] in search of further escapades.

Alert readers may notice other odd and unsettling changes that occur in Su's "translations" of or dialogues with Tao's convictions, for instance, in Poem 8 of the "Untitled" series in Tao's work, one of several summations of Tao's reckoning with his situation at home: "All other men in every way get what suits them. / Inept at life, I have lost the recipe for this. / Since this is the way of it, what else can I do? / Let me enjoy [*tao* 陶] a cup of wine."[354] The last line plays on Tao's name, conflating wine drinking with Tao's highly conscious pleasure in being his unique self, his own man. Aside from this single play on words, the language of Tao's four lines could not be plainer. Contrast Su's showy "matching poem" employing no fewer than six separate allusions in fourteen lines.[355] And whereas Tao downplayed the significance of his poetry, calling it something he did to "amuse himself" (*ziyu* 自娛) on long winter evenings after the effects of wine had worn off ("I write them [the poems] as a way of enjoying life and laughing — they are that and nothing more!"),[356] Su more commonly directed attention to the ennobling disciplines he practiced during his poetic project commemorating Tao.[357]

At the same time, Su is generally more sentimental than Tao. That mature sentimentality is on display in Su's considered response to the twelfth segment of the lengthy "Drinking Poems," which Tao ends with the angry blast:

> When one goes, that must be the end. [往便當己]
> Why be any longer in doubt?
> Go! Go! What more is there to do?
> Vulgar conventions have long cheated us.
> Cast away useless chatter!
> Let me follow the way I have chosen.

Su's answer to Tao raises an idyllic scene of years past, with the man rooted in the child. Since both dreams and drink offer solace, the perfect antidotes for taking the rough-and-tumble of life to heart, one *should* be able to live out one's days in "no harm" and "no doubt," well pleased with one's own genius:

> I dreamed I had entered the lower school
> And called myself a youngster with a topknot.
> I did not note the white hairs [on my head].
> I was still intoning the *Analects* phrases.

> The world of men is based in child's play, yet
> It's all topsy-turvy, and "encumbered like this."[358]
> Only when drunk is one sometimes true,
> Emptied, hollowed out, with no doubts at all, so
> One can tumble from a cart and not get hurt.[359]
> The old farmers do not cheat me.
> I call my boy to prepare paper and brush.
> Such drunken words, I record them in a flash.[360]

Of course, the Buddhist theology that pervades Su's writing (but is largely absent from Tao's) correlates strong attachment to things and people with a less-enlightened nature.[361] A comparison of Tao's "Reading the Classic of the Mountains and Seas" with Su's poem of the same title nicely shows how this colors their respective poems. Tao's poem bespeaks his wonderment at the variety of forms that material things take in life, in books, or in fantasy — the sheer physicality to which Tao clings. Su, by contrast, takes the occasion to lambaste Ge Hong, author of the *Master who Embraced Unadorned Simplicity*, for being "in truth every bit as much attached to things as Kongzi and Yan Hui,"[362] on the grounds that Ge yearned for physical immortality (though Su was not above dabbling in elixirs).[363] A comparison of the first poem in Tao's "Reading the Classic of the Mountains and Seas" cycle with Su's third poem in the "Drinking" cycle strengthens the suspicion that their occasional incongruences reflect Su's ambivalence toward his own love of material objects, with his avid collecting and reputation for connoisseurship vying with his desire to become a better Buddhist.[364] Su himself details his changing attitudes toward material objects, and intensity is even more a subject that Su returns to, somewhat obsessively as he comes to the realization, through successive exiles, that most of life is to be lived without that intensity, and to achieve that is truly to "return to one's nature."[365]

Let us compare the two poems. Tao's Poem 54, on reading the *Shanhai jing*, exults in the thinginess of the things around him, whose lively particulars suffuse the person and persona of the poet:

> Summer's first month, with plants growing tall.
> Around my house, the shrubs are thickening.[366]
> Flocks of birds trust to find refuges there.
> Like them, I love my humble home.

Fields plowed; new seed sown.
There's time once more to read my books, and
My dead-end lane, well off from deep-worn ruts,
Will still draw in my old friends' carts.[367]
With pleasure, I pour the spring wine
And in the garden pick some greens.
A fine rain comes from the east.
A gentle wind accompanies it.
I skim through [汎覽] the *Tale of King Mu*,
Peruse [流觀] pictures in the *Seas and Mountains*,[368]
Glancing up and down, like Fuxi, surveying all time and space.[369]
Were I *not* to take pleasure in this,
What would I do with myself, what would the future be like?[370]

Taking his turn, Su's poem exalts immaterial men who exist outside the realms of life and death, space and time, specifically, Tao Yuanming and Ge Hong:

Even though Tao got but middling longevity,[371]
His elegant aspirations were those of the immortals' Cinnabar Mound.
Far-ranging indeed was Tao Yuanming himself!
Superb! Boundless! Without peer!
His rare writings issued forth until his dying breath.
Didn't the two [Tao and Ge] surmount the flow of life and death?
I myself am traveling to the Nine Plains.[372]
From three eras, Tao, Ge, and I are firm friends.[373]

At least two possibilities lie before us: Su's discomfort with his own complex attitude toward material things caused him to turn a blind eye to the persistent delight in material forms that informs Tao's pleasure-taking. Or, more charitably, Su may have knowingly refused to dwell upon Tao's ardent love for the material, thinking it would raise red flags among Su's peers about Tao's fitness for the superlatives found in Su's accolades ("Superb! Boundless! Without peer!"). Su's last poem in the long cycle ends with "Holding hands with Ge [Hong] and Tao, I say 'Let's return! Let's return!'" and by that time, we sense that Su interprets Tao's theme of "returning home" to mean that the earlier poet did not merely relinquish a career, but more crucially, he "cleansed himself of all worldly dross," aware that "To get involved with the world is truly to scatter one's talent."[374]

Very occasionally, Su's matching poems slide into woodenness or, worse, sanctimoniousness, though in all fairness, Su's matching poems rarely falter badly as *poetry*, with most critics tut-tutting over two or three couplets, at most. But in his determination to put a name to what Tao found inexpressible, Tao's jumble of reactions to his world, Su sometimes subverted the spirit pervading Tao's poems to keep up a cheerful facade. Early in his efforts, Su, when matching Tao's eleventh poem in the "Drinking Wine" cycle in Yangzhou, avoids Tao's casual unconventionality. Of two paragons of virtue, Yan Hui and Rong Qiqi, Tao had remarked acidly,

> They may have left behind an honored name,
> But it cost a lifetime of privation.
> What do we know of them,[375] once dead and gone?...
> To accord with our hearts — surely that is best.
> Though but guests, we value our bodies,
> But in light of Change, their value vanishes.
> So why must it be so dreadful to be buried naked?
> Men must understand the limits of understanding.[376]

Responding to such skepticism, Su Shi dutifully praised his enlightened ruler: "An imperial rescript generous enough to forgive the pile of debts. / The old ones wear a good face/their color returned [in relief]. / I bow low twice, to felicitate our lord, my emperor. / Receiving this, I want no other treasure!"[377] Living in a later age, which continued to debate Wang Anshi's reforms, Su, as a conscientious administrator, in all likelihood did consider the peasant farmers' welfare more.[378] By this sort of substitution, he managed the astonishing feat of converting Tao's mood into pledges of loyal dynastic service.[379] So, we must ask again what would spur Su to diverge so from Tao's models, with Su sometimes verging on rudeness as he "goes beyond" Tao, his avowed model?[380]

A clear pattern emerges with respect to the type and degree of the changes Su made to Tao's body of work. Indisputably, as the years went by, Su's matches began to swerve ever more sharply from Tao's poems, with Su's poems generally careening between two poles: either they grossly idealize Tao, constructing him as a "saint" or "sage" above all care and sorrow (despite abundant counterevidence in Tao's poems) or they lament the deficiencies of Su's own situation, implying that Tao had had it relatively easy in *his* era. Apparently, after Su was exiled in 1094 to Huizhou, where nearly three-quarters of the matching poems were written, the serene front became harder to maintain, although Su did not publicly succumb to outright despair immediately. As his productions circulated in the capital and attracted reviews by hostile censors, Su could not but feel vulnerable, with "no date set for return."[381] Su's matching poems for Tao's "Lauding Impoverished Gentlemen," compiled in 1096, year two of his second exile, show the somber tone that marks his writing once he faced the prospect of permanent exile from the court, most family members, and friends. Tao's "Impoverished Gentlemen" had frankly discussed the stark reality that his chosen path would make it harder to form the sort of true friendship between men of equal cultivation that he desired ("If 'recognizing the tone'[382] [of a true friend] is no longer an option...").[383] As Su's vision sours, he opines that the best response a worthy man can muster is a sort of weary resignation, noting that it has been ever thus; as Tao had said, "I rely on the past, and those many worthies [who experienced similarly dire straits]."[384]

An even blacker mood descended upon Su once he was sent even farther away from court, to Danzhou, where he was deprived of family, except for one son, denied housing, any useful occupation, and ordinary amusements. His resources, mental and financial, were depleted.[385] When Su lauds his island exile as "the finest pleasure outing of my whole life," comparing the pestilence-ridden location to the immortal isle of Penglai or Tao's Peach Blossom Spring, the bleak irony spills out in full force.[386] One of the first poems written in Danzhou looks back in bitter disbelief at the comparative comforts of Su's earlier sojourn in Huizhou, where Su had built his beloved White Crane Lookout retreat. Tao's poem "Returning to My Old Home" shows him saddened by the inevitable changes that passing time brings, but Su's matching poem catalogues loss in uncharacteristically crude terms (a failed joke?): "An impotent man always dreams of "getting it up."[387] / How could I forget the dream of going home? / Yet I dare not dream of my old mountain. / I fear they're raising grave mounds there."[388]

Another poem from the Danzhou years captures Su's mood during this third exile: in describing an essential spirit leaving a ruined garden,[389] Su's own blighted spirit merges with the miasmic vapors thought to shroud the extreme South. The poem ends with the dreary image (almost Gothic) of broken-branched bamboo and cypresses, unprotected by a fence—the very antonym of the safety of home. Bereft of suitable companions and far away from any place he could call home, Su could no longer sustain the tremendous effort needed to assure others of his health and sunny indifference. A passage from Su's old friend Zhuangzi seems relevant here:

> Haven't you heard about the people who are exiled to Yue? A few days after they have left their homelands, they are delighted if they come across an old acquaintance. When a few weeks or a month has passed, they are delighted if they come across someone they used to know by sight when they were at home. And by the time a year has passed, they are delighted if they come across someone who even *looks* as though he might be a countryman.[390]

This same "progressive" arc we can track in Su's matching poems.

Su Shi was hardly the only person to supply Tao Yuanming with new guises. In the most disreputable version of Tao devised by the harshest critics, Tao had figured as the irresponsible drunkard destroying his credit, his family, and his friends alike.[391] (Bo Juyi, no mean toper, for instance, had cast Tao's entire way of life as a joke of little consequence: "Piece after piece advising us to drink / Other than that, he [Tao] has nothing to say.")[392] Yet Su had joined a growing number portraying Tao, in visual representations and in countless literary pieces, via allusion, paraphrase, and commentary,[393] as a superhuman detached from mundane cares and concerns, as superior to ordinary mortals as the phoenix to the crow. One encomium before Su's time had asserted, for instance, "Tao Qian had a lofty and untrammeled air [*fengqi gaoyi*], unlike anything seen before him. It's something that can be appreciated only by the sort of people who know what it is to lie drunk by the north window and fancy themselves men from the time of the primeval ruler Fuxi."[394] While this last sentence draws upon a line from Tao,[395] Tao made no such elevated claims for himself. But, as is so often true in human history, invented traditions could exert incredible power. Thus, those stylized flattering portraits of Tao, multiplying from the Northern and Southern Song, are just as apt to color our reception of Tao's work as the Buddhist "reinterpretations" of Tao Yuanming locating Tao's true "home" in the Pure Land paradise, occluding his delight in his fortunate escape from official life.[396]

All of this prompts a tentative conclusion that flies in the face of beloved literary conventions: to assume a different person's persona demands a virtual metamorphosis—an alchemical miracle, in a well-worn Chinese metaphor[397]—and Su was the first to admit quietly in more than one poem cycle that alchemical experiments invariably fizzled before the final transformation.[398] Forced to adjust not only to a different personality and place, but also to temporary abodes that proved no true home or respite to him, Su had the harder job, always, in that for Tao, returning home was evidently "returning to himself" after being "for so long imprisoned in a cage."[399] Su, on the other hand, had to make sense of several unpleasant and shifting geopolitical and psychological realities simultaneously, not only for himself, but also for his friends, family, and vocal critics, all the while with inadequate information from and about the court. This was doubly arduous, since the rhetoric of the time constrained him to style exiles from the court's favor as valuable "leisure time," to maintain the standard line that the throne in punishing displayed its benevolence.[400]

That said, Su's decision to "match" Tao's corpus of poems had momentous consequences for the lives of both poets, insuring Tao's *post-facto* canonization as gifted secular saint, improving Su's daily life in exile, altering the style of some of Su's writings, and elevating his own posthumous reputation. In fact, Su's repeated imaginative flights to Tao's reclusion resonated so powerfully that after Su's death, popular plays and *ci* poetry generously credited Su with such thoroughgoing enlightenment that he readily declined renewed offers of high rank,[401] his three exiles having taught him the inconsequential nature of worldly status and courtly affairs. Finally, in the popular imagination, Su had truly "in singular fashion, with Tao come home."[402] Such romanticized figures, by definition, lie outside the ordinary because of the intensity and purity they represent in fictional form, whereas, in the end, Tao and even Su came to see in ordinary existence, moments absent intensity, a remarkable reservoir of untold pleasures (see below).

The Pleasures of Return
For most of human history, a huge proportion of resources and rituals has been devoted either to furnishing the homes of the living or to sacralizing the resting places of the dead. How fitting, then, that the "last" pleasure to be catalogued here for early and middle-period China is not a passion for more things, for greater erudition, or for more vivid sensations, but rather a deep longing for a quiet return to abiding and essential relations, however difficult those are to define and to achieve. For Tao (perhaps less so for Su), all forms of return held

the promise of experiential pleasures, because Tao was "thoroughly drunk on each day's riches."[403] This, coupled with Tao's deep love of the quotidian turns of life, made for a clear-eyed and unsentimental approach, which saves him from a simple or simple-minded embrace of wine, women, song, and scenic views,[404] even if Tao could provide no clear path for entering into that realm "beyond words." With Su, one simply observes, as he once did, that "a forced song is no true pleasure." Yet Su reported that such poetic "amusements" as he found in matching Tao brought him intense pleasure, both in the creative moment and when he later shared the poems with intimate friends and his beloved brother Che, despite his restiveness during the exiles of "prolonged leisure."[405] No later portrait of the poets, certainly not mine, or Su's of Tao, can hope to capture or project that complexity fully. At this remove, we cannot know how much the sensuous surface of the two authors' writings contributed to their early readers' appreciation of the phenomenal order, but these poets' avowed belief in such a fundamental pattern has helped many a later admirer.

According to Virginia Woolf, "The reason why poetry excites... one to such abandonment, such rapture, is that it celebrates some feeling one used to have."[406] To read Tao and Su is to experience just this sense of déjà vu. On the surface, Tao and Su espoused the "common wisdom" of their day, which derived fine lyric poetry from authenticity and expressiveness,[407] even as their poems gestured toward a lofty "aesthetic... adequate to preserve the ambiguities of a contingent, historical existence... bereft of ontological certainty."[408] Hence Su Shi's careful forging of an analogy, for example, between his writing and a flowing spring while evoking the age-old trope of "spirit wandering."[409] But since I have attended to the two poets' differences, I would end on a record by Su that almost seems as if it could have been written by Tao himself. Written shortly before Su's death, a temple dedication inquires about the mysterious relation of wisdom to aging, invisible to visible transformation, the intensive to the nonintensive, that preoccupied Tao in his later days.

> The Daoist master Zhang Yijian taught primers.... When I was young, I was his pupil, living for something like three years in the Northern Dipper Courtyard of the Felicitous Heaven Temple. After moving to Hainan, one night in a dream I saw Master Zhang.... The courtyard had been sprinkled and swept clean... as if some dignitary was expected.... One of Zhang's followers was reciting the *Laozi* line "Mystery upon mystery / Entry to a host of marvels."

I piped up, "Since the Mystery is One, there's no more room for a crowd of other mysteries."

The Daoist master laughed and said, "Oneness is already an inferior state. What sublime mysteries can it hold? But if you set your mind on surveying all marvels, no matter how many they are, this can be done."[410]

At that Zhang pointed to the beads of water on the vegetation [in the courtyard]. "Each of these is also a mystery."

When I went back to take a look, two classmates, quick as lightning, accompanied me there. Right then the rosy clouds parted and suddenly dispersed. Startled, I remarked,

"Ha! The marvels are indeed here! I understand Butcher Ding and the plasterer who had plaster dust sliced from his nose by a friend. I believe it now!"

The two boys . . . came up and said,

"You have never yet seen true marvels! And Butcher Ding and the plasterer are not the proper models. . . . Have you ever seen the cicada and the chicken? The cicada climbs a tree and makes its chirring, never knowing to stop itself. The chicken, head bowed, uses its beak [to peck at the dirt], never knowing to look up [to find the cicada]. Such are their fixed natures. But when it comes to the cicada molting and burrowing underground, we have suspended animation: no seeing, no hearing, no starving, no thirst. . . . The silent transformations take place in the murk; the waiting, in a space infinitely small. Even the sage, with all his knowledge and understanding, cannot grasp it. That being so, do you mean to say that finger pointing and daily recitations will be of any help?"

The two boys were leaving, and the Daoist master called out to them, "Wait a bit.[411] Wait for the old master and *then* ask him about it." The two boys, looking back, retorted, "Old men don't necessarily comprehend anything one whit better. Look at the cicada and chicken and ask *them* whether they know how to nourish life or extend the lifespan.". . .

Now a begging letter has come to Hainan, my place of exile, seeking words in calligraphy to commemorate it, so I made use of the words in my dream for this record.

Su had come to realize, belatedly, that the very intensity he had so often celebrated as the fount of creativity did not account for much of human existence, the better part of which is lived going through the motions, without acute awareness, like the cicada undergoing changes underground. Ordinary pleasures is where he leaves us, in company with Tao.

Afterword

While researching and writing this topic of pleasure, I chanced upon a brand new scientific study reporting that "the small, brave act of cooperating with another person, of choosing trust over cynicism, generosity over selfishness, makes the brain light up with quiet joy."[1] This is interesting, if true. It calls to mind Fan Zhongyan's 范仲淹 (989–1052) famous dictum, "Those in power must worry about problems in the realm before the others do and take their pleasures after the others" (先天下之憂而憂, 後天下之樂而樂). This is a dictum worth our consideration, for all in Fan's vision contribute to the community's pleasures, despite class, ethnic, and status distinctions.

In a recent book, the philosopher Hans Sluga speaks of the academy being "still in thrall" to the "normative delusion" that equates the "best" theories with stories that appear "pure" and "logical" and "commonsensical," that is, the stories most likely to incorporate and rely upon presentist misrepresentations of the past.[2] I continue to study early China happily because I have gone through several stages of engagement with my texts. I started with the question "What can be learned about them?" I now ask, "What can be learned from them and, most importantly, through them?"

As Fredric Jameson pointed out long ago in his essay "Pleasure: A Political Issue," what we "really" think is often a residue of one or more older ideologies, too often without recourse to the thoughts of a great many smart people. In consequence, our notions of pleasure may remain little more than attempts at imagining diversions with a sensualist taint from the business at hand.[3] The early and middle-period Chinese thinkers worry about the degradation of culture, the distortions of the heart and mind through spurious appeals to "reason," the ease with which language can be twisted, and the misapplication of institutions. They, too, were looking for some deeper grounding in which the values of "decorum," on the one hand, and physical pleasure, on the other, might be stimulated and fostered.

Xunzi said there are people who create order, but no rules and institutions that create order.[4] Having seen in my own lifetime what cold-blooded cost-benefit analysis has done in wartime and in peacetime to vulnerable populations, I ascribe greater wisdom to the thinkers and poets here, who never promise pain-free or quick solutions for endemic problems. They agreed on the human scale and the human costs, though their ions for living well differed in other respects.

Some people have said that it has taken me nearly two decades to finish the pleasure book because it has been too pleasurable to write it. I agree, but I would add this: one cannot work on a pleasure book when grieving at the state of the world or the loss of a friend. Happily, thanks to several co-conspirators, life seems well worth living now. This book I dedicate to Naomi Richard and to the other generous souls like her in my circle.

Notes

Due to the length of this book, the publisher has decided to make the endnotes and works cited available digitally. They can be downloaded from our web site at: *zonebooks.org/books/133-the-chinese-pleasure-book*. A digial file may also be requested via e-mail by writing to *info@zonebooks.org*.

We apologize for any inconvenience.

List of Names

Premodern Statesmen, Thinkers, and Writers Who People the Main Text

Aulus Gellius (d. ca 180)
Ban Gu 班固 (32–92 CE)
Bao Shuya 鮑叔牙 (sixth century BCE?)
Bo Juyi 白居易 (772–846)
Boli Xi 百里奚 (fl. 650–625 BCE)
Boluan 伯鸞, elder brother of Dai Liang
Bo Ya 伯牙 (legendary figure, Eastern Zhou)
Bo Yi 伯夷 (fl. ca. 1050 BCE)
Cai Yong 蔡邕 (132–92)
Callimachus (ca. 305–240 BCE)
Cao Pi 曹丕 (87–29 June 226)
Chen Nong 陳農 (fl. 26 BCE)
Chen Yu 陳餘, time of Han Gaozu
Chenggong Sui 成公綏 (fl. Western Jin 231–273)
Chao Cuo 鼂錯 (d. 154 BCE)
Chu Pou 褚裒 (Jin)
Confucius/Kongzi 孔子 (tradit. 551–479 BCE)
Dai Liang 戴良 (Eastern Han)
Dou Ying 竇嬰 (d. 131 BCE)
Du Fu 杜甫 (712–770)
Duke Huan 桓 of Qi 齊 (hegemon, 374 to 357 BCE)
Duke of Zhou 周公 (fl. 1050 BCE)

Fan Shi 范式 (d. before 235 CE)
Fan Zhongyan 范仲淹 (989–1052)
Feng Xuan 馮諼, contemporary of the Lord of Mengchang
Flying Swallow Zhao 趙飛燕 (ca. 32–1 BCE)
Gaozi 高子, contemporary of Mencius
Ge Hong 葛洪 (280–ca. 343)
Gongdu 公都, contemporary of Mencius
Guan Fu 灌夫 (fl. 154–ca. 131 BCE)
Guan Ning 管寧 (158–241)
Guan Zhong 管仲 (d. 645 BCE)
Guo Xiang 郭象 (d. 312 CE)
Honorable Zhai 翟公, aka Zhai Fangjin 翟方進 (fl. 130–127 BCE)
Hua Xin 華歆 (157–231)
Huan Tan 桓譚 (43 BCE–28 CE)
Huangfu Mi 皇甫謐 (215–282)
Hui Shi 惠施 (fl. ca. 370–319)
Huiyuan 慧遠, 334–416
Ji Kang 嵇康 (223–262)
Kongzi: see Confucius
Kuang Heng 匡衡 (active ca. 31 BCE)
Li Ling 李陵 (d. 74 BCE)
Li Guang 李廣 (d. 119 BCE)
Li Zhuguo 李柱國 (fl. 26 BCE)
Li Yanzhi 李琰之 (Wei dynasty)
Liu Xiang 劉向 (79–7 BCE)
Liu Xin 劉歆 (53 BCE–23 CE)
Liu Xie 劉勰 (ca. 465–ca. 522)
Liu Yiqing 劉義慶 (403–444)
Liu Zhiji 劉知幾 (661–721)
Liu Zongyuan 柳宗元 (773–819)
Lord of Mengchang 孟嘗 (d. 279 BCE)
Master Gongdu: see Gongdu
Master Wulu 屋廬子, contemporary of Mencius
Mei Yaochen 梅堯臣 (1002–1060)
Mencius 孟子 (act. 320 BCE)

Mi Zixia 彌子瑕 (late sixth century BCE)
Mozi 墨子 (active 400 BCE)
Qu Yuan 屈原 (332–295 BCE)
Ren An 任安 (d. 91 BCE)
Ren Fang 任昉 (460–508)
Ren Hong 任宏 (fl. 26 BCE)
Shan Tao 山濤 (205–283), contemporary of Ji Kang
Shu Qi 叔齊, fl. 1050 BCE
Sima Guang 司馬光 (1019–1086)
Sima Qian 司馬遷 (145–ca. 80 BCE)
Su Che 蘇轍 (1039–1112)
Su Shi 蘇軾 (1037–1101)
Sun Huizong 孫會宗 (fl. 60 BCE)
Sun Jing 孫敬 (Eastern Han)
Tao Yuanming 陶淵明 (365?–427)
Tian Fen 田蚡 (d. 131 BCE?)
Wan Zhang 萬章, contemporary of Mencius
Wang Bi 王弼 (226–49)
Wang Can 王粲 (177–217)
Wang Chong 王充 (27–97)
Wang Shang 王商 (d. 29? BCE)
Wang Xizhi 王羲之 ((303–361)
Wan Zhang 萬章, contemporary of Mencius
Xi Kang: see Ji Kang
Xie An 謝安 (320–385)
Xu You 許由, said to live in many eras
Xunzi 荀子 (d. before 221 BCE?)
Yan Hui 颜回, personal name *Yuan* 淵
Yan Zhitui 颜之推 (531–591)
Yang Hu 陽虎, contemporary of Confucius/Kongzi
Yang Xiong 揚雄 (53 BCE–18 CE)
Yang Yun 楊惲 (d. 54 BCE)
Yin Xian 尹咸 (fl. 26 BCE)
Yuan Qiao 袁喬 (312–347)
Zeng Gong 曾鞏 (1019–83)

Zhang Chang 張敞 (fl. 74–51 BCE)
Zhang Er 張耳 (d. 202 BCE)
Zhang Shao 張劭 (late Eastern Han)
Zheng Qiao 鄭樵 (1104–62)
Zheng Xuan 鄭玄 (127–200)
Zhong Rong 鍾嶸 (468–518)
Zhong Ziqi 鍾子期 (sixth century BCE?)
Zhou Yafu 周亞夫 (d. 143 BCE)
Zhu Mu 朱穆 (100–163)
Zhu Xi 朱熹 (1130–1200)
Zhuangzi, i.e., Zhuang Zhou 莊子/莊周 (mid-fourth century BCE?)
Zichan 子產 (act. 542–522 BCE)

Suggested Readings

For general background, one may consult the following reliable works:

CHAPTER ONE

The ultimate inspiration for this work (however great the divergences) is:

Bauer, Wolfgang. *China and the Search for Happiness: Recurring Themes in Four Thousand Years of Chinese Cultural History*. Translated by Michael Shaw. New York: Seabury, 1976. The original German was entitled about translation: *China und die Hoffnung auf Glück; Paradiese, Utopien, Idealvorstellungen* (1971).

Bennett, Steven J. "Patterns of the Sky and Earth: A Chinese Science of Applied Cosmology." *Chinese Science* 3 (March 1978), pp. 1–26.

Cheng, Anne. *Histoire de la pensée chinoise*. Paris: Éditions du Seuil, 1997.

Csikszentmihalyi, Mark. *Material Virtue: Ethics and the Body in Early China*. Leiden: Brill, 2006.

Harper, Donald. *Early Chinese Medical Literature: The Mawangdui Medical Manuscripts*. Trans. and study by Donald Harper. New York, Kegan Paul, 1998.

Henry, Eric. "The Motif of Recognition in Early China." *Harvard Journal of Asiatic Studies* 47 (June 1987), pp. 5–30.

King, Richard A. H. "Freedom in Parts of the Zhuangzi and Epictetus." In Geoffrey Lloyd and Jenny Jingyi Zhao (eds.). *Ancient Greece and China Compared: Interdisciplinary and Cross-Cultural Perspectives*. Cambridge: Cambridge University Press, 2017.

Kuriyama, Shigehisa. *The Expressiveness of the Body and the Divergence of Greek and Chinese Medicine*. New York: Zone Books, 1999.

Mair, Victor H., trans. *Mei Cherng's "Seven Stimuli" and Wang Bor's "Pavilion of King Terng": Chinese Poems for Princes*. Lewiston, NY: E. Mellen Press, 1988.

Nylan, Michael and Michael Loewe (eds.). *China's Early Empires: A Re-appraisal*. University of Cambridge Oriental Publications 67. Cambridge: Cambridge University Press, 2010.

Riegel, Jeffrey K. "Eros, Introversion, and the Beginnings of Shijing Commentary." *Harvard Journal of Asiatic Studies* 57.1 (June 1997), pp. 143–77.

Rosemont, Henry. *Against Individualism: A Confucian Rethinking of the Foundations of Morality, Politics, Family, and Religion*. Lanham, MD: Lexington Books, 2015.
Sommers, Deborah. "Boundaries of the Ti Body." *Asia Major* 21.1 (2008), pp. 293-394.
Wang, Robin. *Yinyang: The Way of Heaven and Earth in Chinese Thought and Culture*. Cambridge: Cambridge University Press, 2012.

There are several superb new translations of early classics and masterworks. Every reader should examine the new translation of the *Zuozhuan* entitled *Zuo Tradition*, trans. Stephen Durrant, Li Wai-yee, and David Schaberg, 3 vols. Seattle: University of Washington Press, 2016. For another, see Suggested Readings for Chapter 5 on Xunzi below.

CHAPTER TWO

Bagley, Robert, W. "Percussion." In Jenny F. So (ed.). *Music in the Age of Confucius*. Washington, DC: Smithsonian Institution, 2000, pp. 34-66.
———. "The Prehistory of Chinese Musical Theory." *Proceedings of the British Academy* 131 (2005), pp. 41-90.
DeWoskin, Kenneth. *A Song for One or Two: Music and the Concept of Art in Early China*. Ann Arbor: Center for Chinese Studies, University of Michigan, 1982.
Ehrenreich, Barbara. *Dancing in the Streets: A History of Collective Joy*. New York: Metropolitan Books, 2007.
Mittag, Achim. "Change in *Shijing* Exegesis: Some Notes on the Rediscovery of the *Odes* in the Song Period." *T'oung Pao* 79.4-5 (1993), pp. 197-224.
Nylan, Michael. *The Five "Confucian" Classics*. New Haven: Yale University Press, 2001.
Van Zoeren, Steven. *Poetry and Personality: Reading, Exegesis, and Hermeneutics in Traditional China*. Stanford: Stanford University Press, 1991.
Vervoorn, Aat. "Friendship in Ancient China." *East Asian History* 27 (June 2004), pp. 1-32.
———. "Music and the Rise of Literary Theory in Ancient China," *Journal of Oriental Studies (Dongfang wenhua)* 43.1 (1996), pp. 50-70.

CHAPTER THREE

The best translation of the *Mencius* (if not of the *Analects*) is D. C. Lau, London: Penguin, 1970.

Behuniak, James. "Naturalizing Mencius." *Philosophy East and West* 61.3 (Jul 2011), pp. 492-515.
Graham, A. C., *Disputers of the Tao: Philosophical Argument in Ancient China*. Chicago, IL: Open Court, 1989.
Hadot, Pierre. *What is Ancient Philosophy?* London: Belknap, 2002.
Hutton, Eric. "Mencius, Xunzi, and the Legacy of Confucius." In *The Norton Critical Edition of the Analects: A Collection of Essays Plus a Translation*. Edited by Michael Nylan. New York: W. W. Norton, 2014, pp. 166-77.

King, Richard A. H. and Dennis Schilling. *How Should One Live: Comparing Ethics in Ancient China and Greco-Roman Antiquity*. Berlin: de Gruyter, 2011.

Sahlins, Marshall D. *The Western Illusion of Human Nature: With Reflections on the Long History of Hierarchy, Equality, and the Sublimation of Anarchy in the West, and Comparative Notes on Other Conceptions of the Human Condition*. Chicago: Prickly Paradigm Press, 2008.

Shun, Kwong-loi. *Mencius and Early Chinese Thought*. Stanford, CA: Stanford University Press, 1997.

Virag, Curie. *The Emotions in Early Chinese Philosophy*. Oxford: Oxford University Press, 2017.

Wong, David B. "Is there a Distinction between Reason and Emotion in Mencius?" *Philosophy East and West* 41.1 (Jan. 1991), pp. 31–44.

See also: *Polishing the Chinese Mirror: Essays in Honor of Henry Rosemont*. Eds. Marthe Chandler and Ronnie Littlejohn. La Salle, IL: Association of Chinese Philosophers of America and Open Court, 2007.

CHAPTER FOUR

Cua, Antonio S. "The Ethical Significance of Shame: Insights of Aristotle and Xunzi." *Philosophy East and West* 53.2 (April, 2003), pp. 147–202.

Elstein, David. "Beyond the Five Relationships: teachers and worthies in early Chinese thought." *Philosophy East and West* 62.3 (Jul 2012), pp. 375–91.

Fraser, Chris. "Xunzi versus Zhuangzi: Two Approaches to Death in Classical Chinese Thought." *Frontiers of Philosophy in China (Beijing)* 8.3 (Sep 2013), pp. 410–27.

———. "Moral Reasoning in Aristotle and Xunzi." *Journal of Chinese Philosophy* 29.3 (Sep 2002), pp. 355–84.

Geuss, Raymond. *Philosophy and Real Politics*. Princeton: Princeton University Press, 2008.

Graham, A. C. *Disputers of the Tao: Philosophical Argument in Ancient China*. Chicago, IL: Open Court, 1989.

Hutton, Eric L., ed. *The Dao Companion to the Philosophy of Xunzi*. New York: Springer, 2016.

Kline, T. C., III. "The Therapy of Desire in Early Confucianism: Xunzi." *Dao: A Journal of Comparative Philosophy* 5.2 (Summer 2006), pp. 235–46.

Nylan, Michael. "Xunzi: An Early Reception History, Han through Tang." In *The Dao Companion to the Philosophy of Xunzi*. Edited by Eric L. Hutton. New York: Springer, 2016, pp. 395–433.

Sluga, Hans. *Politics and the Search for the Common Good*. Cambridge: Cambridge University Press, 2014.

CHAPTER FIVE

Two works by A. C. Graham are classics in the field, though I dispute parts of their arguments: *Disputers of the Tao: Philosophical Argument in Ancient China*. Chicago, IL: Open

Court, 1989, and *Reason and Spontaneity: A New Solution to the Problem of Fact and Value.* London: Curzon, 1985. To these we may now add, Moeller, Hans-Georg, and Paul D'Ambrosio. *Genuine Pretending: On the Philosophy of the Zhuangzi.* New York: Columbia University Press, 2017. The two best translations are (1) Burton Watson's *The Complete Works of Zhuangzi.* New York: Columbia University Press, 2013 (reprint of a 1968 translation, with romanization updated); and (2) Brook Ziporyn's *Zhuangzi: The Essential Writings, with Selections from Traditional Commentaries.* Indianapolis: Hackett, 2009.

One may consult with pleasure the following:

Mair, Victor, ed. *Experimental Essays on Chuang Tzu.* Honolulu: University of Hawai'i Press, 1983.

King, Richard A. H. "Freedom in Parts of the Zhuangzi and Epictetus." In Geoffrey Lloyd and Jenny Jingyi Zhao (eds.). *Ancient Greece and China Compared: Interdisciplinary and Cross-Cultural Perspectives.* Cambridge: Cambridge University Press, forthcoming.

Klein, Esther. "Were there Inner Chapters in the Warring States? A New Examination of Evidence about the *Zhuangzi.*" *T'oung Pao* 96 (2010), pp. 299-369.

Liu Xiaogan. *Classifying the Zhuangzi Chapters.* Michigan Monographs in Chinese Studies. Ann Arbor: University of Michigan Center for Chinese Studies, University of Michigan Press, 2003.

Poo Mu-chou. *In Search of Personal Welfare: a view of ancient Chinese religion.* Albany: State University of New York Press, 1998.

Sivin, Nathan. "The Myth of the Naturalists," *Medicine, Philosophy and Religion in Ancient China. Researches and Reflections* (Aldershot: Variorum, 1995), pp. 1-33.

Wu Kuang-ming. *The Butterfly as Companion: Meditations on the First Three Chapters of the Chuang-Tzu.* Albany: State University of New York Press, 1990.

Yearley, Lee H. "The Perfected Person in the Radical Chuang-tzu." In Victor Mair (ed.), *Experimental Essays on Chuang-tzu.* Honolulu: University of Hawai'i Press, 1983, pp. 125-39.

———. "Ethics of Bewilderment." *Journal of Religious Ethics* 38.3 (2010), pp. 436-60.

CHAPTER SIX

Bagley, Robert W. "Anyang Writing and the Origin of the Chinese Writing System." In Stephen D. Houston (ed.). *The First Writing: Script Invention as History and Process.* Cambridge: Cambridge University Press, 2004, pp. 190-249.

Cherniack, Susan. "Book Culture and Textual Transmission in Sung China." *Harvard Journal of Asiatic Studies* 54.1 (1994), pp. 5-125.

Greatrex, Roger. "An Early Western Han Synonymicon: The Fuyang Copy of the *Cang Jie Pian.*" In *Outstretched Leaves on His Bamboo Staff: Essays in Honour of Göran Malmqvist on his Seventieth Birthday.* Edited by Joakim Enwall. Stockholm: Association of Oriental Studies, 1994, pp. 97-113.

L'Haridon, Béatrice. "La recherche du modèle dans les dialogues du Fayan de Yang Xiong (53 av. J.-C.-18 apr. J.C.): écriture, éthique, and réflexion historique à la fin des Han occidentaux." Ph.D. thesis, Institute National des Langues and Civilisations orientales, 2006, 2 vols.

Knechtges, David. R. and Ban Gu. *The Han shu Biography of Yang Xiong (53 B.C. to A.D. 18)*. Tempe: Center for Asian Studies, Arizona State University, 1982.

Nylan, Michael, *Yang Xiong and the Pleasures of Reading and Classical Learning in Han China*. New Haven, CT: American Oriental Society, 2011.

———. trans. of Yang Xiong, *Exemplary Figures / Fayan*. Seattle: University of Washington Press, 2013.

CHAPTER SEVEN

Brotherton, Elizabeth. "Beyond the Written Word: Li Gonglin's Illustrations to Tao Yuanming's 'Returning Home.'" *Artibus Asiae* 59.3-4 (2000), pp. 225-63.

de Pee, Christian. "Purchase on Power: imperial space and commercial space in Song-dynasty Kaifeng 960-1127." *Journal of the Social and Economic History of the Orient*. 53.1-2 (2010), pp. 149-84.

Egan, Ronald. *Word, Image, and Deed in the Life of Su Shi*. Cambridge, MA: Harvard University Press, 1994.

Feuillas, Stéphane. "Un lieu à soi?: Construction de l'espace et de soi chez Su Shi (1037-1101)." In Jean-Jacques Wunenberger and Valentina Tirloni (eds.), *Esthétiques de l'espace: Occident et Orient*. Paris: MIMESIS, 2010, pp. 27-46.

Fuller, Michael A. *Drifting among Rivers and Lakes: Southern Song Dynasty Poetry and the Problem of Literary History*. Cambridge, MA: Harvard University Press, 2013.

Murck, Alfreda. *Poetry and Painting in Song China: The Subtle Art of Dissent*. Harvard-Yenching Institute Monograph Series 50. Cambridge, MA: Harvard University Press, 2000.

Nelson, Susan E. "What I Do Today Is Right: Picturing Tao Yuanming's Return." *Journal of Sung-Yuan Studies* 28 (1998), pp. 61-90.

———. "Tao Yuanming's Sashes: or, The Gendering of Immortality." *Ars Orientalis* 29 (1999), pp. 1-27.

———. "'The Thing in the Cup': Pictures and Tales of a Drunken Poet." *Oriental Art* 46.4 (2000), pp. 49-61.

———. "Catching Sight of South Mountain: Tao Yuanming, Mount Lu, and the Iconographies of Escape." *Archives of Asian Art* 52 (2000-2001), pp. 11-43.

———. "The Bridge at Tiger Brook: Tao Qian and the Three Teachings in Chinese Art." *Journal of Oriental Studies* 50.1 (2002), pp. 257-94.

Index

Page numbers in italics represent illustrations.

ABEL-RÉMUSAT, JEAN-PIERRE, 314.
Academicians, 278, 296, 305, 306–308.
Activist editing, 275, 281–82, 285, 305.
Aged, treatment of, 144, 168–69, 295.
Ai (grieved by a loss), 35.
Ai (love or care for), 44.
Aidi (Han), 279.
An (secure/ease), 35, 41.
Analects: allusions to, 342, 377 n.5, 382 n.2, 389 n.1, 394 n.9, 402 n.2; commonalities with Mencius, 165; Confucius of, 97; on fellow feeling (*shu*), 136; on flows of *qi*, 50; on friendship, 64, 114; on human motivations, 312; on *ren*, 156; "steadfastness in adversity," 331, 354, 388, 410, 415; teachings on righteous rule, 147; "words are merely for communication," 113; and Yang Xiong's *Exemplary Figures*, 272.
Ancestor worship, 23. *See also* Sacrifices.
Annals (Chunqiu), 31, 271, 275.
Anxiety, 35, 40–41, 51, 209.
Apatheia, 244.
Archaic Script corpus, 305.
Archives, 96, 274, 277–78. *See also* Libraries.
Aristotle, 87, 121, 213.
Army Regulations for the Colonel, 306.
Art, 50, 207–208. *See also* Music.
Art of War, 306.
Artisans, 216–19; Woodworker Qing and Butcher Ding, 215, 216–21, 224–25.
Asceticism, 52, 239.
Ataraxia, 244.
Attitudinal pleasure, 36–37.

Aulus Gellius, 269.
Authenticity: in Mencius, 149; in *Odes* and *Changes*, 64; Su Shi's notion of, 359, 363, 367, 391; Tao Yuanming's notion of, 321, 329, 348, 414; of texts, 282; in *Zhuangzi*, 256–57.
Authorship, 274–76, 282, 285, 294.
Autonomy, 39–40, 54, 148, 219, 235, 324, 347.

BACHELARD, GASTON, 254.
Balancing the faculties, 217, 218, 239, 248, 254.
Balazs, Étienne, "Political Philosophy and Social Crisis at the End of the Han," 29.
Bamboo slips, documents on, *200*.
Ban Gu, 313; *Han Histories*, 272, 298, 300; "Tables of Men, Ancient and Recent," 300.
Banquets, 23, 99–100, 103–104, 121, 207; in tomb mural, *102*; in *Zuozhuan*, 98, 107. *See also* Drinking sessions.
Bao Shuya, 88–90.
Barthes, Roland, 84.
Bauer, Wolfgang, *China and the Search for Happiness*, 20.
Bells: chariot, 69; in early depictions of musical performance, 66, *68*; in ritualized music, 67; tones of, 63, 85; Xunzi on, 79; Zeng hou yi tomb bell set, 76, 77, 85.
Bergson, Henri, 255.
Berry, Wendell, 133.
Bhutan, Gross National Happiness index, 37.
"Binding" (*jie*), 85.
Birds: calls of, 69, 120, 241, 242, 257; *luan*, 302; Peng, 232–33, 262; in poems of Tao Yuanming, 327, 329, 331, 332, 333, 340, 341, 346–47, 374, 382, 400, 402, 406, 418; in Yang Xiong *fu*, 286.

441

Bo Juyi, 359, 401 n.4, 422.
Bo Ya and Zhong Ziqi, 69, 90.
Bo Yi, 94, 343, 351, 388, 403.
Body: circulatory flows, 50–51; theories of, and sensory experience, 36–37; in *Zhuangzi*, 227–28, 243. See also *Qi*.
Boli Xi, 70.
Boluan, 109.
Books: Kuang Heng and, 265–66; in Yang Xiong's view of pleasure, 19–20, 263–65, 271, 290, 312–14. See also Libraries; Manuscript culture.
Bronzes: with depiction of musical performance, 66, *68*; inscriptions, 117, 302–304; Western Zhou wine vessel from Guodun Shan, *303*.
Buddhism, 30, 418.
Butcher Ding, 215, 216–21, 224–25, 226, 425.

CAI YONG, "CORRECTING CONTACTS," 125–26.
Callimachus, 20, 269.
Cang Jie primer, 302, 310.
Cao Pi (Emperor Wen of Wei), 108–109; "Discourse on Writing," 118; letter to a friend, 64.
Careerism, 22, 95, 101, 122, 125, 296, 371; of Su Shi, 369, 371; views of Tao Yuanming, 323, 335, 343, 363, 414, 419; in *Zhuangzi*, 216, 231, 246.
Ceremonial aspects of authority, 28.
Chang Qu, *Record of the Lands South of Mount Hua*, 272.
Changes classic (*Yijing*): on friendship, 64, 118; trust and time in, 118; "Xici" tradition," 28, 383 n.7; and Yang Xiong's writings, 272, 312.
Cheerfulness, 39, 328.
Chen Nong, 278.
Chen Yu and Zhang Er, 129, 131.
Cheng, Emperor of Han (Chengdi), 271, 294, 308; palace libraries under, 278–81; social welfare measures, 170.
Cheng, King of Zhou, 267.
Chenggong Sui, "Rhapsody on Whistling," 73.
Chimes, 66, 77, 79, 304; in depiction of musical performance, *68*.
Chu Pou, 123.
Classic of Filial Piety, handscroll illustrating, *134*.

Classics: allusions to, 290; meanings of, 57; as medium of communication with sages and worthies, 293; rhetoric of pleasure in, 26–27. See also Five Classics learning; Su Shi: classical allusions by; Tao Yuanming: classical allusions obyYang Xiong: passion for classical learning.
Cloud imagery, *38*, 232, 346.
Commitments (*zhi*), 47, 53, 55, 67, 111–12, 127, 159–60, 181.
Community, 23, 26, 61, 99, 133; in Mencius, 137, 138, 148, 166; in Tao Yuanming, 319, 328; in Xunzi, 77–78, 195–97. See also Community banquets; Drinking sessions.
Community banquets, 23, 99–100, 103–104, 121, 207.
Confucius. See *Analects*; Kongzi/Confucius.
Conjugal love, 120; Chinese wedding manual, *38*.
Connoisseurship, 22.
Constants, 146; constant *xin*, 243; Five Constant Social Relations, 52, 74.
Consuming and sustaining pleasures, 24, 26, 33, 52, 147.
Copying texts, 200, 268, 290–91, 293, 297. See also Manuscript culture.
Cosmic eggs, *38*.
Csikszentmihalyi, Mihalyi, 36.
Cycles of the Yellow Emperor, 306.

DA DAIJI, 309.
Dai Kui, 343.
Dai Liang, 109.
Dan (indulge in or be addicted to), 44; *dan Dao*, 312.
Dance: in the *Analects*, 64; and Butcher Ding's technique, 219, 220; at drinking sessions, 100, 101; early depictions of, 66, *68*; paired with music, 65, 69, 77, 78, 79, 81, 82; out of pleasure, 131, 173, 312.
Davidson, Donald, 224.
Davis, A. R., 376 n.1.
De (virtue, grace and charisma), 177, 187, 204.
Death: hatred and, 131; Tao Yuanming's view, 329, 333, 335–39, 357; Zhuangzi's view, 19, 236–37, 250–52.
Declaration of Independence, 37.
Delayed gratification, 184–86.
Delight, 42–44.

INDEX

Desires: in Mencius, 148-49, 175; mimetic, 47, 53, 198; No Desires and Refined Desires advocates, 54; in pleasure rhetoric, 46-47, 52-55, 324; in Xunzi, 175, 176, 178, 179, 182-86, 194-95, 198; in *Zhuangzi*, 244.

Dialogue genre, 137; in *Mencius*, 135, 136, 139, 140, 148, 151, 153, 160, 173; in Yang Xiong's *Exemplary Figures*, 265, 269, 286, 294; in *Zhuangzi*, 214.

Dichotomies, 31, 72, 115, 132, 249, 258-59; pleasure-pain, 40, 41-42.

Display culture, 28, 29-30, 196-97.

Documents, 31, 275, 282; "Be Not Idle" chapter, 267; "Yueming" chapter, 86.

Dong (motions or feelings), 45, 48, 67-69. *See also* Emotions.

Dou Ying, friendship with Guan Fu, 95-96.

Dreams, 239-41.

Drinking sessions, 98-101, 103-107; in classical texts, 107-108; entertainment at, 104, 106; principles of sociability at, 105-106; status conventions in, 104; Xunzi's account of, 206-207.

Drums, 66, 67, 79.

Du Fu, 323, 365, 387 n.5.

Du Lin, 308.

EASTERN HAN ESTATES, 28-29, 30.

Egan, Ronald, 364, 415.

Eight Sounds, 67.

Eight Winds, 67.

Emotions, 45, 48; music and, 67-69, 71-72, 74-75, 78, 80, 84, 85.

Epicurus, 41.

Epigraphy, 284, 305, 308-10. *See also* Philology.

Equitable distribution, 23.

Eroticism, Chinese wedding manual illustration, *38*.

Erya, 306, 308, 309-11.

Ethics: in ancient Greece, 10; Chinese, 10, 117, 131, 156-57, 276; Western, 17, 97, 144, 172.

Evaluations (*si*), 47.

Ever-normal granaries, 170.

Experiential pleasures, 34, 41, 46-48, 424.

FAME, 228, 335-36, 338, 354.

Family: and friendship, 61, 87, 119, 344; for Su Shi and Tao Yuanming, 348-49, 369, 383, 387 n.6, 421.

Fan painting: "Appraisals for Paintings on Fans" by Tao Yuanming, 411; Su Shi and, 412.

Fan Shi, 117.

Fan Zhongyan, 427.

Farmers, 29, 141, 163, 181, 191-92, 202, 384, 420; and distribution of farmland, 168-70; gentleman, 130, 346. *See also* Peasants.

Fayan. *See* Yang Xiong: *Exemplary Figures*.

Feng Xuan, 122.

Feng Yan, 383 n.5.

Fingarette, Herbert, 84.

Fitzgerald, F. Scott, 282.

Five Classics learning: and the *haogu* movement, 278, 306-308, 311-12, 316; and office-holding, 270; Yang Xiong and, 276, 292, 305-308.

"Five Conducts," 185.

Five Constant Social Relations, 52, 74.

Fleeting pleasures, 35, 177.

Flying Swallow Zhao, 271.

Food and wine, 35, 50. *See also* Banquets; Drinking sessions.

Forgetting, 254-55, 340-41.

Formalized exchanges, 23, 27-28, 96-98, 100, 206-207. *See also* Gift giving.

Forster, E. M., 211.

Four Books, 31, 276.

Friendship: of Bo Ya and Zhong Ziqi, 69-70, 90; and career, 122, 125; in the classics, 60, 64-66, 94-95, 111, 114, 118-20; contrasted with hierarchical relations, 66, 98, 119; of Dou Ying and General Guan Fu, 95-96; and drinking, 99-100, 105-106; and family, 61, 87, 119, 344; and funerals, 108-10; of Guan Ning and Hua Xin, 129; of Guan Zhong and Bao Shuya, 88-90; intimacy and sociability, 61, 98-99, 111-12; letters severing, 123-24, 126-28, 129-31; letters to friends, 112-13; "making friends in history," 362; in modern philosophy and classical Chinese rhetoric, 132; music and, 18, *58*, 59-63, 65-66, 69, 72-73, 84-86, 132-33; nature of, 87-88; in novels of Wendell Berry, 133; "old," 118; parting from friends, 18, 73, 128; portraits of, *320*, 407, 407 n.7; "preverbal," 112-13; and relational pleasures, 34-35, 328; role of election in, 115-16; and romantic love, 116, 120; and

443

the ruler-minister relationship, 119–20; severing of, 117, 121–31; and social relations, 66, 116–18, 119–21; Tao Yuanming and, 328, 334–35, 341–42, 344, 350, 352, 359, 370; terms for, 65–66, 117–18, 120, 126; Xunzi on, 60, 119; Yang Xiong on, 117; of Yang Yun and Sun Huizong, 129–30; Zhuangzi on, 90, 91–92, 120; of Zhuangzi and Hu Shi, 90–91, 113. *See also* Intimate friendship.

Fu rhapsodies: contrast with Tao Yuanming's poetry, 324; criticized by Yang Xiong, 284, 291, 298, 306; display *fu*, 269, 272, 284, 324; of frustration, 343–44; "*Fu* on Reading" by Shu Xi, 314; and Mencian dialogues, 173; "Moved by Good Men's Failures to Meet Good Fortune" by Tao Yuanming, 343–44; "Returning to the Fields" by Zhang Heng, 313; of Xunzi, 173; of Yang Xiong, 173, 263, 269, 271, 272, 285–86, 287–88, 294, 306.

Fukui Shigemasa, 307.

Fully Present Man (*zhi ren*), 216, 217, 218, 225–26.

Funerals, 108–10. *See also* Mourning rituals.

GAN (RESONANT FEELING), 166.

Ganying (sympathetic resonance), 45. *See also* Resonance theory.

Gaozi, 159, 167.

Gaozong, colophon for "Illustrations of the Classic of Filial Piety," *134*.

Gaozu (Han), 81.

Ge de qi suo (everyone in his proper place), 53.

Ge Hong, 269, 314, 366, 407, 409, 418, 419.

Gift giving, 27, 96–98, 199–201; manuscripts and, 283, 290–91, 294.

Gong Liu, 153.

Gongdu, Master, 159.

Gongsun Hong, 395 n.14.

Good life, 10, 37, 87, 100, 159, 182, 313, 328, 341.

Goody, Jack, 32, 275.

Gordon, Adrian, untitled photograph, 2017, *234*.

Grain, 141–42, 154, 170.

Great Decrees, 235.

Gross National Happiness index, 37.

Guan Fu, friendship with Dou Ying, 95–96.

Guangwu emperor, 311.

Guan Ning, and Hua Xin, 129.

Guan Zhong, and Bao Shuya, 88–90.

Gui (coming home), 317–19, 350, 352. *See also* Tao Yuanming: on the pleasures of returning home.

Guo Xiang, commentary of, 237, 261.

Guodian: "Black Robe" chapter, 51; "Chengzhi wenzhi" manuscript, 171–72.

Guodun Shan, wine vessel from, *303*.

HAIHUNHOU GRAVE GOODS, *303*.

Hall, Donald, *Lifework*, 36.

Han Feizi, 171.

Hanshu: bibliographic treatise, 279; biography of Yang Xiong, 299; contrasted with *Shiji*, 74; on court music and court rites, 74–75; "Treatise on Rites and Music," 82.

Hao (be fond of), 44.

Haogu (loving antiquity) movement: court sponsorship of, 266, 298; historical background of, 315–16; and pre-Qin bronzes, 302, *303*; and revival of ancient institutions, 296–97; and textual traditions, 270, 278, 297–98, 305, 306–11; and Yang Xiong's writings, 268, 314.

Happiness, 35–36, 37–40, 39.

Harmony (*he*): describes music and friendship, 85; *he er butong* (in harmony, yet not identical), 61; music and, 77–78, 150.

Hatred, 131.

Hedonism, 41–42.

Hejian, King of (Liu De), 277–78.

Hierarchical relationships, 66, 98, 119.

Home, 317–19, 321, 346; and roots, *318*. *See also* Tao Yuanming: on the pleasures of returning home.

Homer, 57.

Honor and glory (*rong*), 27–28.

Hopkins, Gerard Manley, 324–25.

Hou Hanshu, 109.

Hu Shi, 90–91, 91, 113, 255.

Hua Xin, and Guan Ning, 129.

Huainanzi, 266, 306.

Huan, Duke of Qi, 88–89, 90, 143, 238, 295, 408.

Huan Tan, 81, 90, 111, 114, 272.

Huangfu Mi, 314.

Hui, King of Liang, 138–39, 140–43.

Huiyuan, 347; in "Three Laughing Masters at Tiger Ravine," 320.

Huizi, 243–46, 258.

INDEX

Human nature: desires in, 155, 176; Gaozi on, 159; Mencius on, 18, 56, 159-65, 167, 312; music and, 52, 73, 74; in pleasure rhetoric, 25, 26, 45; Xunzi on, 19, 165, 176, 178-79, 189-90, 198, 207. *See also* Second nature.
Hutcheson, Francis, 36.

IMPACT-RESPONSE MODEL, 31-32.
Impulses: animating, 60, 225; antisocial, 42; contradictory, 177, 198, 205; evaluative, 180; impulsive consumption, 26, 35, 47, 50, 54; and moral potential, 157, 189; to seek pleasure, 176, 184, 186; spontaneous and unreflective, 47, 181, 221, 325. *See also* Human nature.
Individuality, 114-15.
Insecurity, 17, 35, 40-41, 51.
Integrity and wholeness (*cheng*), 138, 158, 177, 203, 211, 229; Xunzi on, 203, 204-205, 208.
Intimate friendship, 111-16; severing of, 121-31. *See also* Friendship.
Is-ought problem, 32.

JAMESON, FREDRIC, 194; "Pleasure: A Political Issue," 427.
Jannings *hu*, 68.
Jansen, Thomas, 127.
Ji An, 387 n.5.
Ji jie (raillery), 86.
Ji Kang: "On the Absence of Emotions in Music," 72; execution of, 128; "Letter to Shan Tao," 126-28.
Jiang, Lady, 153.
Jiaoji (social intercourse), 97. *See also* Social relations.
Jiao xin (relational pleasures), 328. *See also* Relational pleasures.
Jie and Zhou (tyrants), 294, 298.
Jin, Prince of, 107.
Jing (quintessential *qi*), 49, 75.
Jingdi (d. 141 BCE), 95.
Jingzhou Academy, 269, 313.
Jinpenling, 280.
Jinshu, 123.
Joy, 35, 40.
Junzi (noble man/aristocrat), 203-204. *See also* Xunzi: view of the noble man.
Juyuan, Eastern Han document from, 200.

KAI (TO BE STIRRED OR THRILLED), 44.
Kang (unruffled), 44.
Kant, Immanuel, 97, 144, 172, 178.
Ke (approval), 178.
Klein, Esther, 261-62.
Kongzi/Confucius: as author and editor, 271, 275-76, 282; biography of, 272; capacity for longing, 163; demeanor in leaving courts, 125; disciples of, 120, *134*, 286, 287, 342, 377 n.5, 382 n.2, 398 n.2, 399, 404 n.5; as example of integral wholeness, 203; on drinking rites, 105; and friendship, 97-98; on Guan Zhong and Bao Shuya, 88, 90; injunction to "reanimate the old" (*wengu*), 292; at leisure, 100, 342; and music, 63, 70-71, 72; "praise and blame," 297, 299; and preservation of the classics, 307; "rectifying names," 311; on sagely behavior, 328; in Tao Yuanming's poems, 342, 384 nn.8-9, 402 n.1; Yang Xiong on, 266, 288; in the *Zhuangzi*, 221-22, 255, 256. *See also Analects*; Yan Hui.
Konstan, David, 116.
Kuang, Music Master, 288.
Kuang Heng, 265-66.
Kunlun Peak, 408.

LABOR SERVICE, 146, 315.
Language and logic, 241-47.
Lao Dan, 233, 257.
Laozi, 306, 389 n.2, 425.
Laughter, 255-56.
Le (pleasure-seeking, pleasure-taking, imparting pleasure), 17, 34; antonyms of, 35, 40-41; contrasted with *xi* (delight), 42-44; graph shared with "music," 63, 72, 74, 80, 132, 150, 176; paired with *an* (to secure), 35; translations of, 35-41; used with noun object of consequence, 35; verbal use of, 17-18.
Legalism, 178.
Legan wenhua (culture alive to pleasure), 36.
Li (profit), 139, 179, 192. *See also* Profit.
Li Deyu, 283.
Li Gonglin, 359; "Tao Yuanming Returning to Seclusion," 359, *360*.
Li Guang, 343.
Li Ling, 93-94.
Li Yanzhi, 314.

Li Zehou, 36.
Li Zhuguo, 279.
Liang Kai, "Gentleman of the Eastern Fence," 357–59, *358*.
Liangshu, 118.
Libraries: imperial, 20, 270, 278–81, 282, 284, 304; private, 270, 274; site of Tianlu ge palace library, *264*; transition from archives to, 274, 277–78.
Liezi, 258, 281.
Liji (Record on Rites): on drinking ceremonies, 104; friendship in, 111, 119; "Notes on Learning," 119; "Record on Music," 79–80, 132.
Liu De (King of Hejian), 277–78.
Liu Songnian, "Listening to the *Qin*," *58*.
Liu Xiang: editions of classics, 266, 278–81; employment in palace library, *264*, 279; and the *haogu* movement, 266, 278, 297, 304–305; on importance of *Erya*, 308, 309.
Liu Xie, 272.
Liu Xin: denied access to Yang Xiong's drafts, 294; edition of *Shanhai jing*, 281; and the *haogu* movement, 266, 278, 296, 304–305, 308; letters exchanged with Yang Xiong, 272; rivalry with Yang Xiong, 272, 278, 297, 304; *Seven Summaries*, 279; son of, 309.
Liu Yiqing, *New Account of Tales of the World* (*Shishuo xinyu*), 107–108, 300.
Liu Yu, 323.
Liu Zhiji, 299, 300.
Liu Zongyuan, "Account of Song Qing," 122.
Liyi (duty and appropriate action), 99.
Logic: Mencian, 138–39, 146; of Xunzi, 176, 178, 184, 192, 203; in *Zhuangzi*, 91, 213, 216, 224, 241, 243–47.
Lu, Duke of, 154.
Lü Lihan, 127.
Lu Xiujing, *320*.
Lu Zhaolin, 314.
Lüshi chunqiu, 79, 80, 306.
Lyrics (*wen*), 83.

MA HEZHI, "Illustrations of the Classic of Filial Piety," *134*.
Malraux, André, *The Temptation of the West*, 33.
Manuscript culture: activist editing and authoritative texts, 281–84, 305; bamboo or wooden slats, 199, *200*; compilation and editing process, 262; copying texts, 290–91, 293, 297, 307; manuscripts as gifts, 283, 290–91, 294; manuscripts as status objects, 199, *200*, 201; preconditions for, 273–74; reading as reciting, 289–91; risks of texts, 294; Tao Yuanming and, 346; tomb sculpture of figures collating texts, *280*; in Western Han, 276–82; and the world of Yang Xiong, 265–66, 270–71, 273–76, 284, 289–90.
Master-disciple relationship, 87, 120; handscroll illustrating, *134*. See also Kongzi/Confucius: disciples of.
Matching poems, 364–66, 371. See also under Su Shi.
Matisse, Henri, 218.
Ma Yuan, 407 n.5; *Riding a Dragon*, 359.
McLean, Karen, untitled photograph, Southern Arizona, 1980s, *251*.
Mei Yaochen, 366.
Melodies (*qu*), 65, 69, 79, 80–83, 128.
"Melody of Guangling," 128.
Memories: homecoming and, 321; of pleasure, 26, 44–45, 46, 341, 352.
Mencius: advice to King Hui of Liang, 138–43; advice to King Xuan of Qi, 143–46, 153–54, 171; advice to Lord Wen of Teng, 168; advice to rulers, 144, 148, 152–55, 166; "basics" of, 145–46; Book 1 on pleasure, 135–39, 149, 154–55; commonalities with *Analects*, 165; compared with Xunzi, 165–68, 175, 192, 194; on compensation of rulers, 162; contrasted with Zhuangzi, 214; on floodlike *qi*, 155–56; "friends in history," 362; on gift giving, 96–97; and Han *fu*, 173; on his own sageliness, 164; inner/outer distinctions, 167; on love of money and women, 153; on moral potential, 148, 156–57, 161–62, 166, 172; on music, 149–51, 152; on profit and rightness, 138–40; on sensory perception, 47; on sharing, 151–52, 154–55, 162, 168–69, 312; and social welfare, 170–71; teachings on pleasure and rulership, 18–19, 137–38, 140–41, 149–52, 153, 161–63; tribalism of, 172; on the true king, 144, 147, 148, 150, 152–55; use of dialogue form, 136, 137, 140; use of word *xin*, 328; view of human nature, 18, 56, 148, 159–65, 167, 312;

on wholeness, 138, 158-59; and Yang Xiong, 173, 292; on yearnings, 163.
Meng Jia, 341.
Mengchang, Lord of, 122.
Meyer, Dirk, 275.
Mi Zixia, 124.
Mian (thinking fondly), 341.
Ming (charge or writ), 24.
Ming (light or clarity), 250.
Mingdi (Han), 311.
Mohists, 150, 199, 240. *See also* Mozi.
Motion, 45, 48, 67-69.
Mourning rituals, 98, 108-10, 197.
Mozi, 77-78, 80, 296. *See also* Mohists.
Mu, Duke of Zhou, 408.
Music: accompanied by dancers, 65, 69, 77, 78, 79, 81, 82; appreciation of, 149; cosmic dimensions of, 60, 79-80, 85-86; court performance of, 66, 77-78, 81, 82, 150, 196; and cultivation of personal character, 72; depictions on ancient bronzes, 66, 68; and drinking, 99, 104, 106; early literature on, 67; elegant versus popular, 81, 149, 150; and the emotions, 67-69, 71-72, 74-75, 78, 80, 84, 85; focus on lyrics, 83; and governance, 61, 63-64, 74, 80; graph shared with "pleasure," 63, 72, 74, 80, 132, 150, 176; intimate, 66, 82-83, 84; and intimate friendship, 18, 58, 112, 114; invention of, 52; in letter from Yang Yun to Sun Huizong, 130-31; and listening, 84-85, 86; and the loss of true friends, 73; lost classics and old music, 65, 81-82, 83; and order, 79-80; pitch standards, 85; power to induce awareness, 70-72; professional musicians, 75, 82, 104; relationship with friendship, 59-63, 65-66, 69-70, 72-73, 84-86, 132-33; resonance theory in, 60, 225; rhythm and harmony in, 77-80, 85-86, 150; and ritual, 65, 66, 67, 72, 74-75, 81, 194; role in moderating pleasure, 74; as shared pleasure, 150-51; and social relations, 74, 77; "soundless," 83, 114; and status, 75; work songs, 69; Xunzi's view of, 77-79, 177; of Zheng and Wei, 83. *See also* Musical instruments.
Music Bureau, 75, 81-82.
Musical instruments, 65-67, 80, 82, 83; bells, 63, 67, 77, 79, 85; chimes, 66, 77, 79, 304; drums, 66, 67, 79; early depiction of, 66,

68; "ocarina and flute," 61; winds and strings, 67, 79, 83; Xunzi on, 79; Zeng hou yi tomb bell set, 76, 77, 85; zithers, 69, 70-71, 79, 82, 128.
Musical phrasing (*yuezhang*), 83.
Mystery Learning (*xuanxue*) movement, 269, 313.

"NAMES" AND "ACTUALITIES," 29-30.
Natural disaster relief, 170.
Nelson, Susan, 357.
Ni (indulge in or be addicted to), 44.
Nineteen Old Poems, 337, 401 n.7.

ODES: allusions to, 334, 382 n.1; authorship of, 275; commentaries, 312; in cultivated behavior, 63; on drinking rites, 107, 108; on friendship, 64; "The Guests Take Their Seats," 107; "Hewing Wood," 120-21; "Jigu," 118; Kongzi as editor of, 282; on the parks of King Wen, 141; on pleasures of music and friendship, 60; redaction of, 282, 293; "What harm would there be in curbing our lord?" 152; Zheng Qiao on musical phrasing in, 83.
Old Text corpus, 305.
Ouyang Xiu, 365.
Overindulgence, 35, 49-50, 194.
Ox Mountain, 161.

PAIN, 17, 35, 40, 41-42.
Parks, 140-41, 151.
Paternalism, 171.
Peach Blossom Spring, 355-56, 421.
Peasants, 29, 186, 420. *See also* Farmers.
Periods of Disunion, 30.
Perkins, Maxwell, 282.
Philology, 271, 301-12.
Pines, Yuri, 119-20.
Pingdi (Han), 310. *See also* Wang Mang.
Pitch standards, 85.
Plato: image of chariot, 48; *Protagoras*, 223.
Pleasure: association with motion, 45, 48, 67-69; books and, 19-20, 263-66, 268-71, 290, 312-14; consuming and sustaining, 24, 26, 33, 52, 147; physiology of, 45; desires in, 46-47, 52-55, 324; prolongation of, 33; relational, 34-35, 36, 42, 328; of returning home, 319-22; rhetoric of, 24-27, 34, 45-51,

56; and ritual, 193; solitary, 88, 151; and statecraft, 51–56, 147; temporal aspects, 36, 44–45; vocabulary of, 44; Western conceptualization of, 17, 35. See also Friendship; *Le*; Music; *and under* Mencius; Su Shi; Tao Yuanming; Xunzi; Yang Xiong; *Zhuangzi*.
Pleasure parks, 18, 140–41, 151.
Parting, 128, 350, 357.
Pollock, Sheldon, 57.
Power and pleasure, 11, 22–23, 25–27, 52–55, 79–80, 135–36. See also Rulers.
Praise and blame, 125, 297–300.
Pre-Qin writing, 297, 299, 302–306, 309.
Professional musicians, 75, 82, 104.
Profit, 111, 138–39, 179, 192, 201, 287.

QI (SPIRIT OR VITAL ENERGY): and Hopkins's "instress" and "inscape," 325; leakage of, 49–51, 52; Mencius on, 155–56, 157; motion of, 48; music and, 52, 67, 75; refined, 49; resonant exchange of, 48–49, 226; in *Zhuangzi*, 216, 236, 239, 241.
Qi (state), 88–90. See also Huan, Duke of Qi.
Qin, First Emperor of, 298, 301.
Qin you (dear friend), 94, 126. See also Intimate friendship.
Qing (inclinations/feelings), 53, 72.
Qu Yuan, 306.

RAN QIU, 414.
Reading, 265–66, 269–71, 288, 289–90, 312; as consolation for Su Shi, 362, 364; practice recommended by Yang Xiong, 304–307, 311, 313–14; Shu Xi's "*Fu* on Reading," 314. See also Manuscript culture.
Reclusion: of Ji Kang, 127–28; in poems of Su and Tao, 20, 323, 335, 343, 363, 382 n.3, 399 n.3, 403, 407, 411 n.2, 423; of Xu You and Bo Yi, 94; in *Zhuangzi*, 215, 221, 236, 244.
Record on Rites (*Liji*): on drinking ceremonies, 104; friendship in, 111, 119; "Notes on Learning," 119; "Record on Music," 79–80, 132.
Relational pleasures, 34–35, 36, 42, 328. See also Friendship; Music.
Ren (humaneness), 156–57, 164.
Ren An, letter to, 92–94.
Ren Fang, 118.

Ren Hong, 279.
Resonance theory, 18, 37, 45–49, 59, 224–25, 323–24.
"Return to antiquity" (*fugu*), 82. See also *Haogu* movement.
Reverie (Bachelard), 254.
Richter, Antje, 112.
Ritual: drinking ceremonies, 206–207; mourning, 98, 108–10, 197; music and, 65, 66, 67, 72, 74–75, 81, 194; and pleasure in Xunzi, 193.
Ritual vessels. See Bronzes.
Rong Qiqi, 420.
Rulers: and the common good, 168; income of, 53; as models of virtue and generosity, 30, 196; and music, 65, 74, 80, 150; pleasure-taking by, 51–52, 137–41, 151–52; policies to promote pleasures, 33; providing for subjects, 141–44, 146; relations with ministers, 66, 119–20; relations with subjects, 51, 168; as textual scholars, 315–16; tours of, 152; as "true kings," 144, 147, 148, 150, 152, 153, 154–55; as unifiers, 24, 27. See also Power and pleasure.
Ryles, Gilbert, 229.

SACRIFICES, 23, 44, 117.
Sages: Kongzi on, 328; Mencius on, 164, 192; Seven Sages of the Bamboo Grove, 357; Xunzi on, 191–92; Yang Xiong on, 201; Zhuangzi on, 243–44, 250, 262.
Schadenfreude, 42.
"Science of Happiness" project (UC-Berkeley), 39.
Second nature, 47, 114, 181, 189, 199, 222, 257.
Security, 19, 30, 35, 41, 62, 312. See also Insecurity.
Self-indulgence, 24, 34, 42, 257. See also Overindulgence.
Self-understanding (*zhi ji*), 115.
Sender-medium/percept receiver model, 31.
Seneca, 87.
Sensory organs, 46–48.
Sensory pleasure, 34, 36–37, 47. See also Relational pleasures.
Seven Kingdoms Revolt, 95.
Seven Sages of the Bamboo Grove, 357.
Sex, 36, 50, 62, 100–191, 107–108, 159, 160, 185; conjugal love, 38, 120; and *qi*, 38, 49–50.

Shan (goodness in a given situation), 158.
Shan Tao, 126–28.
Shang, last king of, 141.
Shanglin Park, 170.
Shanhai jing (Classic of the Mountains and Seas), 281, 381 n.11; "Reading the *Classic of the Mountains and Seas*" matched poems by Su and Tao, 406–409, 415, 418–19.
Sharing, 151–52, 154–55, 162, 168–69, 312.
Shen (divinity), 177.
Shen (refined *qi*), 49.
Shen ming (divine insight), 75.
Shennong, 296.
Shi you (colleague in office), 118.
Shi zhen (damage to the true self), 49.
Shiji (Historical Records; Sima Qian): allusions to, 387 n.5; biographies, 28; contrasted with *Hanshu*, 74; and the letter to Ren An, 92, 94; and manuscript culture, 276; on music, 74; record of friendship between Dou Ying and Guan Fu, 95; on sacrifices of ardent lovers, 93; Yang Xiong and, 285, 296, 300, 306. *See also* Sima Qian.
Shishuo xinyu (New Account of Tales of the World; Liu Yiqing), 107–108, 300.
Shitao, *Reminiscences of the Qinhuai River*, 361.
Shu (fellow feeling), 136.
Shu (physical ease), 44.
Shu Qi, 343, 351, 388, 403.
Shu Xi, "*Fu* on Reading," 314.
Shun, 162, 163, 165.
Shuo (interpretive readings), 307.
Shuowen jiezi (Xu Shen) 111, 310, 312.
Shuoyuan, 107, 120.
Si (long for/to ponder), 177, 190.
Sima family, 126–28.
Sima Guang, 365.
Sima Qian: and authorship, 275–76; on friendship, 70, 94–95; "Letter to Ren An," 92–94; on music, 70, 74; and Yang Xiong, 300. *See also Shiji*.
Sima Xiangru, 324.
Singsong girls, 62.
Six Arts, 279, 309.
Sluga, Hans, 427.
Social relations, 35, 50, 61, 78, 97–99, 111–12, 118–19, 155, 283.
Social welfare measures, 170, 171.
Socrates, 100; Socratic dialogues, 137.

Soga Shohaku (Miura Sakonjirō), "Three Laughing Masters at Tiger Ravine," *320*.
Solitary pleasures, 88, 151.
Song Bian, 184.
South Mountain, 332–33, 356, 378 n.1, 390.
Spectacles, 196–97. *See also* Display culture; Music: court performance of.
Statecraft theory, 51–56.
Stoicism, 252.
Su Che, 319, 363, 365, 369, 416, 424.
Su Shi (Su Dongpo): "After an Old Poem" matching poem, 399; appreciation of Tao's poetry, 323; attitude toward material things, 418–19; Buddhist ideology, 418; *ci* celebrating imperial pardon, 362; classical illusions by, 377 n.5, 381 n.11, 382 nn.1–2, 383 n.7, 387 nn.5,7, 389 nn.1–2, 403 n.3; across "clouds and seas," 282; construction of Tao Yuanming, 20, 319, 357, *358*, 359, 365–66, 367–68, 370, 371, 420; contrasted with Tao Yuanming, 321, 322, 369–71, 418–19; court service and exile, 20, 322–23, 362, 363–64, 368–69, 371, 421–23; "On Drinking" matched poem, 389, 391, 393, 395; drinking poems, 365, 372, 414, 417–18, 420; "Early Spring" matched poem, 383, 385; "Excursion to Xie Brook" matched poem, 377, 413–14; family of, 369, 387 n.5; and fan painting, 412; illness and death, 369; "Lauding Impoverished Gentlemen" matched poem, 403, 405, 413, 421; letters of, 346, 365, 416; matched poems with Tao Yuanming, 20–21, 319, 322, 359–63, 368, 371–72, 410 n.1, 413, 424; as medical author, 364; "Miscellaneous Poems" matched poem, 401; motivations for matching Tao's poetry, 363–68; "Moving House" matched poem, 379, 381; and ordinary pleasures, 426; pleasure in poetry, 424; "Reading the *Classic of the Mountains and Seas*" matched poem, 407, 409, 415, 418; returning home poems, 364, 369–70; "Returning to My Old Home" matched poem, 387, 421; "Returning to the Orchards and Fields to Dwell" matched poem, 375, 416; sentimentality of, 417; "Stopping Wine" matched poem, 397; temple dedication written shortly before his death, 424–26.

Sumptuary regulations, 53, 197–98, 201.
Sun Huizong, letter from Yang Yun, 130–31.
Sun Jing, 313.
Sustaining pleasures. *See* Consuming and sustaining pleasures.
Swordsman's Treatise, 288, 306.
Sympathetic response, 30, 45, 61. *See also* Resonance theory.
Symposia, 99–100.

TAI WANG, 153.
Tao Yuanming (Tao Qian): abandonment of bureaucratic career, 20, 368; admiration for *Zhuangzi*, 323; admiration for Yang Xiong, 269, 272, 314; "After an Old Poem," 398; "Appraisals for Paintings on Fans," 345, 411–12; "Appraising Shang Zhang and Qin Qing," 345; appreciation of, 323, 365; appreciation of material things, 418–19; biography of Meng Jia, 341; classical allusions by, 342, 354, 394 n.9, 402 nn.1–2; compared with Su Shi, 321, 322, 369–71, 418–19; "Composed on the first day, fifth month, in answer to Registrar Dai's poem," 349; contrasted with display *fu*, 324; contrasted with Seven Sages of the Bamboo Grove, 357; depicted in "Three Laughing Masters at Tiger Ravine," *320*; "On Drinking," 350, 388, 390, 392, 394; "Drinking Alone in the Rainy Seasons," 349; "Drinking Poems," 331–33, 340, 356, 362, 365, 414–15, 417; "Early Spring, in the *Guimao* Year, Yearning for the Ancients," 382, 384; "Endless Rain, Drinking Alone," 343; "Excursion to Xie Brook," 350, 354, 376, 413–14; on fame, 330, 335–38, 344, 345, 354, 414; "For My Sons," 340–41; on forgetting, 340–41; "Form, Shadow, and Spirit," 336–39, 349, 350–51; "Grief for My Late Cousin Zhongde," 352; home and family life, 348–49; and Hopkins's "instress" and "inscape," 324–25; imagery of, 346–49; "Lauding Impoverished Gentlemen," 345, 402, 404, 413, 421; "Living in Retirement, Composed on the Double Ninth Holiday," 335, 349, 356; love of wine, 331–33, 334, 339, 343, 422; "Miscellaneous Poems," 400; "Moved by Good Men's Failures to Meet Good Fortune," 343–44, 410; "Moving House," 370, 378, 380; "Offering for My Late Cousin Jingyuan," 352; "Offering for My Late Sister, Madame Cheng," 352; old age and death of, 350–51, 352–55; ordinary pleasures of, 330–31, 424, 426; paintings depicting, 346, 357–59, *358*, *360*, 422; on the pleasures of friendship, 328, 334–35, 341–42, 344, 350, 352, 359, 370; on the pleasures of returning home, 319–22, 323, 326–28, 340, 344, 346–48, *360*; "Poem of Resentment in the Chu Mode," 336, 351; "Reading the *Classic of the Mountains and Seas*," 406, 408, 415, 418–19; "Record of Peach Blossom Spring," 355–56, 421; "Reply to Adjutant Pang," 332–33, 344; "Reply to a Poem by Clerk Hu," 357; "Reply to Senior Officer Yang," 351; "Reproving My Sons," 348; return as theme, 357, 369, 419, 423, 424; "return" poem sent to Magistrate Ding in Chaisang, 332–33, 344; "Returning Birds," 347; "Returning Home," 326–27, 328, 348, 354, 359, 364; "Returning to My Old Home," 350, 355, 386, 421; "Returning to the Orchards and Fields to Dwell," 374, 416; "Sacrifice to Myself," 352–54 ;"Seasonal Round," 342; "Soul," 329; "steadfastness in adversity," 331, 354, 388, 410, 415; "Stopping Clouds," 341–42; "Stopping Wine," 396; "Tale of Master Five Willows," 354–55; "To My Grandfather's Cousin," 349; "A Tree in Bloom," 349, 350; unease with impermanence, 336, 349–50; "Untitled" series, 417; use of word *mian*, 341, 383 n.6; use of word *xin*, 328–29; views on life and death, 329–30, 333, 335–40, 352, 356, 424; "Written after Reading History," 323. *See also* titles of matched poems under Su Shi.
Taxation, 53, 169, 300.
Textual authority, 274, 281, 283–84. *See also* Manuscript culture.
Tian Fen, 95.
Tian Xiaofei, 346.
Tianlu ge palace library, *264*.
Timi Ming, 107.
Togetherness, 166, 258.
Tomb figurines, from Jinpenling, 280.
Tomb murals, "Feast with the Married Couple," *102*.
Tong (pain), 35. *See also* Pain.

True kingship, 144, 147, 148, 150, 152–55.
True Way Learning, 165, 166.

UELSMANN, JERRY N., untitled photograph, 1982, *318*.
Unifiers, 24, 27.
"Universal Declaration of Human Rights" (UN), 172.

VERVOORN, AAT, 118–19, 120.
Vices of disproportion, 57.
Virtue, 30, 177, 187, 196, 204.

WAN (PLAY), 44, 129, 268.
Wan Zhang, 96–97.
Wandering, 232, 253, 262, 424.
Wang Anshi, 420.
Wang Bi, 269.
Wang Can (Zhongxuan), 108–109.
Wang Chong, 300.
Wang Fu, *Qianfu lun*, 122.
Wang Mang, 273, 310, 312.
Wang Shang, 343.
Wang Shumin, 262.
Wang Xizhi, 73, 370; letters to friends, 112–13.
Wang Yi, 127.
"Warming up the old," 57, 312.
Wei (in danger), 35.
Wei, Lord of, 124.
Wei Sheng, 96.
Well-field system, 168.
Wen, Emperor of Wei. *See* Cao Pi.
Wen, King of Zhou, 70–71, 140–41, 169, 267.
Wen, Lord of Teng, 168.
Wenxin diaolong, 127.
Whistling, 73.
White Crane Lookout, 381 n.10, 387 n.6, 421.
Wholeness. *See* Integrity and wholeness.
Wilde, Oscar, 33.
Wittgenstein, Ludwig, 242.
Wolfe, Thomas, 282.
Wood, Allen W., 172.
Woodworker Qing and Butcher Ding, 215, 216–21, 224–25, 226, 425.
Woolf, Virginia, 424.
Worthy men, 55, 123, 126, 229, 233, 235, 242, 255.
Wu, King of Zhou, 267.
Wu Zixu, 228.
Wude (lacking charisma), 41.

Wudi (Han), 81–82, 92–93, 278, 297, 301.
Wulu, Master, 160.
Wuwei (act without fixed goals or polarizing effects), 228, 232.

XI (DELIGHT), 42–44.
Xiang (enjoy), 44.
Xiang gan (mutually attracting and affecting), 114.
Xiang le (pleasurable mutuality), 48.
Xiao He, 28.
Xiaoxue (elementary learning), 309.
Xie An, 73.
Xie Lingyun, 314.
Xijing zaji (Diverse Records of the Western Capital), 309.
Xin (appreciate/be heartened), 44, 328–29.
Xin (heart), 45–48, 49, 243.
Xing (human nature), 165. *See also* Human nature.
Xing (second nature), 47. *See also* Second nature.
Xu Shen, *Shuowen jiezi*, 111, 310, 312.
Xu You, 94.
Xu Zhi, 109–10.
Xuan, Emperor of Han, 131, 170, 298.
Xuan, King of Qi, 143–46, 153, 157, 171.
Xuanxue (Mystery Learning) movement, 269, 313.
Xue (study or learning), 302.
Xun Yue, 312.
Xunzi: aesthetic theory of, 178; borrowings from *Zhuangzi*, 261; compared with Mencius, 165–68, 175, 192, 194; compared with Zhuangzi, 214; contrast with Mozi, 80; on creating order, 428; "Enrich the Realm" chapter, 199; on fear and anxiety, 178, 182–83; on friendship and order, 119; *fu* of, 173; on human nature, 19, 165, 176, 178–79, 189–90, 198, 207; on the ideal ruler and realm, 198–202; on integrity and wholeness, 19, 203, 204–205, 208, 211; as Legalist, 178; "Letting Go of One-Sidedness" chapter, 261; Liu Xiang's version of, 281; on music, 52, 60, 74, 77–79; "On Music and Pleasure," 201; paradox of risking death on the battlefield, 51; pleasure theory in, 176, 178, 179–83, 186–87, 190, 192–93, 201–203; pupils of, 211; rejection of moral intuition,

56; on rites and music, 177, 184, 196; "On Ritual," 194–96; on ritual exchanges and village drinking ceremonies, 206–207; on sociopolitical policy, 194–95, 201; and sumptuary regulations, 53, 198, 201; as systematic thinker, 175; theory of desire, 175, 176, 178, 179–80, 182–86, 194–95, 198; view of the noble man, 175–78, 187–88, 190–91, 192, 197, 203–206, 207, 208–11.

YAN (SATISFIED), 44.
Yan Hui: mentioned by Su Shi, 399, 418; in poetry of Tao Yuanming, 374 n.3, 382, 398 n.2, 420; suffering of, 203, 343; teacher-student relationship with Kongzi, 120; in Yang Xiong's writings, 286–87, 289, 292; in *Zhuangzi*, 255.
Yan Lingfeng, 269–70.
Yan Zhitui, 75.
Yang Hu, 98.
Yang Xiong: allusion and word play, 294, 298; authorial voice of, 265, 285–86; biographies of, 272–73; *Canon of Supreme Mystery* (*Taixuan jing*), 269, 272, 289, 300; categories of texts, 292–94; commentaries on, 269–70; compared with Confucius, Mencius, and Xunzi, 284, 292; *Correct Words* (*Fangyan*), 272, 310; criticism of *fu*, 284, 291, 298, 306; *Exemplary Figures* (*Fayan*), 173, 263, 265, 269, 272, 282–83, 284–85, 286–89, 294–95, 298–99, 299–300, 306; on friendship, 117; on good rule, 295; and the *haogu* movement, 278, 296–97, 302, 306–308, 311; on the importance of finding a good teacher, 291–92; influence of, 269–70; inspired by Xunzi, 178; library work, 264, 284, 294, 304; and Liu Xin, 272, 278, 294, 297, 304; and manuscript culture, 270–71, 273–76, 282–84, 289–90, 294; and music, 81; passion for classical learning, 19–20, 263–65, 268, 271, 290, 292–93, 297, 312–14; and philology and epigraphy, 270, 301–306, 308–11; "play with the Ancients," 298–99; as praise-and blame historian, 297–301; prefaces of, 285; preoccupation with Five Classics, 276, 281, 282, 286–87, 288, 292, 305–307, 308; and pre-Qin script, 297, 299, 304–305; readership of, 289; rejection of "full awareness," 56; "the sage takes pleasure in being a sage," 210;

selection of pre-Qin canon, 306; and Sima Qian, 300; suicide attempt, 294; support for Wang Mang, 273; and Tao Yuanming and Su Shi, 324; writer of *fu*, 173, 263, 269, 271, 272, 285–86, 287–88, 294; writing style of, 265, 268–69, 271–72, 285–86.
Yang Yun, letter to Sun Huizong, 129–31.
Yang Zhu, 54.
Yao and Shun, 165, 192, 236, 408.
Yearley, Lee, 215.
Yearning, 163, 173, 295. See also Desires.
Yellow Emperor, 296.
Yi (duty), 139.
Yi (sated), 44.
Yi (unruffled), 44.
Yi, Marquis of Zeng, bell set of, 63, 76, 77, 85.
Yijing. See *Changes* classic.
Yin (to go to excess), 44, 49.
Yin jin (on good terms), 44.
Yin Xian, 279.
Ying (impulses), 47.
Ying Shao, *Fengsu tongyi*, 109–10, 123.
Yiya (chef), 288.
"Yongyuan qi wu bu" (Eastern Han official document), 200.
You (anxious, worried, concerned), 35, 40–41, 51.
You (friendship), 65, 117, 120; terms using, 117–18, 120, 126. See also Friendship; Intimate friendship.
You yu ren (to be seduced by others), 53.
Yu (amuse or be amused), 44.
Yu (driving), 48.
Yu (witless), 53.
Yuan An, 345, 404.
Yuan Qiao, 123.
Yuan Taotu, 295.
Yuan Xian (Zisi), 404 n.5, 404 n.7.
Yuanyou faction, 365, 369.
Yue (music), 80. See also *Le*: graph shared with *yue*; Music.
Yue (think well of), 44.

ZENG GONG, 365.
Zeng hou yi tomb bell set, 63, 76, 77, 85.
Zeng Xi, 342.
Zengzi, in "Illustrations of the Classic of Filial Piety" handscroll, 134.
Zhai of Xiagui, 122.

452

Zhang Chang, 302–304.
Zhang Er, and Chen Yu, 129, 131.
Zhang Heng, 313.
Zhang Shao, 117.
Zhanguo ce, 281.
Zhanguo era treatises, 22–23.
Zhao, Flying Swallow, 271.
Zhao Dun, 107.
Zheng Qiao, 83.
Zheng Xuan, 114.
Zhi (will or commitment), 48, 53.
Zhi ji (self-understanding), 115.
Zhile/yue (maximum pleasure/ultimate music), 80.
Zhi ren (fully present man), 216, 217, 218, 225–26.
Zhi sheng (construct a life worth living), 113.
Zhi yin (know the tone), 69–70, 72, 109.
Zhou, Duke of (Zhougong), 266, 292, 295, 297, 313; album leaf illustration of, 267; "Be Not Idle" chapter of the *Documents*, 267.
Zhou Yafu, 95.
Zhou Yangkui, 411 n.2.
Zhu Kangshu, 346.
Zhu Mu: "On Severing Unofficial Relations," 125; "On Upholding Tolerance and Magnanimity," 124–25.
Zhu Xi, 31, 82, 83–84, 276.
Zhuang Bao, 149.
Zhuangzi: agnosticism and embrace of uncertainty, 19, 56, 222–24, 231, 258–59; belief in inherent goodness of life, 250–53; curiosity of, 252–53, 258–59; and Hu Shi, 90–91, 113; as imagined author and fictional protagonist, 213, 261; mourning for his wife, 237–38. See also *Zhuangzi*.
Zhuangzi: allusions to, 387 n.5, 414; assessed by Yang Xiong, 306; "Attaining Life and Vitality" chapter, 221–22; "Autumn Floods" chapter, 262; "balancing the faculties of the heart," 217, 218, 239, 248, 254; bleak assessment of the human condition, 230, 231–32; commentaries and interpretations, 237, 239, 261; compared with Mencius and Xunzi, 120, 171, 178, 181, 214; dating and composition of, 261; on death, 236–38, 240, 250–52; dreams in, 239–41; on duty to family and ruler, 235–36; element of time, 218; on exile, 422; on forgetting, 254–55; Fully Present Man, 19, 216, 217, 218, 225–26; "Inner Chapters," 214, 261–62; interpretations of, 19, 20, 214–15, 220, 235–36, 237, 252, 255, 261; on language and logic, 241–47; on laughter, 255–56; "Let it be!" and "Let it be spring!" 253; "Old Fisherman" chapter, 256; opening image of Kun fish and Peng bird, 232–33, 262; "Outer and Mixed Chapters," 262; on pleasure, 178, 213, 252, 258; portrayal of friendship, 90, 91–92; "quieting the mind," 217; on reputation, 228; and resonance theory, 224–25; "return to basics/the source," 214, 249, 257–58; "Seeing All as Equal" chapter, 241, 351; story of the piglets, 237, 356; style of, 214; "Supreme Pleasure" chapter, 215, 223, 227–31, 237, 239–40; Supreme Swindles, 226; Su Shi and, 422; Tao Yuanming's admiration for, 323, 414; on the "true," 256–57; view of life and vitality, 216, 220, 221, 229, 247–48, 258–59; view of the worthy man, 233–35; "Wandering in a Daze" chapter, 262; Woodworker Qing and Butcher Ding, 215, 216–21, 224–25, 226, 425. See also Zhuangzi.
Zhuozhuan, 103, 150, 296.
Zichan of Zheng, 21–22.
Zide (self-possession), 62, 115.
Zigong, 90, 257, 404 nn.5,7.
Ziqi of South Wall, 248.
Ziran (self-propelling), 55.
Zithers, 69, 70–71, 79, 82, 128.
Zuowang (sitting and forgetting), 254–55.
Zuozhuan, 98, 107, 120.

Zone Books series design by Bruce Mau
Typesetting and image placement by Julie Fry
Printed and bound by Maple Press